Handbook of Clay Minerals

Edited by **Callum Lloyd**

R CALLISTO REFERENCE

New York

Published by Callisto Reference,
106 Park Avenue, Suite 200,
New York, NY 10016, USA
www.callistoreference.com

Handbook of Clay Minerals
Edited by Callum Lloyd

International Standard Book Number: 978-1-63239-381-4 (Hardback)

Contents

Preface

This book provides a concise and in-depth overview of clay minerals. Clay is a versatile material which has varied applications depending upon its properties and composition. Clay minerals are inexpensive and sufficiently available on earth. They have 1 nm silicate layers in almost all the sediments and are eco-friendly because they are natural minerals. This book discusses basic properties and role of clay minerals in deposits. It also deals with the importance of clay in the soils. This book also covers various aspects of research in this domain, especially to the identification of clay structure and its alteration for different applications.

The information contained in this book is the result of intensive hard work done by researchers in this field. All due efforts have been made to make this book serve as a complete guiding source for students and researchers. The topics in this book have been comprehensively explained to help readers understand the growing trends in the field.

I would like to thank the entire group of writers who made sincere efforts in this book and my family who supported me in my efforts of working on this book. I take this opportunity to thank all those who have been a guiding force throughout my life.

Editor

Clay Minerals in Deposits

Clay Minerals from the Perspective of Oil and Gas Exploration

Shu Jiang

Additional information is available at the end of the chapter

1. Introduction

The clay minerals e.g. kaolinite, smectite, illite, chlorite, etc. are ubiquitous in the targeting rocks of oil and gas exploration. During the early age (1940s) of worldwide oil exploration, clay minerals were studied to predict the quality of organic rick source rock and generation mechanism when scientists tried to investigate the origin of oil and gas (Grim, 1947, Brooks, 1952). Then the clay minerals analysis was used as a tool in terms of environmental determination, stratigraphic correlation and hydrocarbon generation zone identification to find exploration target interval, which was preliminarily and generally summarized by Weaver in 1960. By the 1970s, the clay minerals began to be widely studied for diagenesis and reservoir quality prediction due to the application of petrological analysis and quantitative mineralogical analysis by X-ray diffraction (Griffin, 1971; Pettijohn, 1975; Heald and Larese, 1974; Bloch et al., 2002). Since 1980s, the clay minerals analysis has been used to determine the hydrocarbon emplacement time and petroleum system analysis (Lee et al., 1985). These intermittent clay minerals research progresses are the result of exploration demands of conventional reservoirs (sandstone and carbonate rocks) at different times.

Despite their increasing importance in fundamental geological research and oil industry, clay minerals prove difficult to study in the past. Their sheet structure results in features that can only be resolvable at the sub-micron scale. They are also subtly variable in chemical composition (Fe, Mg, K, Al, etc) and can be confused with each other and other silicates. the recent innovative analytical tools and modern analysis techniques, e.g., QEMSCAN (Automated Mineralogy and Petrography), FIB/SEM (Focused Ion Beam/Scanning Electron Microscope), EDS (Energy-dispersive X-ray spectroscopy), etc., have the capability of quantitative and qualitative characterizing nano-pore and mineralogy of fine grained shale rocks (Lemmens et al., 2011), which creates new era of studying clay minerals for facilitating unconventional (shale) reservoir exploration.

Even though there were the numerous sporadic reports about the application of clay minerals in the oil and gas exploration. So far, relatively little work has been documented on the detailed summary of clay minerals from the perspective of oil and gas exploration. This paper is to systematically summarize the important role of clay minerals in oil and gas exploration from many points of view: basin tectonic evolution, depositional environment, thermal history and maturation history of organic matter in the source rock, hydrocarbon generation, migration and accumulation process, diagenetic history and reservoir quality prediction. The traditional and cutting-edge analytical tools and techniques are also be introduced to identify and characterize the clay mineralogy, rock fabrics property and micro- to nano-scale pores both conventional and unconventional oil and gas exploration.

2. The uses of clay minerals in oil and gas exploration

2.1. Indication of tectonics and sedimentation

During the evolution of petroliferous sedimentary basin, the clay minerals contained in the rocks undergo a series of changes in composition and crystal structure in response to tectonics and sedimentation. The amount and type of clay minerals are a function of the provenance of clastic minerals and of diagenetic reactions at shallow and greater depth in different tectonic and sedimentary settings. Clay minerals can be used to infer tectonic/structural regime, basin evolution history and the timing of various geologic events. This may even provide useful tool in helping to unravel the histories in tectonically complex area, e.g., Schoonmaker et al. (1986) found that the depth distribution of illite/smectite (I/S) compositions showed an irregular, zig-zag trend with depth. This trend is probably the result of multi-stage reverse faultings resulted from the compressional tectonic movement. I/S data were also used to infer several kilometers of uplift and subsequent erosion of the section. The depositional facies appears to be an important factor controlling the abundance of clays in the sediments. Fluvial facies generally possesses higher clay mineral abundance. Well-sorted clean aeolian sands typically have a low clay abundance (<15%).

2.2. Indicator of hydrocarbon generation and expulsion

For oil and gas exploration, we need at least to confirm the exploration area has potential source that generates the oil and gas. This drives geologists to study the potential source rocks (usually organic rich shales) to understand if the organic matter in the source rock can generate hydrocarbons at a given depth in a specific geologic time and when the generated hydrocarbons reach the expulsion peak. Organic geochemistry is the main discipline for studying oil and gas generation and expulsion. However, clay mineralogy is also important for evaluation of these parameters since clay minerals and organic matters usually coexist in the sedimentary rocks and the ultrafine clay minerals are sensitive to the changes in the rocks accompanying the hydrocarbon generation and expulsion processes. Association of clay minerals and organic matter in shales is a significant factor in petroleum genesis. Grim (1947) emphasized the likelihood that the clay minerals in shales concentrated organic constituents by adsorption to form abundant source material, and subsequently acted as catalysts in petroleum generation (Brooks,1952) .

Many authors report the transformation of clay minerals during diagenesis is from montmorillonite to mixed-layer montmorillonite/illite to illite (Hower et al, 1976) and changes in the ordering of Illite/smectite (I/S) are particularly useful in studying the hydrocarbon generation because of the common coincidence between the temperatures for the conversion from random to ordered I/S and those for the onset of peak oil generation. Percentage of expandable layers in Illite/smectite decreases sharply where the Tmax (from Rock-Eval pyrolysis) of S2 hydrocarbon production peak and indicator of thermal maturation-production index (PI) [PI=S1/(S1+S2)] indicate it is oil generation zone (Burtner and Warner,1986). The use of mixed-layer illite/smectite (I/S) as a geothermometer and indicator of thermal maturity is based on the concepts of shale diagenesis that were first described in detailed studies of Gulf Coast (Powers, 1957; Hower et al., 1976; Hoffman and Hower, 1979). The good agreement between changes in ordering of I/S and calculated maximum burial temperatures or hydrocarbon maturity suggests that I/S is a reliable semi-quantitative geothermometer and an excellent measures of thermal maturity (Waples, 1980; Bruce, 1984; Pollastro, 1993). The clay mineral association even can be used to evaluate the hydrocarbon generation degree, e.g., the presence of illite-smectite-tobelite demonstrates that oil generation has taken place and absence of tobelite layers shows that the rock has not been heated sufficiently to generate large amounts of oil (Drits et al., 2002).

The significant changes of clay minerals during burial and their relations with diagenetic stages, temperature, organic matter maturity, hydrocarbon generation and expulsion can be summarized in Figure 1. During early diagenesis, the maturity of source rock indicated by vitrinite reflectance (Ro) is low and low percentate of illitic beds in illite-smectite mixed-layer clay minerals, e.g. Ro= 0.5% approximately corresponds to around 25% illite presence. The Clay minerals mainly experience loss of pore water and little oil is generated during this period. 25 to 50% illitic beds in illite-smectite mixed-layer clay minerals correspond to major oil generating zone (Ro= 0.5 to 1.0%). When more than 75% illitic layers are present in illite-smectite mixed-layer clay minerals, cracking of hydrocarbons form dry gas (Ro> 1.5%). This general trend can be used to predict if the source rock is able to generate hydrocarbon in an area. For example, the smectite alters to illite at temperature of 80 to 120°C, which corresponds to the oil generation peak at the same temperature range. Figure 2 presents data from Liaodong Bay area in Bohai Bay Basin in Northeast China to this aspect showing the change in maturity of organic matter and reaction progress in the smectite to illite transformation, which indicates that the rapid increase in illite and decrease in smectite (montmorillonite) in I/S correspond to rapid oil generation.

The reaction of smectite to illite in these clays also can indicate the producing high pore-fluid pressures (Powers, 1967) and expulse hydrocarbons from the shales (Burst,1959; Bruce, 1984). This can be demonstrated in Figure 2 that the overpressure development interval corresponds to the transformation of smectite to illite and hydrocarbon generation zone (Figure 2C, D, E).

Burial depth /km	Hydrocarbon product	Temper-ature (°C)	Diagenetic stage	Vitrinite reflectivity (Maturity)	Ordering of Illite/Smectite (I/S)	Percent illite in I/S mixed layer
1	Biogenic Gas	30	Diagenesis			
2		60		— 0.5	Onsite	— 25%
3		90	Katagenesis		R=0 Illitization increase	
4	Oil	120		— 1.2	R=1	— 50%
5	Thermogenic Gas	150		— 2.0		— 75%
			Metagenesis		R=3	

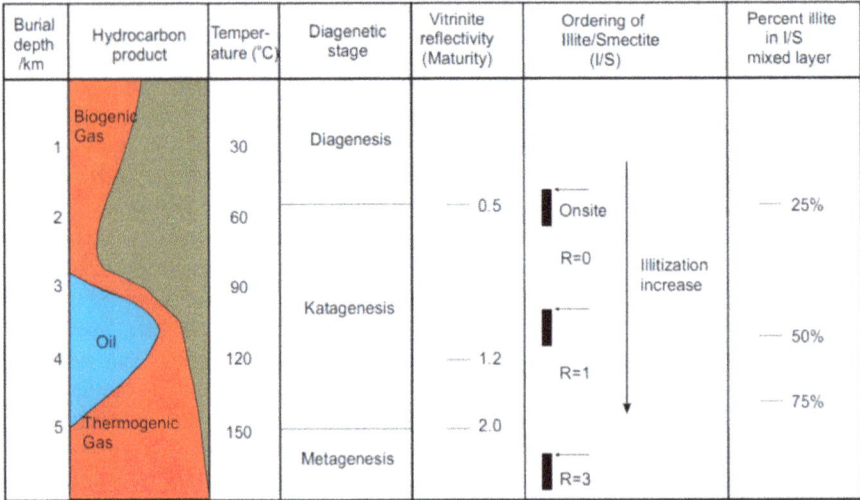

Figure 1. Generalized relationship between temperature, hydrocarbon generation, diagenesis, source rock maturity (vitrinite reflectance), changes in mixed-layer illite/smectite. Figure and data summarized from Foscolos et al (1976), Hoffman and Hower (1979), Waples (1980), Tissot and Welte (1984).

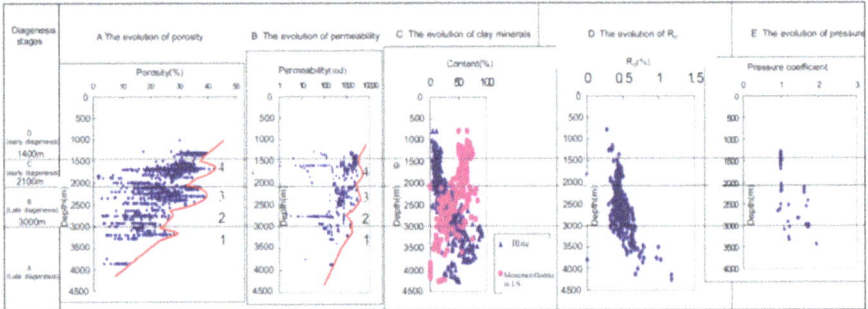

Figure 2. Five plots showing the relationships between diagenetic stages, porosity (A), permeability (B), clay minerals evolution (C), vitrinite reflectance (Ro) (D), and pressure (E) in Liaozhong depression, Liaodong bay sub-basin, Bohai Bay Basin, Northeast China. The secondary porosity zones are numbered upward from 1 to 4.

2.3. Indices for hydrocarbon migration and accumulation

It is critical to establish that hydrocarbon formation and migration occurred after the formation of the trap (anticline, etc.) that is to hold the oil. There is still very little known about the manner in which hydrocarbons formed in argillaceous source rocks migrate and accumulate in porous reservoirs. Some evidence exists, however, that the clay mineral-kerogen complex plays a role in modifying hydrocarbon compositions during migration.

Some time ago, Legate & Johns (1964) used gas chromatography to measure the affinity of montmorillonite clays for hydrocarbons of differing polarity, and suggested that during the migration of petroleum chromatographic effects might modify their composition. Young & Melver (1977) developed the chromatographic technique further and convincingly showed that they could in numerous instances predict oil compositions following migration, where the clay-kerogen complex was the chromatographic agent. "Organo-pores" proposed by Yariv (1976) could be migration paths for hydrocarbons in argillaceous source rocks.

A number of investigators (Powers, 1967; Burst, 1969) have focused attention on the late-stage dehydration which accompanies smectite to illite transformation during burial diagenesis. This firstly suggests that the replacement of kaolinite by illite or direct precipitation of illite indicates fluid flow where the chemical potential of the fluids is in disequilibrium within the reservoir sandstone. The existence of secondary illite does indicate aqueous fluid flow and thus can be used as indices of fluid movement and hence signal the possible hydrocarbon migration. Secondly, it indicates that the water release could create a flushing action responsible for the migration of petroleum hydrocarbons from the source rock through the migration paths to nearby reservoirs. Also, the water liberation can build up abnormal pressures in less permeable sediments, which can provide migration dynamic for hydrocarbons (Figure 2 C, E).

Abnormal Illite distribution has been used as an index to determine if certain rocks/strata/areas are a hydrocarbon migration pathway and its conducting capability (Zeng and Yu, 2006, Jiang et al., 2011). If there shows abnormal illite distribution, it indicates the hydrocarbon migration happened. The illite abnormal distribution of three wells from three different structure zones in Liaodong Bay Sub-basin of Bohai Bai Basin in Northeast China in Figure 3 suggests hydrocarbon migration happened in these three areas represented by three wells, but the conduiting capabilities are different in the three areas based on different abnormal magnitude of illite content. At the same depth of these three wells, illite content of well JZ25-1s-1 is the highest and the illite content of well JZ21-1-1 is the lowest, which indicates the hydrocarbon migration in the JZ25-1s-1 well area is the most active and the JZ21-1-1 area is the relatively least active area regarding to hydrocarbon migration (Figure 3). This result is consistent with current oil discoveries: The Liaoxi uplift (represented by well JZ25-1s-1) to the west of Liaodong Bay contributes to the most reserves in the Liaodong Bay sub-basin. Tan-Lu strike-slip area (represented by well JZ23-1-1) is emerging as the second largest hydrocarbon migration and accumulation area (Jiang et al., 2010, 2011). Almost no oil and gas discoveries in the rest area of Liaozhong depression (represented by well JZ21-1-1) away from strike-slip zone and Liaoxi uplift so far due to poor hydrocarbon migration pathway and poor conduiting capability.

The smearing of clay minerals can also prohibit the hydrocarbons' further migration and facilitate the hydrocarbon accumulation. When the soft clays are smeared into the fault plane during movement and they will provide an effective seal . In many cases, the presence of clay types and their proportions can even indicate if there is oil and gas accumulation.

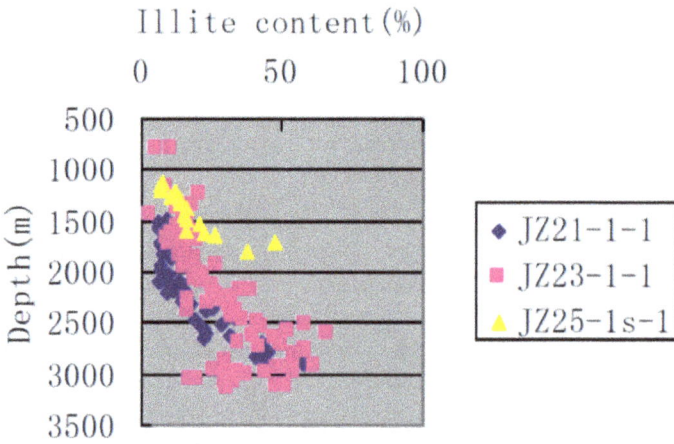

Figure 3. The illite content distribution versus depth from three wells in Liaodong Bay area, Bohai Bay Basin.

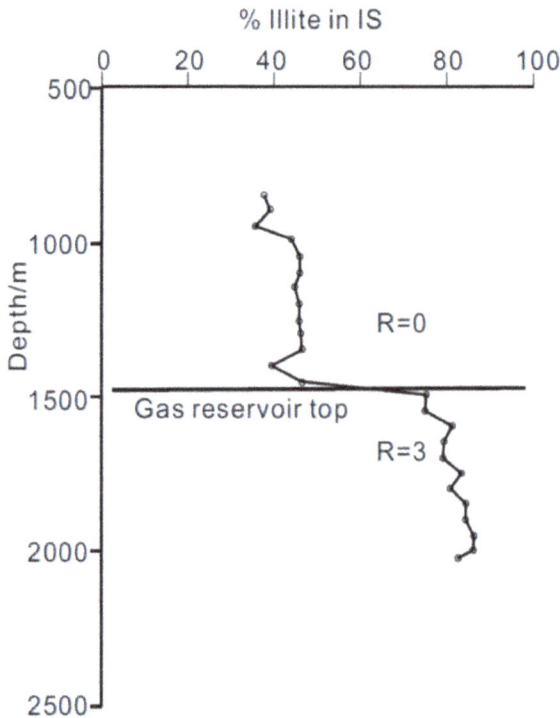

Figure 4. The abrupt changes in the percent illite in I/S and the ordering of I/S (R) in well Dongfang 6 in Dongfang Gas Field, Northern South China Sea.

Webb (1974) recorded that the Cretaceous sandstones of Wyoming generally contain abundant authigenic kaolinite where water saturated, but little if any authigenic clay is found where the sandstones are hydrocarbon saturated. Investigation on clay mineral and its relationship with gas reservoirs show that the clay mineral percentage and ordering of I/S can indicate the hydrocarbon reservoir, e.g., the high content of illite in I/S and higher ordering of I/S change indicate the gas reservoir interval in Dongfang 6 well from Dongfang gas field in Northern South China Sea (Figure 4).

2.4. Significance of clay mineralogy for reservoir Quality prediction

Porosity and permeability are the most important attributes of reservoir quality. They determine the amount of oil and gas a rock can contain and the rate at which that oil and gas can be produced. Most sandstones and carbonates contain appreciable fine-grained clay material including kaolinite, chlorite, smectite, mixed layer illite-smectite and illite. These clay minerals commonly occur as both detrital matrix and authigenic cement in reservoir sandstones. The reservoirs initially have intergranular pores that are main space for oil and gas accumulation. When the reservoirs are deposited, their primary porosity is frequently destroyed or substantially reduced during burial compaction. The clay minerals are usually assumed to be detrimental to sandstone reservoir quality because they can plug pore throats as they locate on grain surface in the form of films, plates and bridge and some clay minerals promote chemical compaction. Not only in sandstone reservoir, the clay content also greatly accelerated the rate of porosity loss in limestone reservoir (Brown, 1997). Generally, the porosity loss is mainly caused by the diagenetic process including mechanical compaction, quartz and K-feldspar overgrowths, carbonate cementing and clay mineralization. Especially, the diagenetic clay minerals play a very important role in determining the reservoir quality.

Authigenic clays from diagenesis in the sandstones studied occur as illite, illite-smectite and kaolinite. They form cements around the detrital minerals. During the period of intermediate to deep burial diagenesis, Ilite and illite-smectite clays are the first cements. These early-formed clay films play an important role in reducing reservoir porosity and permeability during burial diagenesis. For example, pore-filling illite formed mainly at the expense of kaolinite.The illitic clays usually occur as pore-bridging clays to reduce the pore space and block the fluid movement by reducing permeability. For clay minerals that replaced rigid feldspar minerals are easily compacted and can be squeezed into pore throats between grains. This will also greatly influence the decrease of reservoir quality.

For oil and gas exploration, we expect the occurrences of high-quality reservoirs. Even though the porosity and permeability of reservoir generally decrease with the increase of burial depth due the diagenetic processes as state above, other diagenetic processes may enhance porosity through the forming of secondary porosity including fractures, removal of cements or leaching of framework grains, preexisting cements and clay minerals, limited compaction and/or limited cementation. The dissolution of authigenic minerals that previously replaced sedimentary constituents or authigenic cements may be responsible for a significant percentage of secondary porosity. Some micropores are found in various clays regardless of whether the clay is authigenic or detrital in origin. Also, the existence of clay minerals does not always mean to reduce the reservoir quality, it may be good phenomenon to indicate good reservoir quality, e.g., coats of chlorite on sand grains can preserve reservoir quality because they prevent quartz cementation (Heald and Larese, 1974; Bloch et al., 2002; Taylor et al., 2004). Sometimes, the higher content zone of kaolinite is indicative of higher porosity. The reason is that porosity is created when the acid dissolves feldspar to produce kaolinite (Jiang et al., 2010). These all show the positive aspect to clay authigenesis.

The secondary porosity development and its relationship with clay minerals evolution has been investigated in many basins (Bloch et al., 2002; Taylor et al., 2004; Jiang et al., 2010). Let's use Liaodong Bay Sub-basin in Bohai Bay Basin in Northeast China as example again. There clearly exist four secondary porosity development zones for the Tertiary strata, whose depth intervals are 1600-1800 m, 2000-2500 m, 2700-2800 m and 3200-3300 m, respectively (Figure 2A). These intervals are named 1 upward to 4 informally. Their corresponding permeability zones have relative higher values (Fig.2B). The secondary and third secondary porosity zones have relatively larger scale. Correlation between porosity, clay minerals and Ro demonstrates that the secondary porosity zones are related to the rapid transformation of the clay minerals and hydrocarbon generation (Ro>0.5%) (Figure 2A, C, D). The relation between zones of secondary porosity and pressure distribution illustrated that No.3 secondary porosity is just right below the top surface of overpressure. This is probably because that the overpressure can retard compaction and avoid the excessive porosity reduction.

2.5. Petroleum emplacement chronology

Petroleum emplacement chronology is one of the frontier research subjects in both petroleum geology and isotope geochronology. Determining the oil or gas emplacement ages has important implications for oil or gas genesis and resource prediction. Typical relative chronology for oil or gas migration, emplacement, and accumulation is established by petrology, basin tectonic evolution, trap formation, and hydrocarbon generation from the source rock (Kelly et al., 2000; Middleton et al., 2000). So far, the illite K-Ar and $^{40}Ar/^{39}Ar$ dating technique hold significant promise in establishing absolute constraints on the emplacement age of oil and gas.

Since the middle of the 1980s, authigenic illite K-Ar dating has been applied to determine the ages of petroleum migration in the North Sea oil fields and Permian gas reservoirs in Northern Germany (Lee et al., 1985; Liewig et al., 1987, 2000; Hamilton et al., 1989). The dating is based on the hypothesis that "illite is commonly the last or one of the latest mineral cements to form prior to hydrocarbon accumulation. Because the displacement of formation water by hydrocarbons will cause silicate diagenesis to cease, K-Ar ages for illite will constrain the timing of this event and also constrain the maximum age of formation of the trap structure" (Hamilton et al., 1989). Wang et al. (1997) investigated oil or gas emplacement ages in the Tarim Basin by this technique.

Recently, illite $^{40}Ar/^{39}Ar$ dating was considered better than traditional K-Ar dating. Among the advantages of $^{40}Ar/^{39}Ar$ dating over traditional K-Ar methods are that stepwise heating can distinguish contributions from authigenic illite and detrital K feldspar by interpreting their gas release characteristics. The K-Ar dating and total fusion $^{40}Ar/^{39}Ar$ dating, however, yield a meaningless mixing age of the authigenic illite and detrital K feldspar, e.g., gradually rising age spectra are obtained by $^{40}Ar/^{39}Ar$ laser stepwise heating of the illite samples from the Tertiary reservoir sandstones in the Huizhou sag, Pearl River Mouth Basin in South China Sea. The youngest ages at the first steps are interpreted as being caused by contributions from authigenic illite, suggesting that the petroleum emplacement occurred after 11 Ma. The high plateau ages in the high-temperature steps that are rather variable between the seven samples are interpreted as being caused by contributions of detrital K-feldspar in the sandstones (Yun et al., 2010).

2.6. Significance for petrophysical property study

The reservoir petrophysics e.g. porosity, permeability, water saturation and hydrocarbon saturation are the most important properties that define and control qualitatively and quantitatively the reservoir performance. The minerals present in the reservoir especially the clay mineral (Moll, 2001) can play the utmost role, which affects both the reservoir capacity and production because the grain size of clay minerals is generally very small and result in very low effective porosity and permeability, thus any presence of clay in a reservoir may have direct consequences on the reservoir properties (Said et al. 2003).

Characteristics of clays that strongly affect their electrical behavior are clay composition, internal structure, the tremendous surface to volume ratios of most clays and the charge imbalance along the surface of clay minerals. All these clay mineral manifestations have an impact upon the interpretative petrophysical parameters by well logging responses. In order to better understand the well logging response for petrophysical analysis, the type of clay minerals must be taken into account in reservoir evaluation, e.g., Potassium presence in the reservoir can increase radiation on Gamma Ray logs. Sometimes, the log response can indicate the hydrocarbon saturation and clay content, e.g., the high resistivity zone of resistivity log corresponds to intervals with low water saturation, a more restricted distribution of diagenetic clay (mainly chlorite) and the low resistivity zone corresponds to intervals with more widely distributed diagenetic clay and variably reduced permeability (Nadeau,2000).

3. Traditional clay mineral characterization methods and their applications for conventional oil and gas exploration

The traditional methods e.g. XRD (X-ray Diffraction), petrographic microscope, XRF (X-ray Fluorescence) and SEM (Scanning Electron Microscopy) have been widely used for clay minerals characterization in conventional siliciclastic and carbonate reservoirs for many years. X-ray diffraction (XRD) is used to provide information on the rock mineral composition and type of clay minerals and their content. Petrographic microscope can identify the reservoir mineralogy composition, pore types, authigenic clays and cements. Figure 5A illustrates that the Tertiary lacustrine turbidite reservoir in Jiyang Sub-basin of Bohai Bay Basin is mainly composed of quart (Q), K-feldspar (Fs) and calcite based on thin section observation. The secondary porosities include the intergranular and intragranular pores caused by dissolution of kaolinite, feldspar and carbonate cement. Scanning electron microscopy (SEM) provides wide range of information about the morphology, mineral composition, distribution and paragenesis of the neoformed authigenic clay minerals, mechanically infiltrated clays, transformational clays and pedogenic mud aggregates. Since SEM has a very large depth of field, and can thus yield a three-dimensional image useful for understanding the structure of a sample. It will help understand the clay mineralogy and their effect on the porosity, permeability and other reservoir characteristics. Figure 5B shows secondary pores resulted from dissolution of authigenic kaolinite exist in the similar turbidite sandstone reservoir as that in Figure 5A in Jiyang sub-basin based on SEM observation.

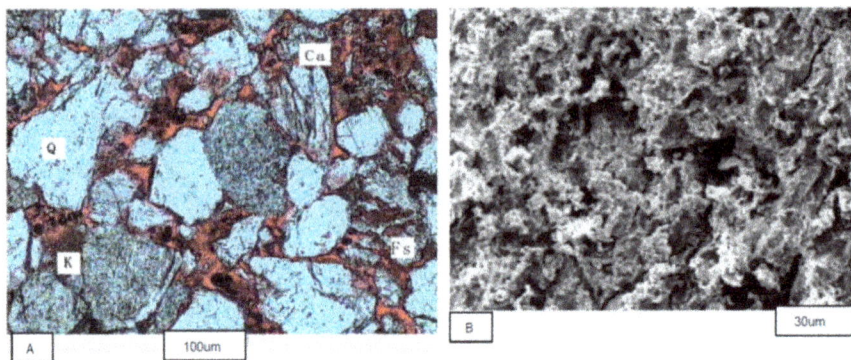

Figure 5. Photomicrography A illustrates secondary pores resulted from dissolution in feldspar (Fs) and kaolinite (K) and some intergranular secondary pores by carbonate cement dissolution based on thin section observation. The sample is from Tertiary lacustrine turbidite sandstone at the depth of 3012m of Well Niu-110; Photomicrograph B is a Scanning Electron Microscope (SEM) image showing secondary pores from dissolution of kaolinite the sample is from Tertiary turbidite sandstone at the depth of 2985.3m from well Niu-35. Both the two wells are both located in in Jiyang Sub-Basin, Bohai Bay Basin, Northeast China.

4. New techniques and their applications for unconventional oil and gas exploration

Over the past decade, interest in shale gas and shale oil reservoirs increased due to commercial success of gas-shale plays in North America. In contrast to conventional oil and gas reservoirs (sandstone and carbonate), these new identified reservoirs typically have very fine-grained rock texture (dominant grain size ≤62.5 μm), low porosity (≤10%) and very low permeabilities (in nanodarcy range). These rocks used to be considered as only source rocks with high organic content (≥ 2% weight fraction Total Organic Carbon, TOC), but now they have been found as reservoir rocks through horizontal drilling and hydraulic fracturing. Gas and oil reserves in the tight shales are huge, US Energy Information Administration (EIA) released a major report in 2011 that there exists potential 6,622 trillion cubic feet (Tcf) of gas contained in shales around the world. Despite the commercial importance of shale formations, their physical properties especially porosity, pore-size distributions and clay mineral fabrics are still poorly understood. Porosity measurements in shales are complicated because of the very fine-grained texture, small pore sizes, extremely low permeability, and the strong interaction of water with clay minerals, which are often important component in these rock types. Shales exhibit dual-porosity structure and have a more complex pore-structure than the sandstones and limestones. One of the biggest challenges in estimating oil and gas transport and storage properties of shales has been a lack of understanding of clay type, clay content, free-gas content, porosity and their relationships. Brittleness of shale has an impact on proppant embedment and maintaining hydraulic-fracture connectivity to the wellbore. High clay-rich shales usually have low Young's Modulus and, by extension, low brittleness index and difficult to frac. So clay minerals play key role in shale gas and shale oil exploration.

Since shales are really fine and tight, estimating reservoir quality in gas shale requires a thorough understanding of pore structure and pore connectivity. MicroCT is a proven technique to resolve pore parameters with a resolution in the order of 1 micrometer. NanoCT technology has resolution down to 200nm but even that may not be enough for gas shale. Gas shales are known to contain finely-dispersed porous organic matter within an inorganic matrix. The porosity within the organic phase has pore and pore throat dimensions typically below 100 nanometers and even down to just a few nanometers, so new techniques are required to characterize the clay minerals in shales. The recent new clay mineral characterization methods include but not limited to FTIR (Fourier transform infrared spectroscopy), QEMSCAN (Automated Mineralogy and Petrography), FIB (Focused Ion Beam), EDS (Energy-dispersive X-ray spectroscopy), etc.

The advancements of special analytical techniques have made significant progress in clay mineral imaging, mineral identification and quantifying by using FEI company's QEMSCAN® , The clay mineral identification and quantification based on the QEMSCAN® EDS spectral analysis method is a reliable alternative to conventional methods. Furthermore, the automated SEM-EDS solution approach provides additional textural information, and the resulting mineral maps can be used to differentiate, in the case of conventional sandstone and unconventional tight shale reservoir rocks, pore linings, from granular mineral alteration products, from intergranular cements, and sedimentary laminations. Figure 6 illustrates QEMSCAN® mineral and texture maps of representative shale samples from China. The Silurian shale (Figure 6A) shows no bedding and has relatively low content of quartz (14.5%) and Cambrian shale (Figure 6B) shows bedding and high content of quartz (62%), dominant clay minerals of two samples are illite and no smectite detected, which indicates that among the two samples the Cambrian shale (Figure 6B) is more easier to frac to produce gas since it has high brittle quartz content and no expandable smectite.

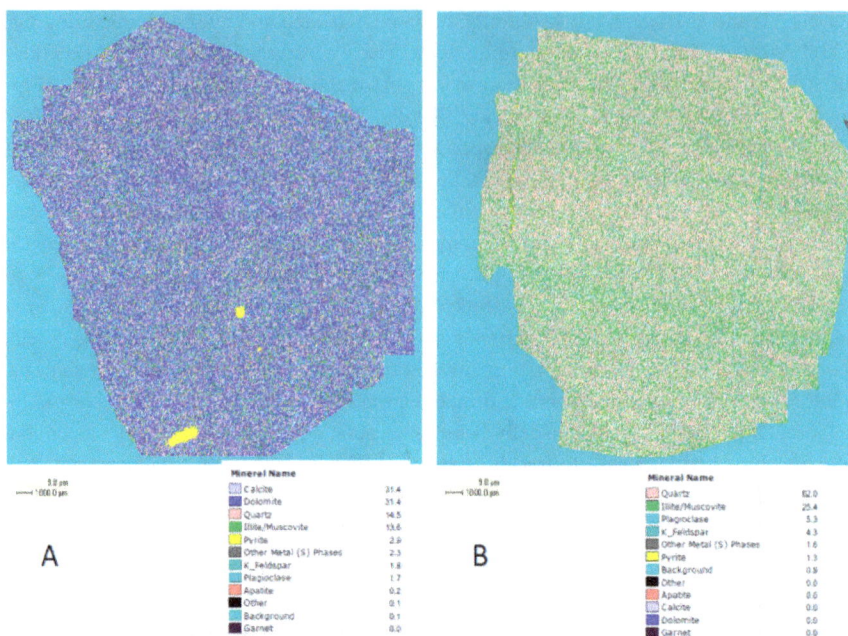

Figure 6. Slices from QEMSCAN showing the rock fabrics and quantitative mineral composition. A is from Silurian shale in Chongqing, China and B is from Cambrian shale in Guizhou, China.

The SEM/EDS has also recently been widely used to study shale reservoirs since SEM has a high resolution and EDS's chemical element analysis can help identify mineral composition precisely through the combination of fabric and chemical composition analysis. Figure 7 illustrate this method to identify that a Silurian shale sample from Sichuan, China is mainly composed of quartze, albite, dolomite, illite, chlorite and kerogen by using the SEM/EDS method.

For the intuitive visualization of the nano-pore network and rock fabric architecture, FIB/SEM is the only technology so far with nanometer resolution in 3 dimensions to reveal the reservoir architecture of broad ion polished shale. The high resolution of a Scanning Electron Microscope (SEM) combined with the precise cutting capability of a Focused Ion Beam (FIB) enables rendering of 3D reconstructions with resolution of a few nanometers (Lemmens et al., 2011). The FIB is capable of removing a controlled amount of material to create a subsequent 2D section parallel and aligned with the previous one, with inter-section spacing of the order of 10nm, and having resolution of a few nanometers in the section plane. Figure 8 is an example of FIB/SEM slice of a Silurian shale reservoir in Sichuan Basin in China, which renders the intra-organic (kerogen) nano-scale pores and illite presence in the kerogen. These nano-scale pores can store huge amount of gas in the basin. In this way, after careful combination of the subsequent slices, a 3D model with nanometers resolution can be obtained. Figure 9 shows an example of 3D reconstruction for a US Paleozoic shale reservoir (Zhang and Klimentidis, 2011).

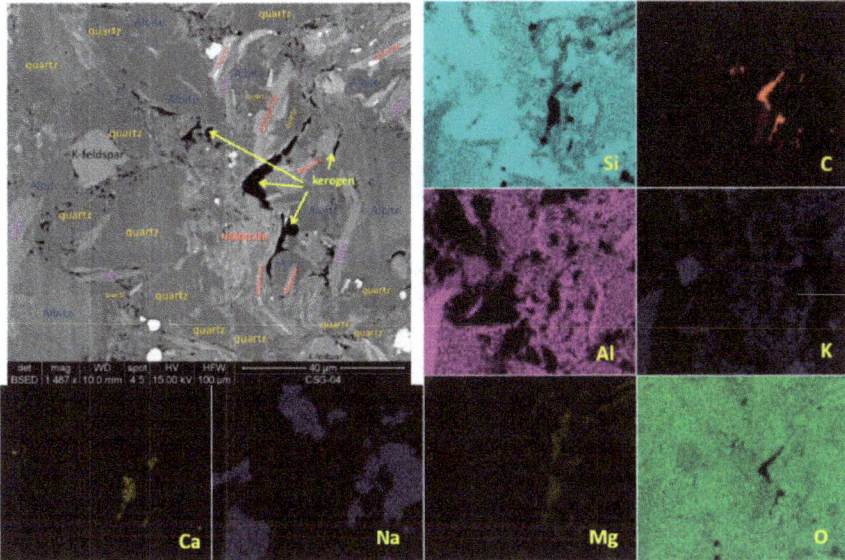

Figure 7. Mineral composition identification based on Scanning Electron Microscope (SEM) and Energy-dispersive X-ray spectroscopy (EDS). The sample is from Silurian shale in Pengshui, Sichuan, China. The upper left and the rest of the slices have the exact same view area under microscope even though their scales are different.

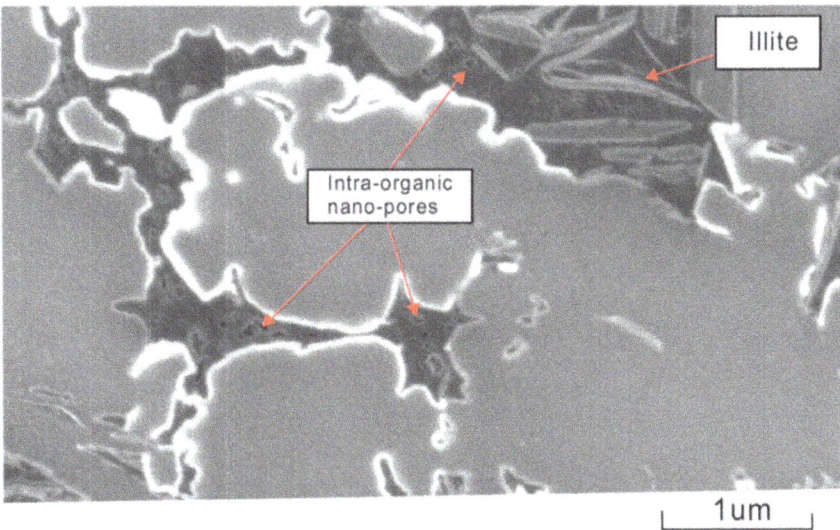

Figure 8. The FIB/SEM slice of Silurian shale reservoir from East Sichuan Basin, China. The sample depth is 2164.8m below surface.

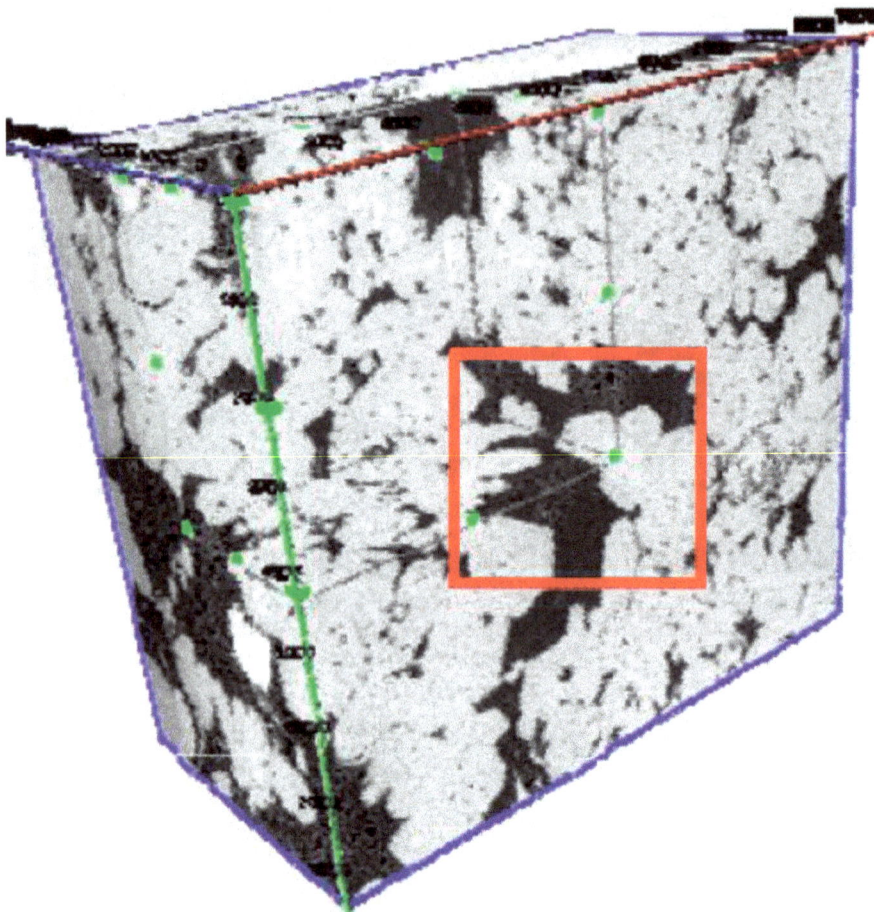

Figure 9. 3D reconstruction of a US Paleozoic shale core sample from a series of FIB/SEM imaging slices (from S. Zhang and R.E. Klimentidis, 2011). The red rectangle area with about 200nm in width shows the nano-pores developed in the kerogen.

5. Summary

The clay minerals are important compositions in source rocks and reservoir rocks that can generate and store oil and gas respectively. The presence of clay minerals strongly influences the physical and chemical properties of conventional sandstone, carbonate and unconventional shale.

Regionally, the clay minerals can be used to interpret and understand such perspectives as the basin evolution on tectonics, sedimentation, burial and thermal history, to infer the sedimentary environment and to correlate strata, etc.

For clay minerals in source rocks, they are important for quality evaluation of the hydrocarbon generation, expulsion and migration. Clay minerals help concentrate organic matter by adsorption and subsequently act as catalyst to generate petroleum. The transformation of montmorillonite to illite and increasing ordering of I/S can indicate the hydrocarbon generation and expulsion events.

For clay minerals in reservoir rocks, their presence has an important impact upon reservoir properties such as porosity and permeability and upon those measured physical data that are used to evaluate reservoir quality. Geologists use clay minerals information to decipher the burial diagenetic process and reveal the pore type and pore evolution. Even though they are usually considered to be detrimental to reservoir quality because they can plug pore throats and can be easily compacted, other diagenetic processes may enhance porosity through the forming of secondary porosity through providing porosity by clay dissolution, creating micropores in clays and coating of chlorite on grains to prevent quartz cementation.

The recent emerging shale oil and gas exploration requires state-of-art imaging and characterization techniques to study the application of clay minerals in the exploration of this unconventional resource. The modern innovative QEMSCAN® and FIB/SEM/EDS have been playing key roles in the identification and quantitative characterization of clay minerals, which help define the best brittle reservoir interval and avoid exploration failure by choosing the compatible drilling and hydraulic fluids.

Author details

Shu Jiang

Energy & Geoscience Institute, University of Utah, Salt Lake City, USA

6. References

Bloch, S., R. H. Lander, and L. M. Bonnell, 2002, Anomalously high porosity and permeability in deeply buried sandstone reservoirs: Origin and predictability: AAPG
Bulletin, v. 86, p. 301–328.
Brooks, B.T., 1952, Evidence of Catalytic Action in Petroleum Formation Ind. Eng. Chem., 44 (11), pp 2570–2577
Brown, A., 1997, Porosity variation in carbonates as a function of depth: Mississippian Madison Group, Williston Basin, in J.A. Kupecz, J. Gluyas, and S. Bloch, eds., Reservoir quality prediction in sandstones and carbonates: AAPG Memoir 69, p. 29--46.
Bruce, C. H., 1984, Smectite dehydration--Its relation to structural development and hydrocarbon accumulation in northern Gulf of Mexico: Amer. Assoc. Petrol. Geol, Bull. 68, 673-683.
Burst, Jr.,J.F.,1959, Post diagenetic clay mineral environmental relationships in the Gulf Coast Eocene. Clays & Clay Minerals, 6, 327-341.

Burtner and Warner,1986, relationship between illite/smectite diagenesis and hydrocarbon generation in lower cretaceous Mowry and skull creek shales of the Northern rocky mountain area, Clays and Clay Minerals. Vol. 34, No. 4, 390-402,

Drits, V.A., Lindgreen, H., Sakharov, B.A., et al., 2002, Tobelitization of smectite during oil generation in oil-source shales. Application to north sea illite-tobelite-smectite-vermiculite. Clays and Clay Minerals, v. 50, no.1, p. 82-98

Foscolos, A. E., Powell, T. G., & Gunther, P. R., 1976, The use of clay minerals and inorganic and organic geochemical indicators for evaluating the degree of diagenesis and oil generating potential of shales. Geochirn. Cosrnochirn. Acta, 40, 953-966.

Griffin, G.M., 1971, Interpretation of X-ray diffraction data, in Carver, R.E., ed., Procedures in Sedimentary Petrology: New York, Wiley, p. 541–569.

Grim, R. E., 1947, Relation of clay mineralogy to origin and recovery of petroleum: Am. Assoc. Petroleum Geologists Bull., v. 31, no. 8, p. 1491-1499.

Hamilton, P. J., S. Kelley, and A. E. Fallick, 1989, K-Ar dating of illite in hydrocarbon reservoirs. Clay Minerals, v. 24, p. 215−231

Heald, M. T., and R. E., Larese, 1974, Influence of coatings on quartz cementation: Journal of Sedimentary Petrology, v. 44, p. 1269–1274.

Hoffman, J. and Hower, J.,1979, Clay mineral assemblages as low grade metamorphic geothermometers: Application to the thrust faulted disturbed belt of Montana: in Aspects of Diagenesis, P. A. Scholle and P. S. Schluger, eds., SEPM Spec. Publ. 26, 55-79.

Hower, J., Eslinger, E. V., Hower, M. E., and Perry, E. A., 1976, Mechanism ofburial metamorphism ofargillaceous sediment: Mineralogical and chemical evidence: GeoL Soc. Amer. Bull. 87, 725-737.

Jiang, S., Wang, H., Cai, D.,et al., 2010, The secondary porosity and permeability characteristics of Tertiary strata and their origins, Liaodong Bay Basin, China. Energy, Exploration & Exploitation. 28(4): 207-222

Jiang, S., Wang, H., Cai, D., et al., 2011, Characteristics of the Tan-Lu Strike-Slip Fault and Its Controls on Hydrocarbon. Accumulation in the Liaodong Bay Sub-Basin, Bohai Bay Basin, China. Advances in Petroleum Exploration and Development. Vol. 2, No. 2, 2011, pp. 1-11

Kelly, J., J. Parnell, and H. H. Chen, 2000, Application of fluid inclusions to studies of fractured sandstone reservoirs. Journal of Geochemical Exploration, v. 69, p. 705−709.

Lee, M., J. L. Aronson, and S. M. Savin, 1985, K/Ar dating of time of gas emplacement in Rotliegendes sandstone, Netherlands. AAPG Bulletin, v. 69, p. 1381−1385.

Legate, C. E. and Johns, W. D.,1964, Gas-chromatographic examination of several systems from clay minerals and organic materials: Beitr. Mineral Petroyraphie. 10, 60-69.

Lemmens, H.J., Butcher, A.R. and Botha, P.W.S.K., 2011, FIB/SEM and SEM/EDX: a new dawn for the SEM in the core lab? Petrophysics, 62 (6):452-456

Liewig, N., N. Clauer, and F. Sommer, 1987, Rb-Sr and K-Ar dating of clay diagenesis in Jurassic sandstone oil reservoir, North Sea. AAPG Bulletin, v. 71, p. 1467−1474.

Leiwig, N. and Clauer,N. 2000, K-Ar dating of varied microtextural illite in Permian gas reservoirs, northern Germany. Clay Minerals, 35, 271-281

Middleton, D., J. Parnell, P. Carey, and G. Xu, 2000, Reconstruction of fluid migration history in northwest Ireland using fluid inclusion studies. Journal of Geochemical Exploration, v. 69, p. 673–677

Moll, W.F., Jr., 2001, Baseline studies of the clay minerals society source clays: Geological origin. Clays and Clay Minerals, V. 49, pp. 374–380.

Nadeau,P.H., 2000, The Sleipner Effect: a subtle relationship between the distribution of diagenetic clay,reservoir porosity, permeability, and water saturation. Clay Minerals, 35, 185-200

Pettijohn, F.J., 1975, Sedimentary Rocks, 3rd Edition: New York, Harper & Row, 628 p.

Pollastro, R.M., 1993, Considerations and applications of the illite/smectite geothermometer in hydrocarbon-bearing rocks of Miocene to Mississippian age. Clay and Clay Minerals, Vol.41, No.2, 119-133.

Powers, M.C.,1957, Adjustment of clays to chemical change and the concept of the equivalence level: Clays & Clay Minerals, Proceedings of the Sixth National Conference, 309-326.

Powers, M. C. 1967, Fluid-release mechanisms in compacting marine mudrocks and their importance in oil exploration. Am. Assoc. Petrol. Geol. Bull. 51: 1240, 54

Said, A., Abdallah, M. and Alaa E., 2003, Application of well logs to assess the effect of clay minerals on the petrophysical parameters of Lower Cretaceous Reservoirs, North Sinai, Egypt. EGS Journal, V. 1, No. 1, pp. 117-127.

Schoonmaker, J., Machenzie, F.T. and Speed, R.C., 1986, Tectinic implications of illite smectite diagenesis, Barbados Accretionary Prism. Clays and Clay Minerals. 34 (4):465-472

Taylor, T. R., R. Stancliffe, C. I. Macaulay, and L. A. Hathon, 2004, High temperature quartz cementation and the timing of hydrocarbon accumulation in the Jurassic Norphlet Sandstone, offshore Gulf of Mexico, U.S.A., in J. M. Cubit, W. A. England, and S. Larter, eds., Understanding petroleum reservoirs: Towards an integrated reservoir engineering and geochemical approach: Geological Society (London) Special Publication 237, p. 257–278.

Tissot, B.P. and Welte, D.H., 1984, Petroleum; Natural gas; Geochemical prospecting; Geology; Prospecting, Springer-Verlag (Berlin and New York) , 699P

Waples, D. W., 1980, Time and temperature in petroleum formation: Application of Lopatin's method to petroleum exploration. Amer. Assoc. Petrol. Geol. Bull. 64, 916-926.

Wang, F. Y., P. He, S. C. Zhang, M. J. Zhao, and J. J. Lei, 1997, The K-Ar isotopic dating of authigenic illites and timing of hydrocarbon fluid emplacement in sandstone reservoir: Geological Review, v. 43, p. 540–546.

Weaver, C.E., 1960, Possible uses of clay minerals in search for oil. Bull. Am. Assoc. Petrol. Geologists 44, 1505-1518.

Webb,J.E.,1974, Relation of Oil Migration to Secondary Clay Cementation, Cretaceous Sandstone, Wyoming AAPG Bulletin, v. 58, p. 2245-2249

Yariv, S., 1976, Organophilic pores as proposed primary migration media for hydrocarbons in argillaceous rocks. Clay Sci. 5: 19-29

Young, A., McIver, R. D., 1977, Distribution of hydrocarbons between oils and associated fine-grained sedimentary rocksPhysical chemistry applied to petroleum geology. Am. Assoc. Petrol. Geol. Bull. 61: 14 07-36

Yun J B, Shi H S, Zhu J Z, et al., 2010, Dating petroleum emplacement by illite 40Ar/39Ar laser stepwise heating, AAPG Bull, 94: 759-771

Zeng, J. and Yu, C.,2006, Hydrocarbon migration along the Shengbei fault zone in the Bohai Bay basin, China: The evidence from geochemistry and fluid inclusions. Journal of Geochemical Exploration, 89(1-3), 455-459.

Zhang, S. and Klimentidis, R.E., 2011, porosity and permeability analysis on Nanoscale fib-sem 3d imaging of shale rock, Society of Core Analysis Meeting, SCA2011-30, 1-12

Documentation, Application and Utilisation of Clay Minerals in Kaduna State (Nigeria)

Oluwafemi Samuel Adelabu

Additional information is available at the end of the chapter

1. Introduction

The significance of solid mineral resources has been of profound value to man since time immemorial. Clay minerals appear not to be the most valuable among the minerals of the earth surface, yet they affect life on earth in far reaching ways. Nigeria in sub-Saharan Africa (surface area: 923,768 km²) is a country with considerable wealth in natural resources, with a record of over 30 minerals of proven reserves [1]. As far back as 1903 and 1904, geological survey in Nigeria evolved when the Mineral Surveys of the Southern and Northern Protectorates of Nigeria were established under the British colony. The Mineral Surveys carried out broad reconnaissance of mineral resources of the two Protectorates with the prospect of using the raw materials for industries in Britain. In course of these activities, such deposits as Tinstone, Columbite Limestone, Bitumen, Lead-zinc Ores, Coal, Clays, Iron Ore, Gold, and Marble etc were discovered in various parts of the country [2]. After the colonial era, government parastatals have been set up such as the Nigeria's Ministry of Solid Minerals Development, Raw Material Development Research Council (RMDRC) and the Federal Institute of Industrial Research Oshodi (FIIRO) which all tried to establish a comprehensive data list of basic mineral resources as they occur at various geological locations in appreciable millions of tonnage that supports experimental and industrial uses [3, 4]. In recent research purview, various studies on solid mineral resources using geo-scientific surveys and mineralogical charaterisation considered that the understanding of the nation's mineral potentials is critical for efficient exploration and exploitation towards promoting sustainable economic development as shown in [1, 5-8]. Results have shown that Nigeria's geosphere is enriched with a wide range of both metallic and non-metallic minerals deposited across the states of the nation which are and could still be beneficiated to provide the raw materials for industrial manufacturing among other productive purposes. Noteworthy, clay minerals constitute over 50% of the non-metallic, earthy and naturally-occurring resources abounding throughout Nigeria's sedimentary basins and on the basement [9]. In [5], it was observed that extensive investigation has been carried out on the liquid mineral endowment of the country, while little has been done to solid mineral endowment of which clay is prominent and as a result, adoption of solid mineral on industrial scale is scanty.

Mararaba-Rido and Kachia areas of Kaduna State are among the largest reserves of clay deposits in Nigeria with over 5.3 million tons [9]. Despite the vast potentials, clay minerals are still grossly underutilized and the few pockets of existing clay-based industries have primarily harnessed the raw for the production of ceramic wares and structural products. A growing number of investigations carried on the solid industrial minerals in Nigeria have been broad based and generic with consideration for geological survey and mineral characterization [see 1,4,6,8]. Besides, documented studies on clay minerals in selected areas of Nigeria tend to focus more on the mineral characterisation and with little emphasis on the economic potentials or usage of the minerals as such in [5]. This study had considered the industrial potentialities in addition to the properties study of clay mineral using Kaduna State of Nigeria as a case study. The qualities of clay found determine its application and suitability for ceramic products such as in bricks, ceramic wares, and refractory. The findings of the study were gathered through field surveys with documentation of relevant information on clay reserves, mineral locations, and the economic significance of the minerals. This includes detailed evaluation of report findings from three clay-based industries at Mararaba-Rido, Jacaranda and Maraba areas in Kaduna State, Nigeria. The result shows a significant usage of clay mineral as a principal raw material for ceramic manufacturing such as structural, refractories, and whitewares products. Clay minerals hold high material value to industries in Kaduna utilizing them for ceramic purposes towards socio-economic and industrial development. This supports the main policy thrust of the economic reform program of the Nigerian government which is targeted at mobilizing national capability in converting the country's endowments into utility products and services for the common man [10].

2. Background

2.1. Documentation on clay minerals in Nigeria

The most abundant, ubiquitous, and accessible material on the earth crust is clay [11]. Reference [5] observed that a great emphasis is placed on exploiting the abundant solid minerals endowments in Nigeria with a view to diversifying the economic base of the country, improving Gross Domestic Product (GDP) and industrial activity. One of these endowments with tremendous potential for economic utilization is clay. Clay deposit is spread across the six geo- political zones of the country [12]. Clays have their origin in natural processes, mostly complex weathering, transport, and deposition by sedimentation within geological periods [13].

The abundance of the clay minerals in Nigeria supports its rich and historic traditional pottery industry that dates from the Stone Age. Archeological evidences from the ancient pottery areas of Nigeria such as Iwo-Eleru near Akure in Ondo State, Rop in Plateau state, Kagoro in Kaduna State and Afikpo in Ebonyi state proved that as far back as the late stone age, the occupants of these areas made productive used of clay for pottery [14]. The composition of clayey and organic materials such as straws made into adobe brick, served as a ubiquitous building material widely used for building weather-friendly houses in the vast rural domains. Modern industrial uses of clay for ceramics and bricks now obtained in notable parts of the country including Kaduna, Northern Nigeria.

Clay is simply defined as earth or soil that is plastic and tenacious when moist and that becomes permanently hard when baked or fired. It consistsof a group of hydrous alumino-silicate minerals formed by the weathering of feldspathic rocks, such as granite. Individual mineral grains are microscopic in size and shaped like flakes. This makes their aggregate surface area much greater than their thickness and allows them to take up large amounts of water by adhesion, giving them plasticity and causing some varieties to swell (expandable clay). Common clay is a mixture of kaolin, or china clay (hydrated clay), and the fine powder of some feldspathic mineral that is anhydrous (without water) and not

No	Mineral	Site Location	State	Estimated Reserve (tonnes)	Remark
1	Kaolin	Kankara	Katsina	20,000,000	Residual
		Major porter, Jos	Plateau	19,000,000	''
		Oshide	Ogun	…………..	''
		Iseyin	Oyo	…………..	''
		Ifon	Ondo	…………..	''
		Ozubulu	Anambra	769,000	Sedimentary
		Illo	Sokoto	…………..	Residual
		Darazo	Bauchi	10,000,000	…………..
		Kpaki; Pategi	Niger	…………..	…………..
		Igbanke; Ozonnogogo	Edo	…………..	Sedimentary
2	Ball clay	Abeokuta	Ogun	…………..	Black
		Auchi; Ujogba	Edo	…………..	Black; Cream
		Nsu	Imo	…………..	Cream
		Giru	Kebbi	…………..	…………..
3	Common clay	Mararaban-Rido	Kaduna	5,500,000	grey
4	Feldspar	Okuta	Ogun	…………..	Potash
		Lanlate	''	…………..	''
		Egbe	Kwara	…………..	''
		Bari	Niger	…………..	''
		Okene	Kogi	…………..	''
		Gwoza	Borno	…………..	''
		Oshogbo	Osun	…………..	''
		Ijero	Ekiti	…………..	Soda

No	Mineral	Site Location	State	Estimated Reserve (tonnes)	Remark
5	Quartz/ Silica	Pankshin; Shabu	Plateau	27,962;	White; Sand
		Biu	Borno	2,540,000	White
		Ijero	Ekiti	....	Sand
		Lokoja	Kogi	4,000,000	''
		Ughelli	Delta	....	''
		Badagry	Lagos	....	''
		Epe	''	....	''
		Igbokoda	Ondo	....	''
		P/ Harcourt	Rivers	....	''
				....	
6	Talc	Shagamu	Ogun	....	....
		Kumunu	Niger	40,000,000	....
		Ilesha	Oyo	....	....
		Okolom	Kogi	....	....
		Zonkwa	Kaduna	....	....
6	Bentonite	Geshua	Yobe	....	....
		M/Belwa	Adamawa	....	....
		Esan/Isan	Edo	....	....
8	Limestone	Okpila	Edo	10,161,000	White
		Jakuru	Kogi	68,000,000	''
		Igumala	Benue	30,161,000	....
		Mfamoging	C/river	26,000,000	Grey
		Nkalagu	Enugu	720,000,000	''
		Ewekoro	Ogun	7.1 Billion	Clayey
		Arochuku	Imo	101,000,000	....
		Shagamu	Ogun	....	Grey
		Isekulu	Delta	....	....
		Sokoto	Sokoto	....	....
9	Dolomite	Osara	Kogi	2,000,000	White
		Itobe	Benue	1,000,000	''
		Igara	Edo	....	''
		Mura	Plateau	....	''
		Elebu	Kogi	....	''
		Igbeti	Oyo	....	''
		Burum	FCT	8,000,000	''
		Kwakuti	Niger	2,540,000	''
		B/Gwari	Kaduna	....	....

Source: (17) 2000

Table 1. A Table Showing Clay related materials and their various locations in Nigeria.

decomposed. Clays vary in plasticity, all being more or less malleable and capable of being molded into any form when moistened with water. The plastic clays are used for making pottery of all kinds, bricks and tiles, tobacco pipes, firebricks, and other products. The commoner varieties of clay and clay rocks are china clay, or kaolin; pipe clay, similar to kaolin, but containing a larger percentage of silica; potter's clay, not as pure as pipe clay; sculptor's clay, or modeling clay, a fine potter's clay, sometimes mixed with fine sand; brick clay, an admixture of clay and sand with some ferruginous (iron-containing) matter; fire clay, containing little or no lime, alkaline earth, or iron (which act as fluxes), and hence infusible or highly refractory; shale; loam; and marl (16). Tables 1-3 below listed industrial clay-based minerals in Nigeria with information about location, reserve, and geology.

States	Location
Cross-River	Appiapumet and Ofumbonghaone, Ogurude, Ovonum
Akwa Ibom	Nkari, Nlung, Ukim, Ikot-Etim, Eket-Uyo, Ekpere-Obom, Ikot-okoro, Ikwa
Benue	Katsina Ala, Otukpo, Buruku, Gwer West,Gwer, Makurdi
Ebonyi	Ohaukwu, Ezza North, Abakaliki, Ezzi, Afikpo South, Ohaozara
Abia	Isikwuato, Ikwuano, Umuahia Bende, Arochukwu
Enugu	Enugu, Isi-Uzo, Uzo-Uwani, Oji River, Udi
Ekiti	Ara-Ijero, Igbara, Ado, Orin
Ondo	Erusu Akoko, Ikale, Ode-Aye, Ute Arimogija, Ifon
Ogun	Bamajo, Onibode
Plateau	Bassa, Barinkin-Ladi, Mangu, Kanam, Langtang north
Niger	Lavun, Gbako Suleja, Minna, Agaie, Paikoro
Kaduna	Kachia, Mararaba-Rido, Farin-Kassa
Kogi	All over the state
Rivers	Etche Ikwere
Kano	All over the state
Delta	Ethiope East, Isoko South, Ndokwa, South/East/West Okpe, Sapele, Ughelli South, Warri North/South.
Niger	Agaie, Bida, Lavun, Mashegu, Murya

Source: Raw Materials Research and Development Council , 2009

Table 2. Locations of Ball Clay in Nigeria [2009 Update]

State	Location
Cross River	Alige, Betukwe, Mba, Behuabon,
Akwa-Ibom	Ibiaku, Ntok Opko, Mbiafum, Ikot Ekwere,
Abia	Umuahia South, Ikwuano, Isiukwato, Nnochi,
Enugu	Uzo Uwani, Nsukka South, Udi, River-Oji, Enugu North,
Imo	Ehime, Mbano, Ahiazu, Mbaise, Orlu, Ngor Okpalla, Okigwe, Oru,
Benue	Apa, Ogbadibo, Okpokwu, Vandikya,
Anambra	Ozubulu, Ukpor, Anyamelum, Ekwusigo, Nnewi South, Ihiala, Njikoka, Aguata,
Ondo	Abusoro, Ewi, Odo-Aye, Omifun,
Ekiti	Isan-Ekiti, Ikere-Ekiti,
Nasarawa	Awe, Keffi,
Ogun	Ibeshe, Onibode,
Kogi	Agbaja,
Niger	Lavum Gbako, Bida, Patigi, Kpaki,
Kaduna	Kachia,
Plateau	Barkin-Ladi, Mangu, Kanam,
Bauchi	Ackaleri, Genjuwa, Darazo, Misau, Kirfi, Dambam,
Yobe	Fika(Turmi),
Borno	Maiduguri, Biu, Dembua,
Edo	All parts of the State,
Delta	Aniochia South, Ndo Kwu East,
Osun	Irewole, Ile-Ife, Ede, Odo-Otin, Ilesha,
Katsina	Kankara, Dutsema, Safana, Batsari, Ingawa, Musawa, Malumfashi,
Kano	Rano, Bichi, Tsanyawa, Dawakin-Tofa, Gwarzo,
Kebbi	Danko, Zuru, Giro, Dakin-Gari,
Oyo	Iwo, Alakia,

Source: Raw Materials Research and Development Council, 2009

Table 3. Sources and Locations of Kaolin in Nigeria [2009 Update]

2.2. Study area and method

Kaduna State is located at the centre of Northern Nigeria (Figure 1). It is situated on the southern end of the High Plains of northern Nigeria, bounded by parallels $9^03'N$ and $11^032'N$, and extends from the upper River Mariga on $6^005'E$ to $8^048'E$ on the foot slopes of the scarp of Jos Plateau [18]. The bedrock geology is predominantly metamorphic rocks of the Nigerian Basement Complex consisting of biotite gneisses and older granites. In the southeastern corner, younger granites and batholiths are evident. Deep chemical weathering and fluvial erosion, influenced by the bioclimatic nature of the environment, have developed the characteristic high undulating plains with subdued interfluves [18]. In some places, the interfluves are capped by high grade lateritic ironstone especially in the Northwest.However, soils within the "fadama" areas are richer in kaolinitic clay and organic matter, very heavy and poorly drained characteristics of vertisols.

Kaduna State is endowed with minerals which include clay, serpentine, asbestos, amethyst, kyannite, gold, graphite and siltimanite graphite, which is found in Sabon Birnin Gwari, in the Birnin Gwari local government. The soils and vegetation are typical red-brown to red-yellow tropical ferruginous soils and savannah grassland with scattered trees and woody shrubs. The soils in the upland areas are rich in red clay and sand but poor in organic matter. In [15], Kaduna area is noted as a historic home of the Nok culture, the earliest producer of terracotta sculptures in the whole of sub-Sahara African, dating over 2,000 years ago (Figure 2). This reference has provided an index to age-long clay mineral heritage; besides serving as a mirror to civilization with which the modern man has been able to find out more about himself and the environment at such point in recorded history [19]

In recent times, apart from traditional purposes, the vast deposit of clay has basically served as raw material for pottery and red bricks production with a handful industrial presence. The fieldwork survey identified three prominent clay-based industries striving to survive the threats of unfavorable economic factors. The clay industrial sites were examined in relation to productive means of utilising the raw materials. The industries included Kaduna Clay Bricks at Mararaba-Rido and Jacaranda Pottery both located around Kaduna South and Maraba Pottery Center in Maraban Jos, Kaduna, Nigeria. The firsthand knowledge of the various productive uses of Kaduna's rich clay reserves, however, indicated prospects for industrial expansion if the mineral is properly explored and harnessed.

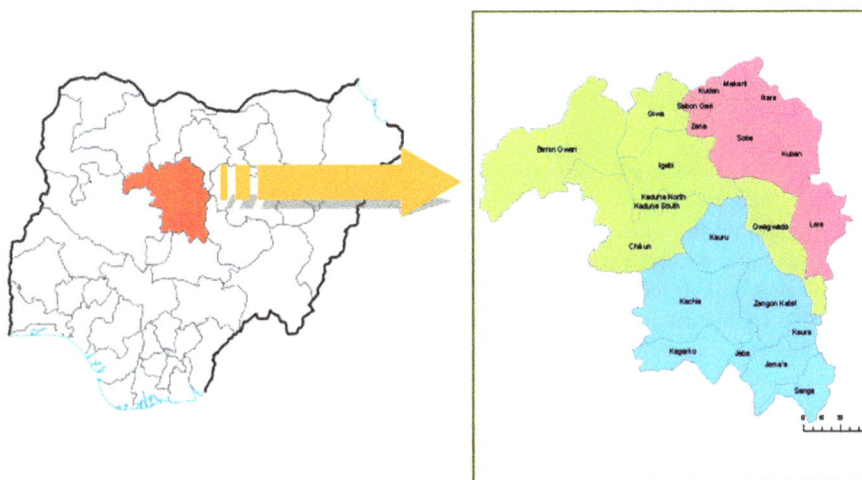

Figure 1. Nigerian map showing location of Kaduna state

Source: Jastrow (2006)

Figure 2. Nok sculpture. Fired clay (Terracotta), 6th century BC–6th century CE, Nigeria. H. 38 cm (14 ¾ in.)

3. Clay minerals and applications in Kaduna State, Nigeria

In most parts of the country, native pottery is a vibrant traditional art practice and an established cottage industry for claywares. Clay has served as an indispensable raw material for the production of products varying from red bricks (for building and decorative purposes) and pottery both at industrial and local levels in Kaduna State. The location of existing brickworks, pottery works and other ceramic production is an evidence of workable deposits within the State.

Specifically, the scope of this study surveyed on the application of the clay deposit as found in Mararaba-Rido, Jacaranda and Maraba outskirt areas of Kaduna State. For these places, the two potential qualities of clay which were of utmost importance to its usage include plasticity and the ability to retain form at the intended firing temperature. The generic property of the clay minerals indicate indicate that of a naturally occurring earthenware/ common clay which is suitable for the production of red bricks and potteries which are refractory enough for stoneware temperature. As observed, majority of the clay fire within the brown-red range of colour commonly referred to as 'terracotta' while grayish/ brownish in its green state.

As noted in [17], earthenware clays are made up of a group of low firing clays that matures at the temperature ranging from cone 08 to cone 02 (940°C- 1060°C). The clays contain relatively high percentage of iron oxide and other mineral impurities, which serve as, flux (a substance that lowers the maturing temperature of the clay). Unlike stoneware clay which is almost completely vitreous after, earthenware clay is known to be quite porous with porosity between 5 and 15 percent. Usually, when the clay is subjected to temperatures above 1150°C, it deforms, bloats or blisters.

Hence, the clay suitably serves as raw material for commercial bricks making and for fashioning aesthetic and utilitarian vases besides tablewares.

3.1. Commercial bricks making at Mararaba-Rido

The vast clay deposit found at Mararaba-Rido is harnessed for a commercial production of red bricks which are made available for building and decorative purposes. Kaduna Bricks and Clay Products Limited, a factory sited in this area, is highly mechanized, bearing fully automated and capital intensive plants with tunnel kilns built to manufacture clay bricks at large scale. According to the reliable sources interviewed at the site, the factory is said to be capable of producing an average number of 70,000 *medium* sized bricks per day.

The process adopted in the manufacturing of bricks ranging from medium, normal to decorative types involves the following:

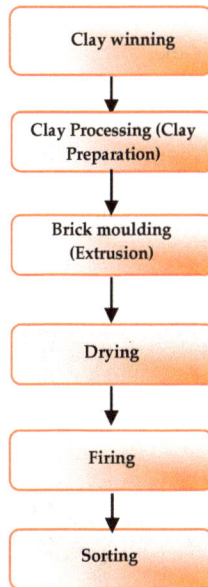

Figure 3. The production process for clay brick industry

- **Clay winning (Quarrying and Transportation)**: This involves the mining of clay from the clay pit or quarry which is situated at about 2km from the main factory. Because the clay material is usually required at bulk quantities, mechanical winning is usually carried out i.e. clay is excavated, transported and dumped at the factory site with the use of drag-line excavator and large dumper truck (Figure 4).

Figure 4. Clay pile at brick factory site for processing

-   **Clay Processing (Clay Preparation)**: The processing phases is a stage where heaps of fairly wet clay is being subjected to crushing, grinding and tempering before it can be suitable for shaping/moulding. At the early phase, fairly wet raw clay is dropped inside a box feeder where the clay is being conveyed through a conveyor belt into the wet-pan grinder. With the aid of two high speed rollers inside the wet grinder, the clay is grinded and mixed with water, and then passed through the screen plate into the double shaft mixer for proper mixing (say tempering). Hence, the clay is conveyed into the vacuum double shaft mixer linked to the extruder where the moulding and shaping take place (Figure 5).

Figure 5. Clay processing plants

-   **Shaping/ Moulding**: the clay with the aid of an extruder is being shaped into various shapes of bricks and cut into standard sizes with the cutting machine set at a particular cutting length. The shape of extruded bricks is determined by the die mould mounted at the extruder mouth (Figure 6).

Figure 6. Brick production unit

- **Drying:** Freshly formed green bricks are systematically packed in procession on palettes and loaded through a cross conveyor and the ascending elevator on a finger car truck. With the green bricks arranged on the finger car truck, they are transported to the drier (Figure 7). In the drier, the bricks are being exposed to hot and cold air for an accelerated drying. The hot air is generated from an oven heater (lintel block) besides the heat siphoned from the tunnel kiln.

Figure 7. Brick drying compartment

- **Firing:** Dried bricks are moved from drier with palettes on wheel barrows to the firing chamber after unloading through a downward elevator connected to the cross conveyor at stationed at the dry side. Hence, dried bricks are inter-sparsely stacked in the firing chamber of the tunnel kiln (Figure 8) and fired to a maturing temperature ranging from 950°C to 1050°C with networks of complex oil burners and fans which help to blow the oil for an accelerated combustion process (air host fixed burner length). Low pour fuel oil (LPFO), also known as black oil mainly serves as the heating fuel.

Figure 8. Tunnel kiln and its inner chamber

- **Dispatching**: Cooled fired bricks are gradually unloaded from the kiln and transported to a designated point around the factory where they are being collected directly by buyers. Prices of bricks products range from N100 (0.63USD) and N38 (0.24USD). Commonly produced brick types include *Medium* size; *Normal* size; decorative types; *Ernest* brick. Refractory bricks (used for the purpose of building furnace, kiln or oven) are also produced but on special demand. In this case, a refractory is composed by blending their clay with other refractory materials like kaolin from Bauchi state, Nigeria.

As noted, Kaduna Bricks and Clay Products Limited is one of the main the main clay-brick producing factories in Nigeria. Most of the brick-making factories that were originally established by the Government are being privatized presently.

3.2. Local pottery production at Jacaranda

The availability of natural earthenware clay at Jacaranda has enabled the production of pottery products varying from decorative and utilitarian vases to tablewares in this area. Situated some few kilometers away from Mararaba-Rido, the clay used in Jacaranda pottery exhibit similar properties in term of plasticity, strength, colour (both at green and fired state). The earthenware clay serves as the basic raw material for the production of pottery articles while some other bodies are derived or composed by blending two or more clay types. For example, a stoneware body will be prepared when tablewares and other articles which require glazing are to be made. Besides, the clay may be enhanced by adding other materials to get a vitrifiable, and more workable clay with less shrinkage.

However, as clearly observed, the pottery center is fully equipped with the basic studio tools and structures required for pottery production. These include kick wheels, throwing kits, kneading table, studio shelves, dewatering tray, clay pits, fuel kilns and kiln furniture. The production processes used at the center for pottery production is described in the following chart:

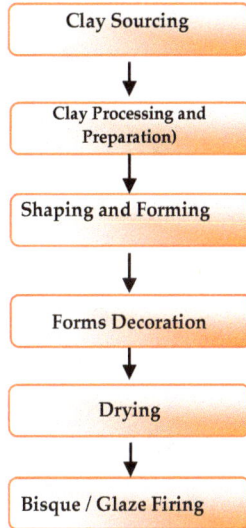

Figure 9. A schematic representation of the pottery production processes and operation

From the information gathered, Jacaranda Pottery is a small-scale ceramic industry which has been in operation since 1982, when it was established by a Briton. Though at present, the ownership of pottery center had been transferred to a Christian organisation with some other assets on the site. This was said to be brought about when the former owner whose residence was in the heart of Kaduna city, got affected by the religious crisis that erupted in the State in 2001 and decided to return to his country. However, the pottery center is still in operation though production activities at the center have not resume to its full capacity as obtained before.

Products include creative ceramic and pottery forms, which serve as ornamental purposes, utilitarian vases and tablewares as shown in Figure 10.

3.3. Maraba pottery, Kaduna

Maraba Pottery is cottage clay industry established in 1985 by set up Danlami Aliyu with the assistance of a British Potter, Michael O'Brien for the purpose of producing local ceramic wares which can meet the needs of local consumers and tourists. The center has also served a skill acquisition center for pottery practice and ceramic studio management. The center was strategically located within the reach of basic raw materials among which are clay minerals carefully collected from nearby sites, blended and manually processed to form stoneware body. The body compositions are basically made from blends of fireclay or ball clay (0-100%) and kaolin (0-70%) Quartz could also be added at 0-30%. Stoneware is generally

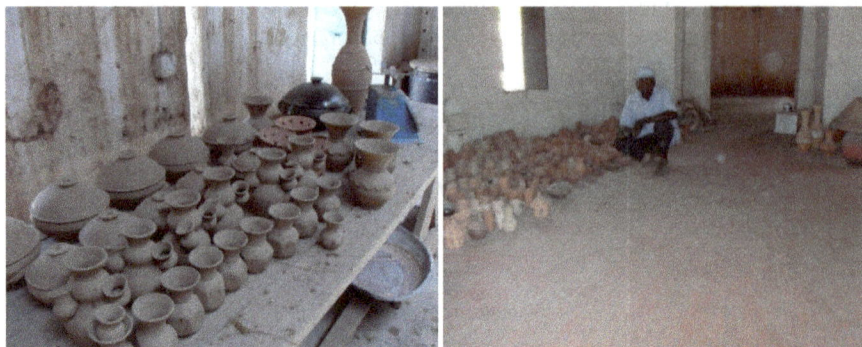

Figure 10. Display of unfired and fired pottery wares at Jacaranda Pottery Centre

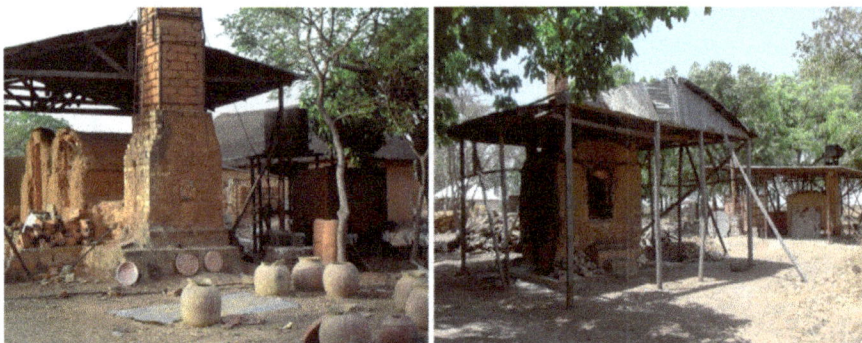

Figure 11. A cross-section of the clay firing facilities (kilns) at Jacaranda and Maraba Pottery Centres

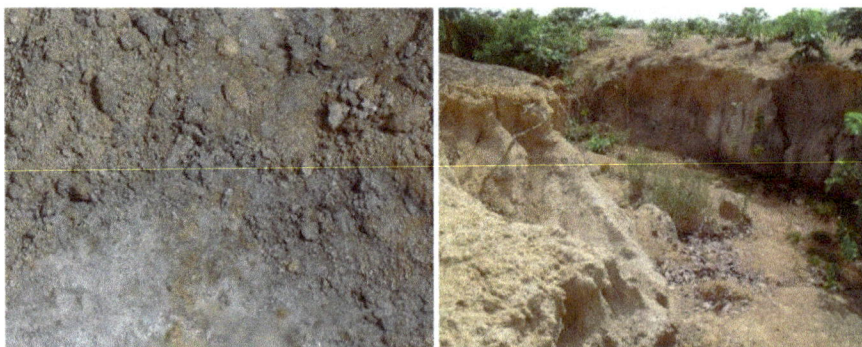

Figure 12. Clay mineral deposition in an eroded mining site

once-fired with firing temperatures which can vary significantly, from 1100 °C to 1300 °C depending on the flux content. The production process adopts simple machines and improvised tools at various stages such clay processing and preparation, clay forming (Figure 13), decoration and firing. The pottery works from Maraba are culturally inspired with items ranging from tablewares, dinnerwares, decorative wares and souvenirs (Figure 14).

Figure 13. Local clay body preparation and clay throwing process at Maraba Pottery Centre

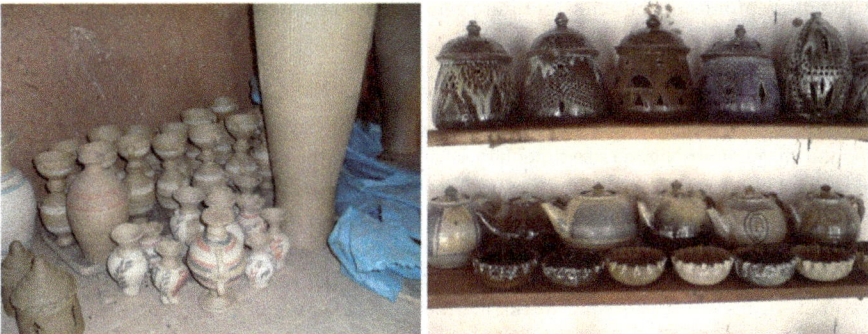

Figure 14. Unfired pottery and glazed ceramic wares on display at Maraba Pottery Centre

4. Discussions

Reference [20] opined that any ceramic industry, be it big or small, simple or complex, is created to serve some certain immediate and long-term needs within a given societies. Considering the usefulness of various ceramic materials/ products, the ceramics industry especially the local-based ones can play a major role in the socio-economic development of their locality and the country at large.

Housing constitutes one of the most important basic needs of life. A number of building materials exist which have proved themselves to be most suitable material for use in a wide variety of situations, and have a great potential for increased use in the future. Clay bricks are one of such products, which make use of available indigenous materials which can be manufactured locally.

Considering costs, locally-made clay bricks are among the cheapest of walling material. Besides, it should be borne in mind that if, as stated at a United Nations Conference, a house is to retain its usefulness, it must be maintained, repaired, adapted, and renovated. Thus, choices concerning standards and materials should consider resource requirements over the whole life of the asset and not merely the monetary cost of its initial production. Durable materials such as clay bricks have a cost advantage in this respect [21].

Significantly, the productive applications of clay from the country's vast mineral reserves will foster the conservation of foreign exchange which otherwise cannot be achieved with overdependence on imported materials. It is therefore notable that the efficient utilisation of locally available clay minerals will contributes to the fulfillment of the national socio-economic development through employments generation and industrialization as buttressed in [22]. Nevertheless, environmental issues should not be ignored while benefiting from the wealth of mineral exploitations. Mining operations should be properly coordinated to avoid the adverse effect of environmental depletion (see Figure 12). Reference [22] also noted that the United Nation General Assembly has implied the integration of economic, social and environmental spheres to meet the need of the present without compromising the ability of future generations need.

5. Conclusion

Previous geoscientific mineral studies have revealed that clays of various kinds and grades abound throughout Nigeria's sedimentary basins and on the basement. The mineral hold a significant importance especially to ceramic (pottery) practices in almost all parts of Nigeria from prehistoric period as also noted in [23]. In all parts of the country, native pottery is a vibrant traditional art form and an established cottage industry for earthenwares. There are various applications of clay use among which ceramics and bricks making are prominently featured in this study. Ceramic works at Abeokuta (Ogun State), Ikorodu (Lagos State), Okigwe (Imo State), Umuahia (Abia State) and Suleja in Niger State produce glazed wares from local kaolin. Refractory clays for refractory bricks have been proven at Onibode near Abeokuta where the refractoriness is very high at about 1,750°C.

Having observed the potentialities of clay minerals in the area of ceramic production, more can be achieved if the raw material can be fully exploited and harnessed. When the local raw materials are explored and exploited, it spurs industrial development and self reliance, thus maximizing the use of local raw materials instead of depending on imported ones with its attendant adverse effect on the economy [20]. A good example in this direction has been projected with the case study. Emphasis should be placed more on research through provision of research fund to the higher institutions, investing on the development of local technologies that utilizes local ceramic raw materials. The mining and geological research industries should be revitalized to rise up to the challenges of assisting towards maximum utilization of these raw materials. More research and developmental institutes should be established and the already existing ones must be properly equipped for effective delivery to researchers.

With the current drive targeted at attaining self-reliance in the local sourcing of industrial raw materials, a new vista can be opened in the previously unexplored areas through mining beneficiation and mineral dressing [24]. This will make it possible to set up profitable ventures for the supply of refined raw materials as feed stock to industries. Furthermore, the empowerment of the small-scale ceramics industry with the ability to compete with foreign products in terms of quality, standard and cost, will better reposition them to contribute immensely to export promotion, employment generation and socio-economic growth of a nation.

Author details

Oluwafemi Samuel Adelabu
Federal University of Technology, Ondo State, Nigeria

Acknowledgement

Special thanks to Mr. Bitrus Lamba and Mr. John at Mararaba-Rido, Kaduna; Mr. Samuel at Jacaranda, Kaduna, Mr. Timothy Olasunboye, who all assisted in gathering relevant information for this study.

6. References

[1] Obaje, Nuhu George (2009) Geology and Mineral Resources of Nigeria. Lecture Notes in Earth Sciences Series, Springer. 120: 221 p.

[2] Records of the Geological Survey of Nigeria (1958) Geological survey in Nigeria Available http://mmsd.gov.ng/Downloads/GSNA.pdf. Accessed 2012 April 13.

[3] Industrial Profile on Ceramic Glaze Material Production (Low and High Temperature Glaze Types) (2009); A Public Presentation on Ceramic Glaze Materials. RMRDC-FIIRO Joint Research and Development Project.

[4] Ministry of Solid Minerals Development (2000). An inventory of solid mineral potentials of Nigeria. Prospectus for Investors, 15pp.

[5] Olokode O.S. and Aiyedun P.O. (2011). Mineralogical Characteristics of Natural Kaolins from Abeokuta, South-West Nigeria. The Pacific Journal of Science and Technology, 12(2): 558-565. Available http://www.akamaiuniversity.us/PJST12_2_558.pdf. Accessed 2012 April 13.

[6] Anifowose A.Y.B & Bamisaye O.A., Odeyemi I.B. (2006)Establishing a Solid Mineral Database for a Part of Southwestern Nigeria. Available http://www.gisdevelopment.net/application/geology/mineral/maf06_ejaculation   . Accessed 2012 April 13.

[7] Ikpatt Clement & Ibanga N. H. (2003). Nigeria's Mineral Resources: A Case for Resource Control. Available

http://www.nigerdeltacongress.com/narticles/nigeria_mineral_resources_a_case.htm. Accessed 2012 April 13.

[8] Orazulike Donatus Maduka (2002) The Solid Mineral Resources Of Nigeria: Maximizing Utilization For Industrial And Technological Growth. Inaugural Lecture Delivered at Abubakar Tafawa Balewa University, Bauchi, Nigeria. 11th September, 2002

[9] Non-Metallic Mineral and Industrial Materials (…) Available http://www.onlinenigeria.com/geology/?blurb=518. Accessed 2012 April 13.

[10] NEEDS (2004) National Planning Commission: Abuja, Nigeria. 144p.

[11] Rado, P. (1988). *An Introduction to the Technology of Pottery*. Oxford: Pergamon Press.

[12] Adegoke, O.S. 1980. "Guide to the Non-Metallic Mineral Industrial Potential of Nigeria". Proceedings of the Raw Materials Research and Development Council. 110-120 pp.

[13] Nosbusch, H., & Mitchell, I. (1988). Clay-Based Materials for the Ceramic Industry . England: Elsevier Science Publisher Ltd.

[14] Fatunsin, A. K. (1992). Yoruba Pottery. National Commission for Museums and Monuments, Lagos: Intec Printers Ltd Ibadan.

[15] Eyo, E & Willet, F. (1982) Treasures of Ancient Nigeria, New York: Alfred A. Knopf.

[16] Microsoft Encarta (2009) Clay. Redmond, WA: Microsoft Corporation. [DVD]

[17] Alasa, S. (2000). Fundamentals of Ceramics. Auchi: Painting and General Art Department.

[18] Nigeria: Physical Setting- Kaduna State (…) Available http://www.onlinenigeria.com/links/kadunaadv.asp?blurb=294. Accessed 2012 April 13.

[19] Opoku, E. V. (2003). Development of Local Raw Materials for the Ceramics Industry in Nigeria. ASHAKWU Journal of Ceramics. 1 (1), 14-17pp.

[20] Alkali, V. (2003). The Impact of Small-Scale Industries on National Development. *Ashakwu: Journal of Ceramics* , 1-4.

[21] UNIDO. (1984). *Small Scale Brickmaking*. Geneva: International Labour Office.

[22] Kashim, I. B. (2011). Solid Mineral Resource Development In Sustaining Nigeria's Economic and Environmental Realities of the 21st Century. Journal of Sustainable Development in Africa. 13 (2), 210-223pp.

[23] Akinbogun, T. (2009). Anglo-Nigeria Studio Pottery Culture: A Differential Factor in Studio Pottery Practice between Northern and Southern Nigeria. The International Journal of the Arts in Society. 3 (5), 87-96pp.

[24] Adelabu, O. & Kashim, I. (2010) Clay mineral: A case study of its potentialities in selected parts of Kaduna State of Nigeria. Proceeding of International Conference on Education and Management Technology (ICEMT), Cairo. 655 – 659pp.

Chapter 3

Claystone as a Potential Host Rock for Nuclear Waste Storage

Károly Lázár and Zoltán Máthé

Additional information is available at the end of the chapter

1. Introduction

Nuclear energy is widely used for production of electricity. There are several advantages of this type of generation of electric power, but severe drawbacks also emerge. These difficulties should be eliminated for appropriate large-scale utilization. One of these difficulties is the formation of long life-time isotopes during the nuclear reactions (fission and capture of neutrons). The life-time of some of these products reaches even the 10^5-10^6 year range. There are procedures, by which these long life time isotopes can be extracted or eliminated (reprocessing and transmutation), but finally, some isotopes are still remaining with significantly decreased activity with life-times in ca. thousand years range.

The reliable and safe deposition of these products should be elaborated. The long-term geological disposal is considered as a suitable option for the isolation of these isotopes from the biosphere during their long life-time. Properties of various types of host rocks were evaluated from this aspect, crystalline (granitic), clayey or even salt types among them.

The clayey host rocks exhibit some andvantages in this respect, namely they may have significant capacity for sorption and ion exhange, both processes considerably may retard the migration of dangerous components. Some of these formations have been characterised in very details – even underground laboratories were established in them for their in situ characterisation (Callovo-Oxfordian formation in France, Opalinus Clay in Switzerland, and Boom Clay in Belgium).

The issue is about to be also addressed in Hungary. C.a. 40 % of electricity is procuded in a nuclear power plant since the 1980's. A significant amount of collected spent fuel has been stored in temporary repositories since then. The procedure for final deposition should be elaborated and implemented within the next 20 – 30 years.

A preliminary screening had been performed, the Boda Claystone Formation has emerged as a potential media for deposition. This formation is an extensive one – samples collected from different parts of it have been characterised. An underground research facility had even been established in it in a depth of 1050 m from 1994 – 1999. A detailed characterization of the formation has been performed based on the data collected from results of measurements performed in this facility.

The intention with the recent chapter is to provide an illustration and to present some aspects of the evaluation on the potential possibilities of the application of clayey geological media for storage of wastes of nuclear origin. Following this short introduction in the second part an overview is given on the types and most important long life-time isotopes, and on the ongoing processes which might occur during the several thousand years of operation time of the waste storage facility. In the third part a short account is presented on the properties of clay minerals which influence their behaviour as an isolation media. Results of some specific evaluations of clays (Callovo-Oxfordian, Opalinus) are also presented in this part to provide an overview and comparison. In the fourth part a more detailed description and characterisation of Boda Claystone is presented. Its formation, components, characterisation of clay minerals is described in more detail. Special attention is devoted to measurements correlated to isolation properties of this media against migration of long life-time isotopes.

2. Nuclear wastes

The production of nuclear wastes is connected primarily to generation of electricity in nuclear power plants. Other sources of nuclear wastes are common as well. For example those radionuclides which had been used for therapeutical or industrial applications. These latter radioisotopes can be generated by neutron or proton irradiation of certain stable isotopes, they usually have short half life-times (less than ca. 30 year), and their amount is comparatively small. The nuclear wastes usally are classified to two main groups: to low-level and to high-level wastes. The criteria for the distinction are the half-life and the amounts of the radionuclides present in them. The main components in the low-level wastes have short half life-times, they can be stored in ground surface facilities, their operation time is expected to elapse for a few hundred years. In contrast, the radioisotopes in the high level nuclear wastes have considerably longer half life-time (up to the range of $10^5 - 10^6$ year). Construction of underground facilities is considered for the final disposal as a reliable mean to maintain an appropriate isolation from the biosphere for the long time interval necessary for the decay of the critical isotopes [1]. One option for the disposal of high level wastes can be the utilisation of clays as isolating media. Prior to discussing some properties of clays it is worth to discuss some properties of the high level wastes in more details with particular respect to their possible interactions with clayey minerals occuring during the long operation time of the disposal facility.

2.1. High level nuclear waste

In conventional nuclear reactors the energy is produced by splitting ^{235}U nuclei with neutron irradiation. In a conventional fuel type the ^{235}U content is enriched to ca. 4 % from the natural 0.6 – 0.8 per cent in the ores. The rest of the fuel is ^{238}U (both isotopes in oxide form). The fission of the heavy U nucleus results in formation of two novel nuclei, and a few neutrons are also emitted which may propagate the process. A small difference exists between the masses of the starting and final products. The corresponding energy equivalent of this difference is the generated heat in the final gross balance. A variety of nuclei are produced in the fission, a number of them is instable and will be stabilized in later processes by emitting β or γ radiation. There are several instable nuclei among these fission products which have considerably long half life-time (Table 1.) Further on, in the nuclear reactors not only fission but capture of neutrons may also proceed. For instance ^{238}U may capture one neutron and the with simultaneous β radiation ^{239}Np forms. This nucleus is also instable and emits one electron by forming ^{239}Pu. This pruduct is still instable but is has a much longer life-time. Plutonium can also be used as a reactor fuel, but for splitting of ^{239}Pu nuclei other type of reactor conditions are optimal [2].

Isotope	Half-time 10^6 year	Rel %	β energy keV
^{99}Tc	0.211	6.1	292
^{93}Zr	1.53	5.4	90
^{135}Cs	2.3	6.9	210
^{129}I	15.7	0.8	198

(Remark: the relative abundance depends on the extent of the burnout of the fuel)

Table 1. Some typical long life-time fission isotopes produced in nuclear reactors

A large variety of other short half life-time isotopes are also formed, they influence the working conditions of the reactor. After a certain period of the usage (3-4 years) the fuel rods are removed from the reactor. They can be stored afterwards without any further utilization. They are highly radiating, a great amount of heat is still generated in them (without exposed them to neutron irradiation). Thus, after operation they are usually kept in a water pool, then in other temporary storage facilities. But, as was shown above, isotopes with long half life-time are present in them, their final safe disposal should be provided in some manner. This type of the straight utilization of the nuclear fuel is called open fuel cycle.

Another option is the reprocessing. The used fuel still contains ^{235}U amounting to ca. ~ 1 %, and it is also worth to extract the newly formed transuranic isotopes (eg. ^{239}Pu) which can be utilized as a nuclear fuel (although in modified type of reactors). During the various steps of the reprocession the amounts of the long life-time isotopes are significantly decreased, but they are still present in significant amounts at the end of the procedure (Figure 1). This version of utilization is named closed fuel cycle. It is seen that the activity is rather high, even after c.a. one million years of storage the tera-Becquerel level is still maintained.

The transmutation is also a further possibility to reduce the amounts of the radiounuclides present in the high level waste. Specific controlled nuclear reactions can be initiated by irradiating different target isotopes with accelerated electrons, protons, etc. and the energy of stabilisation from the ustable target nuclei can also be extracted. This process is very

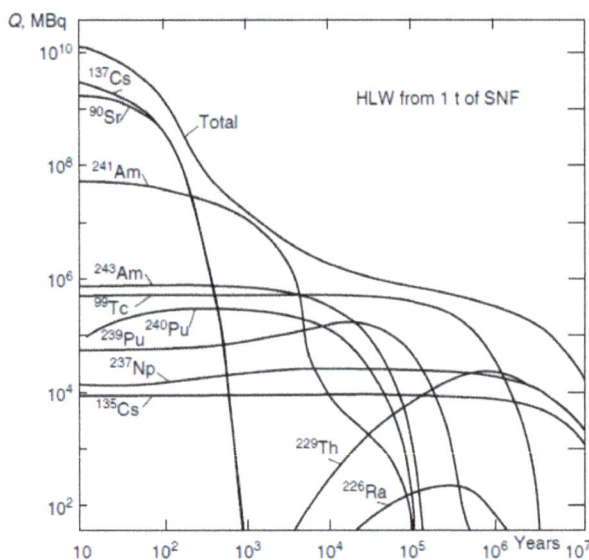

Figure 1. Activity of high level waste remaining after processing of 1 ton spent fuel in dependence of time (reprinted with permission from [1].)

sophisticated, should be performed in different manner with each isotopes separately, thus its practical application cannot be expected in the near future.

This brief overview can be summarised with the conclusions that long life-time isotopes are byproducts in generation of electricity in nuclear power plants. The life time of some of these isotopes may expand to even 10^6 years. The amounts of this isotopes can be decreased by various procedures but they cannot be fully eliminated.

Due to their long life-time these waste isotopes should be isolated from the biosphere. By commonly accepted recent considerations a perspective solution might be the disposal in geological media, i.e. host rocks.

2.2. Disposal and subsequent possible processes in host rocks

Two types of barriers are usually mentioned with respect to the geological disposal, the engineered and the natural ones. The engineered barrier comprises the immediate capsulation (metal cannister), the backfill material (high sorption capacity porous material – e.g. bentonite, etc.) and the strengthened wall of the shaft (made of concrete) in which the cannisters are situated. The natural barrier is the geological media. In the first few hundred years the engineered barrier should sustain the isolation from the long half-life isotopes. During this period the components of this barrier will probably lose their ability for protection due corrosion and other hydrothermal processes, Thus the role of the natural barrier will probably gain emphasised importance later on [3].

Several processes may take place in the working cycle of a disposal site. It should be taken into account that the extent of decaying nuclei is large, thus, a significant amount of heat is generated simultaneously. Further on, most of the design schemes consider deposition sites below the ground level, thus presence some amount of water cannot be excluded either. In addition, due to radiation effects, free oxidizing or reducing radicals may also form, (e.g. •OH from water).

The situation can be illustrated for example with the case of neptunium on a Pourbaix (Eh-pH) diagram [4]. At the closure of the disposal site the environment is exposed to air, the pH of the ground water (in equilibrium with the CO_2 of the air) is c.a. 5.5 (position A in Figure 2.). After the closure the site is isolated, the pH approaches to the mean of the rock (usually it is between 7 and 8). As the time elapses, the concrete sealing around the engineered barrier starts to erode, the pH in the close environment may increase to ~10. Later, very slowly the system may approach the state charcterising generally the whole isolating rock host (Eh -0.1 V, pH 7.5 can be estimated, see Figure 2).

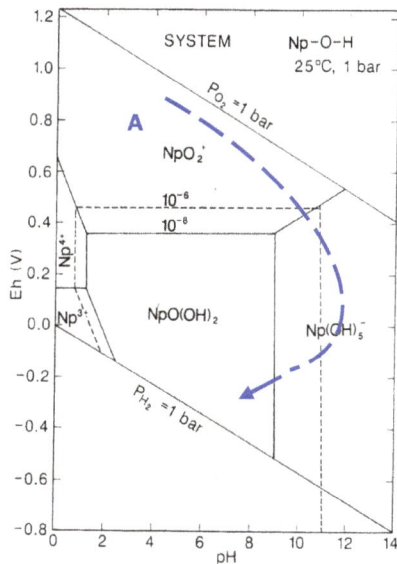

Figure 2. Schematic representation of the change of conditions on the Eh-pH diagram of Np during the life-time of a disposal site. Processes start with the construction, from a stage marked with A. (the base Eh-pH diagram is taken from [4])

It is seen in Figure 2 that Np is present as a cationic species in NpO_2^+ form at the start, which can be sorbed in aqueous media on the components of the rock minerals. However, the speciation changes with the increase of pH, the anionic $Np(OH)_5^-$ component will be dominant later. This component probably will migrate easily due to its negative charge. Having migrated for several hundred years $NpO(OH)_2$ component may be formed at the end, which is neutral, thus it will precipitate.

The simplified example may already provide an impression that the speciation and the pH dependence may be different for the various long half life-time isotopes. The real situation is far more sophisticated: the temperature may reach 80 – 100 °C (depending on the time elapsed, and the depth), the hydrostatic pressure may be in the 50 – 80 bar range (depending on the depth), various interlinked geochemical processes may occur under these conditions which influence the spreading of isotopes in the host rock (for further details see e.g. in [5]).

2.3. Simple practical approaches to estimate the speed of migration

The speed of migration of the various long half life-time components is a crucial factor for the evaluation of the properties of the perspective host rocks. In the first approach the migration of isotopes can be considered as a cyclic repetition of frequent sorption/desorption steps in a slowly flowing aqueous media. The strength of sorption can simply be characterised by the distribution coefficient, K_d.

$$K_d = \frac{I_0 - I_e}{I_e} \cdot \frac{V}{m} \tag{1}$$

where I-s are concentration (activity) values of the studied isotope, I_0 in the starting solution, I_e in equilibrium, after having kept the m mass of sample in contact with V volume of solution.

The K_d values of various rock samples can easily be obtained, the concentration change of the particular ion should only be determined experimentally. Consequently, K_d values are widely used. Further on, the speed of the migration relative to the slowly flowing media can also be estimated by using the K_d values:

$$\frac{v_{rad}}{v_{gw}} = \frac{1}{1 + \left(\dfrac{1-\varepsilon}{\varepsilon}\right) \cdot \rho \cdot K_d}, \tag{2}$$

where v_{rad} is the velocity of migration of the given radionuclide, v_{gw} is the velocity of the flow of the ground water, ε is the porosity, ρ is the density and K_d is the distribution coefficient of the isotope in question.

Equation (2) has been widely used since K_d values can easily be measured. However it should be used with particular precaution. Namely, it should be taken into account that:

- The structure of the powdered rock is different from the real one – usually not all the mineral components are contacted with the aqueous medium in the real case.
- The mass ratios of the liquid solution related to the amount of the solid rock are distinctly different in the K_d determinations and in the real situation. In the former instance the liquid component is applied in large excess (at least ten-fold), whereas the conditions in the reality are the opposite, usually only a few per cent of pore water is present in the rocks.

- Equation (2) provides a relative, retention velocity. It may occur that the flow is practically zero in the rocky media, whilst the simultaneous diffusion takes place still. Under these conditions Eq. (2) cannot be applied at all.

There are other simple experimental methods by which more reliable determination of the crucial constants can be provided. Rock samples preserved in conditions closer to the natural ones are suitable for this type of studies, obtained for example from bore cores. One simple type of these more reliable measurements is the so-called „break-through" type. Two compartments are separated with the studied borecore sample in the measuring cell. One compartment contains the very dilute solution of the studied tracer isotope (in ground water), and the break-through (i.e. the appearance) and the rate of further increase of its concentration is followed in the ground water in the other, opposite compartment. If the trace isotopes are used in really low concentrations the experimental conditions may actually be close to the real, natural ones. In other words the applied small concentration gradient hardly modifies the natural conditions. On the other side, however, rather long time intervals are necessary to perform these studies (several hundred days). From these measurements the effective diffusion constant (D_{eff}) can directly be determined (Eq. 3)

$$\frac{C'(t)}{C_0} = \frac{AD_{eff}}{V'L}t - \frac{\alpha AL}{6V'} ,$$

(3)

where C_0 is the original concentration (activity) of the isotope in a compartment at the start of the experiment, $C'(t)$ is the concentration measured in the opposite compartment, after having the break-through started, A is the cross section of the borecore sample, L is the thickness of the borecore, V' is the volume of the compartment. α is the so called retention factor, its value is related to the delay of the break-through. D_{eff} is proportional to the slope of the increase of concentration of the studied isotope with time in the originally non-active compartment.

D_{eff} is one of the most appropriate parameters which can be used for the characterisation of rocks from the point of view of migration of isotopes expected to proceed in them.

3. Application of clayrocks for the disposal

Clay rocks are widely considered as appropriate media for final waste disposal site. Clay minerals have several advantageous properties for the isolation. They have open, layered structure in which they easily ad- or chemisorb cations.

3.1. Recent investigations on clayrocks

Due to the advantageous properties of clayrocks significant amount of efforts is devoted to investigate them as optional disposal sites. For instance, at the Nuclear Energy Agency (NEA) a working group has been formed devoted to study the scientific bases for stability and buffering capacity of deep geological waste management systems („Clay Club"). This Club organised several conferences, and the results have been published in workshop proceedings [6]. Other specific conferences have also been devoted to this topic (see e.g. [7]).

At present, three clayrock types have been characterised in very detail in Europe. Namely, the Callovo-Oxfordian formation in Bure, France, the Opalinus Clay in Mont Terri, Switzerland, and the Boom Clay in Mol, Belgium [8]. Some basic properties and some characteristic minerals of these clays are summarized and compared in Table 2 [9]. Properties of Boda Claystone are also included for later comparisons.

Clayrock	Age	Depth	Max. temp.	Pore water	Organic matter	Clay fraction	Albite	Calcite
	10⁶ year	m	°C	%	%	%	%	%
Boom	30 - 36	200	16	~ 25	2 - 3	35 - 65	1 - 2	~ 1
Callovo-Oxfordian	155	500	40	~ 7	~ 1	~ 40	~ 1	~ 35
Opalinus	180	500	85	~ 7	~ 1.5	45 - 60	~ 1	4 - 8
Boda	250	1000	220	~ 2	-	35 - 50	4 - 8	3 - 5

Table 2. Certain characteristic data of clayrock types considered as perspective host [9].

Principally, each rock types contain clays in large proportions. Hovewer, in other properties the clays exhibit significant differences. Their ages, the maximal temperatures experienced during formation, water contents, characteristic minerals present in them etc. are distinctly different, attesting their unique history of formation for each of them.

The safety aspects for the construction of disposal sites have also been discussed and evaluated in detail (see eg. for the Callovo-Oxfordian in [10], for the Oplalinus Clay in [11]).

4. Boda Claystone

There had been considerations with respect to the future disposal sites for high level nuclear wastes in Hungary, too. Boda Claystone emerged as potential host. A brief comparison of data shown in the previous Table 2 shows that each clayrock has distincly different properties – each of them should be characterized specifically. Thus, Boda Claystone was characterised in the course of various programmes as well. Some results of these characterisations are summarized below.

4.1. General characterisation

The Upper Permian sedimentary sequence of the Boda Claystone Formation (BCF) is located in Western Mecsek Mountains, southern Transdanubia, SW Hungary. The Mecsek Mts. is part of the Tisza Megaunit comprising the basement of the south-eastern half of the Pannonian Basin [12] (Figure 3). The geological map of BCF is shown in Figure 4.

Following the Variscan orthogeny the continental sedimentation in the Mecsek Mts. began in the Early Permian and lasted until the Lower Triassic. The BCF is part of this about 2000-4000 m thick siliciclastic sequence (continental red beds); its transition with underlying

Cserdi Formation is conformable, while the boundary with overlying strata of the Kővágószőlős Formation is usually sharp (Figure 5) [13].

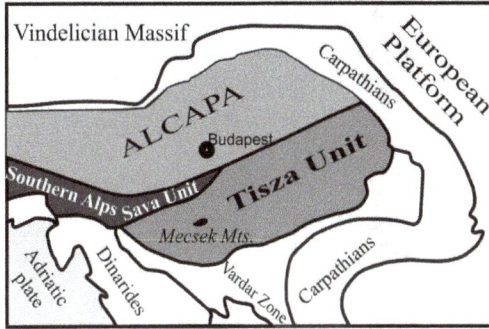

Figure 3. Overview of plate tectonic units in the Pannonian Basin (after [12]).

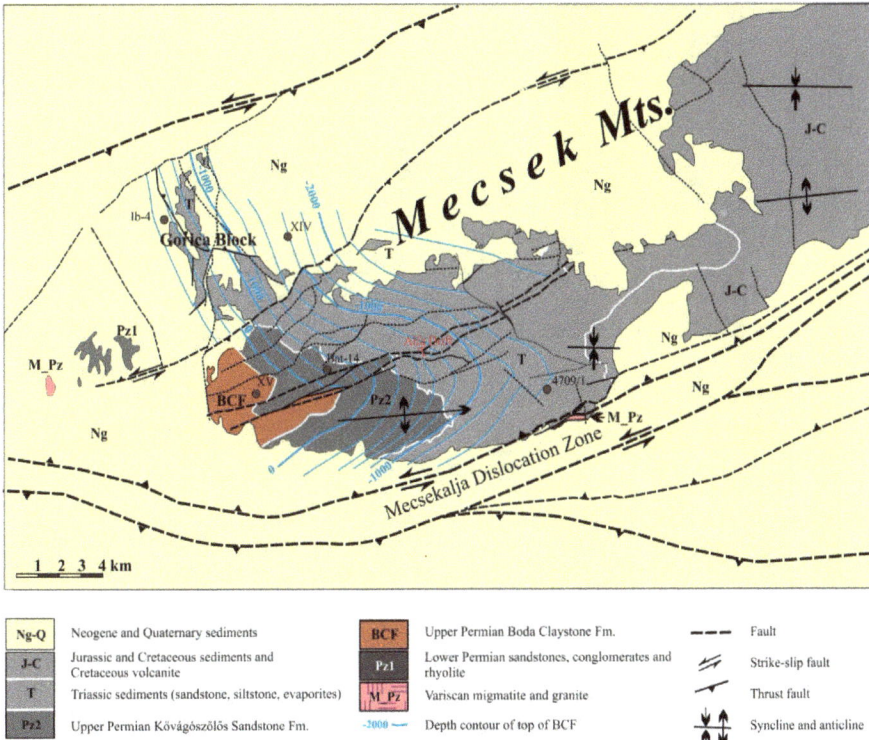

Ng-Q	Neogene and Quaternary sediments	BCF	Upper Permian Boda Claystone Fm.	- - - - Fault
J-C	Jurassic and Cretaceous sediments and Cretaceous volcanite	Pz1	Lower Permian sandstones, conglomerates and rhyolite	Strike-slip fault
T	Triassic sediments (sandstone, siltstone, evaporites)	M_Pz	Variscan migmatite and granite	Thrust fault
Pz2	Upper Permian Kővágószőlős Sandstone Fm.	-2000	Depth contour of top of BCF	Syncline and anticline

Figure 4. Geological map with depth contour of the top of the Boda Claystone Formation and locations of studied objects (boreholes: Bat-14, Ib-4, Alfa Drift) after [13]

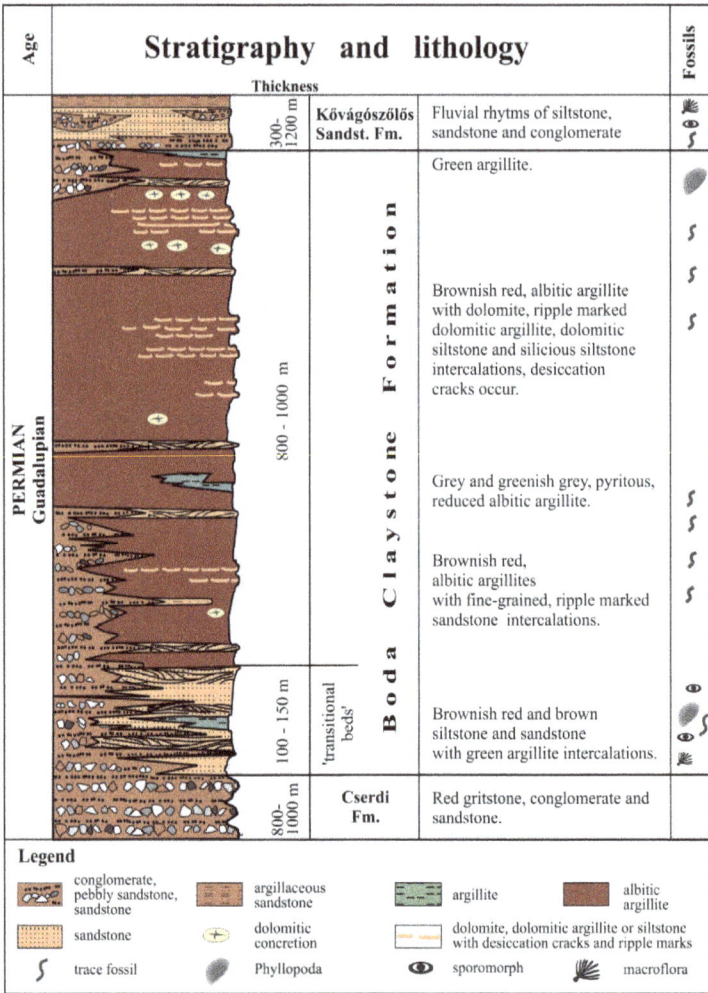

Figure 5. Idealised lithological column of Boda Claystone Formation [13]

On the basis of data of boreholes and geological mapping BCF is known in an area of 150 km². Within this area approximately 15 km² can be found on the surface in W Mecsek Mountains (see Figure 4). Two distribution areas of BCF are known in W Mecsek Mountains: i/ perianticlinal structure of the W Mecsek Mountains; ii/ so called Gorica block. In the Gorica block outcrop of BCF is not known; in this block only the borehole Ib-4 recovers sequence of BCF in significant thickness (between 494,2 and 709 m) (Figure 4).

On the basis of the deep drillings total thickness of BCF is estimated to be about 700-900 m (in perianticlinal structure of the W Mecsek Mountains) whereas according to our knowledge its thickness is smaller in the Gorica block (about 350 m). The sediments of the BCF are dominantly red and reddish brown in color, reflecting the dominantly oxidizing nature of the depositional and early diagenetic environments [14-18] (Figure 6).

Figure 6. Reddish brown albitic analcite-bearing claystone with interbedded aleurolite (Al) and dolomite (D) layers (Ib-4, 591-596 m).

The BCF was deposited in a shallow-water lacustrine environment (playa mudflat, playa lake), under semi-arid to arid climatic conditions. According to our present-day knowledge the middle thickest unit of the BCF has only one reduce interbedding (greyish black albitic claystone containing pyrite and finaly disseminated organic matter), its tickness is about 3-4 m. However, several reduce thin layers (green, greenish-gray claystone, siltstone) can be observed in its lower and upper transitional zones (Figure 5).

On the basis of mineralogical investigation with X-ray diffraction (XRD), differential thermal analysis (DTA), and electron microscopy (EMP), the main rock-forming minerals of the BCF are: clay minerals (absolute dominant are illite-muscovite and chlorite; smectite, kaolinite, vermiculite and mixed-layer clay minerals were identified in considerable amounts), authigenic albite, quartz, carbonate minerals (calcite and dolomite) and hematite [16-19]. In addition, barite, anhydrite, authigenic K-feldspar and detrital constituents (muscovite, biotite, chlorite, zircon, rutile, apatite, ilmenite, Ca-bearing plagioclase) were always also identified in trace amounts. The authigenic albite is present as albite cement (typical of all rock types of formation), albite and carbonate-lined disseminated irregular white voids (typical of albitic claystone) and albite replacement of detrital feldspars in sandstone beds [16-17, 19].

Six main rock types of BCF can be defined based on mineralogical, geochemical and textural considerations, namely, albitic claystone, albitolite, „true" siltstone, dolomite interbeddings, sandstone, and conglomerata [13, 15-17, 19,]. Their mineral compositions are shown in Table 3.

Rock types	clay minerals (wt %)	authigene albite (wt %)	quartz (wt %)	carbonates (wt %)	hematite (wt %)
albitic claystone	20-50	20-50	5-10	~10	7-10
albitolite,	<25	>50	<10	10	5-6
„true" siltstone	approx. 10	>35	>25	approx. 10	5
dolomite interbeddings	10	30-40	5	35-50	5
sandstone	5-traces	25-40 *	20-30	5-20	5-traces

* detrital feldspar + authigene albite

Table 3. Mineral compositions of the main rock types in Boda Claystone Formation

This mineralogical composition is typical of in perianticlinal structure of the W Mecsek Mountains. The principally dominant rock type of the formation is albitic claystone.

BCF recovered from borehole Ib-4 (Gorica block) differs in its mineralogical composition. This succession of BCF contains abundant analcime in significant amounts in addition to the above minerals, in a range between 8 and 25 wt %; (typical examples are shown in the table below).

Samples Ib-4	clay minerals	authigene albit	analcime	quartz	carbonates	hematite
			wt %			
527,2 m	42	8	12	13	18	6
538,7 m	34	7	16	28	10	5
560,64 m	43	16	20	3	10	8

Table 4. Mineral compositions of typical samples taken from Ib-4 borehole in Gorica block

In the Gorica block the same rock types can be defined based on mineralogical, geochemical and textural considerations as in the perianticlinal structure of the W Mecsek Mountains.

The Boda Claystone Formation is divided into three main sections [13]; for the characterization of the units see Figure 5. In the Middle Cretaceous on the basis of thickness of overlying strata in perianticlinal structure of West Mecsek Mts. the BCF was located at least at 3,5 to 4,0 km burial depth. Data determined in area of the perianticlinal structure of the W Mecsek Mountains point to late or deep diagenesis, max. 200-250 °C according to illite and chlorite crystallinity as well as vitrinite reflectance [17]. Higher illite and chlorite crystallinity is determined in core samples of the deep drilling Ib-4 (Gorica block), suggesting that BCF in Gorica block underwent lower grade diagenesis.

There were various stages of exploration and detailed characterisations of BCF. For a period even an Underground Research Laboratory was established and had been maintained in a depth of 1050 m below ground level (1994 – 1998). The results of studies performed there are collected in various reports. The data had been analysed from the aspects of criteria for establishing a waste disposal site, the main conclusions of them are compiled in a Digest [20]. The most important conclusions were presented also at one of the „Clay Club"s conferences [21].

4.2 Specific characterisations - samples

In a further stage of investigations selected Boda Claystone samples were used for measurements performed by applying directly radionuclides to obtain informations on the isolation properties of the rock against long half life-time isotopes.

Sorption and diffusion studies were performed by using fission isotopes and one of the most relavant acitnoide element, uranium. Among others samples from Delta-9, Bat-14 and IB-4 borecores were selected.

Delta-9 deep drilling is located in Alfa Drift at the 1050 m depth below ground level, it is nearly a horizontal borehole; Bat-14 was drilled near the outcrop of BCF, here the formation is covered by 25,5 m thick Quaternary sediments, and Ib-4 borehole starts also from the surface in the Gorica block, and the studied samples were collected from a distant depth 570 m (for their locations see Figure 4). Some specific studies on these samples are presented in the following paragraphs, the results of measurements with radionuclides are presented afterwards.

4.2.1. High resolution SEM with electron diffraction

These measurements were performed on samples from borehole IB-4 (570 m). A SEM/Electrondiffraction image is shown in Figure 7 from a small area of the sample. The corresponding general mineral composition is shown in Table 5.

Figure 7 illustrates the structure of the sample. Very small crystals originated from the secondary processes form the consolidated clay rock. It is also shown that within small distances the composition may change significantly (see e.g the lower right corner – it is more abundant in analcime than the rest). Parallel porosity measurements were also performed from the same regions. It was found that the porosity at the analcime rich bottom right corner is ca. 1.5 %, whereas in the upper part is 3.5 %. In contrast the portion of calcite is larger in the top regions

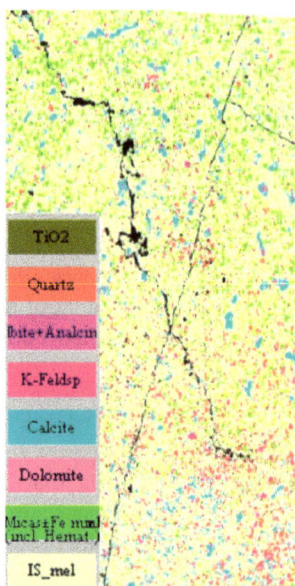

Figure 7. Mineral composition obtained from a SEM & electron diffraction image of a 5 x 2.5 mm section from IB-4 (570 m) sample (IS mel: mixed illite-smectite. The image was recorded in ERM, Poitiers, France)

Mineral	Content (%)
Illite - muscovite	33
Chlorite	2
Smectite	2
Analcime	23
Albite	12
Quartz	11
Calcite	9
Dolomite	1
Hematite	7

Table 5. Average mineral composition of the IB-4 (570 m) sample determined by quantitative XRD analysis

4.2.2. Mössbauer spectroscopy

Oxidation and coordination states of iron can conveniently be studied by Mössbauer spectroscopy. The method can also be used to analyse iron-bearing minerals. (Corresponding Mössbauer parameters of over 400 iron containing minerals are collected in the handbook [22].) Boda Claystone contains also iron-bearing minerals in considerable amounts (hematite, clay minerals), thus samples can also be analysed, and some processes occured in the formation of the rock can be traced.

The ferric iron oxide, hematite exhibits a magnetically split characteristic sextet. Iron in the clay minerals displays two-line doublets in the spectra. Doublets of Fe^{2+} and Fe^{3+} states can clearly be distinguished, their positions are different. Two examples are shown below, in Figures 8 and 9. The effect of weathering can be tracked in Figure 8 in which spectra of samples collected from layers close to the surface (below the 25m Quaternary sediments) are shown for Bat-14 borehole. The very right peak in the spectra belongs to the Fe^{2+} ions in chlorite mineral [22]. It is seen that the intensity of this peak inreases in correspondence with the sampling depth. I.e. clay minerals in the top layers are much abundant in Fe^{3+} whereas in the deeper layers Fe^{2+} starts to dominate in clay minerals. On the other side, the amount of hematite (marked with H in the right side spectra) practically is the same in each sample. Thus it can be concluded that $Fe^{2+} \rightarrow Fe^{3+}$ oxidation had taken place in the layers close to the surface, in other word the process of weathering is reflected in the Fe^{2+}/Fe^{3+} change.

The conditions controlling the formation of minerals can be deduced from specra of Figure 9. Boda Claystone is described in the previous parts of this chapter as a rock which was formed essentially under oxidizing conditions. However a few exceptional strata also exist.

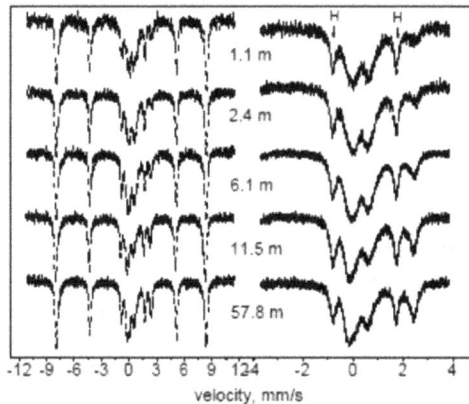

Figure 8. Mössbauer spectra on samples collected from borehole BAT-14. The main iron bearing mineral is hematite (left spectra). Effect of weathering is shown in the right side spectra, approaching to the surface Fe^{3+}/Fe ratio increases in the clay mineral (chlorite)

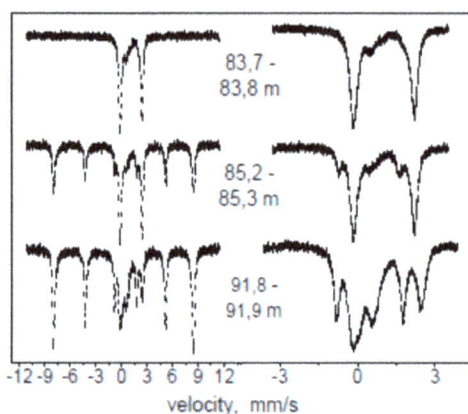

Figure 9. Crossing a reducing zone at 83.7 m in Delta-9 borehole. Iron is present only in Fe²⁺ form in chlorite. Hematite starts to appear in other distances.

For example Delta-9 borehole crosses a few meter thick layer in which reducing conditions controlled the formation of the minerals. This layer is located ca. at the 83 – 84 m distance from the commencement of the borehole (Note: the borehole was drilled at the 1050 m depth.) The overwhelming part of iron is present in Fe^{2+} form in chlorite – hematite is not present at all in the top spectrum of Figure 9. By moving apart a few meters from this layer hematite appears and its presence becames prevailing by moving even farther.

Both examples show that iron ions located in the layered clay minerals are sensitive to the change of redox conditons and they can be used as indicators to monitor the alterations in them. Further description and interpretation of these measurements can be found in [23].

4.2.3. Reduction/oxidation studies from voltammetry of microparticles

The reduction/oxidation properties of rocks and minerals can also be studied by voltammetric measurements. Even small amounts of them can be analyzed by voltammetry of microparticles. Samples shown in Figure 9 were also characterised at the LCPME – CNRS, Nancy, France). For illustraton, Figure 10 displays some cyclic voltammograms obtained on samples collected from Delta 9 borehole.

The results of voltammeric measurements are in good correspondence with the Mössbauer analysis as Figure 10 displays. High anodic current (A1 peak at 0.65 V – relative to standard calomel electrode) could be detected which can be attributed to extended oxidation of Fe^{2+} ions on the sample formed under reducing conditions (collected from 83.7 m position). On the other two samples formed under oxidative environment reduction of iron ions ions can be observed (C1 peak probably corresponds to reduction of ions in the dissolved ions, C2 at –0.45 V to the reduction of ferric ions in the clays and C3 at -0.74 V is for ferric ions in hematite). More extended description of the method, samples and measurements can be found in [24].

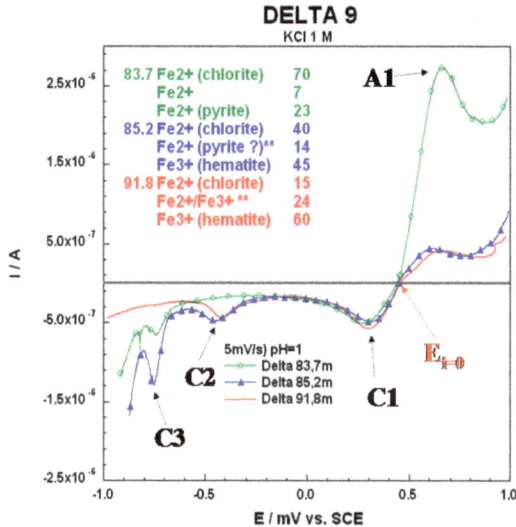

Figure 10. Cyclic voltammograms obtained on samples collected from Delta-9 borehole. The numbers in the top left corner represent the relative amounts of iron bearing minerals extracted from Mössbauer spectra of Figure 9. A1 stands for anodic, C1, C2 and C3 for cathodic peaks, $E_{i=0}$ is the zero potential. (The voltammograms were measured at LCPME Nancy, France)

4.3. Studies with long life-time radionuclides

4.3.1. Sorption measurements

In the first stage of investigations sorption of radionuclides, mostly fission products, were investigated on various types of samples. Namely K_d values were determined for ^{125}I, ^{137}Cs, ^{60}Co and [85]Sr on various Boda Claystone samples collected from boreholes drilled in the 1050 m depth exploratory tunnel. Measurements were performed with 2 g amounts of crushed samples. Radionuclides were added in trace amounts to 20 ml volume of ground water ($10^{-5} - 10^{-3}$ mol/L). K_d values were determined by using Eq. (1). There was a scatter in the results depending of the samples, for [125]I 0.2 -2.0 , for [85]Sr 60 – 120 , for [137]Cs 600 – 5000 and for [60]Co 1200 – 15000 values were determined. Concentration dependence of the sorption was also measured by diluting the tracer isotopes with inactive (natural) isotopes. Sorption of caesium showed a regular behaviour, the constants of a non-linear Freundlich-type isotherm could also be determined. Similar results could not be obtained with Sr^{2+} and Co^{2+}, both ions have low solubility in the applied ground water, they easily form precipitates ($SrCO_3$, and $Co(OH)_2$). Further details and results are described in [25].

It should be mentioned that similar sorption measurements were also performed on samples originated from another facies of Boda Siltstone, rich in analcime component originated from IB-4 borehole (see Figure 7 and Table 5). In these measurements non-radioactive natural isotopes were applied with higher concentrations, their amounts were determined by atomic absorption spectroscopy measurements. The results are in good correspondence with those obtained with radioisotopes [26]. The obtained K_d values might be applied for the further evaluation of the isolation properties of the rock against migration of radionuclides by using Eq. (2).

In relation with Eq. (2) it was already pointed out that principally it decribes the relative retardation of the velocity of migration of radionuclides compared to the velocity of hydraulic flow. The equation is primarily valid for cations, tending to adsorb on negatively charged clay minerals in the aqueous media. Negatively charged species hardly sorb as is reflected for example with the mentioned low K_d value of ^{125}I. Further on, Boda Claystone has small porosity (1-3 %) and the water permeability also lies in an extremely low value region (10^{-20} – 10^{-23} m^2) [27]. Thus, the velocity if hydraulic flow can also be practically negligible. Consequently, there are at least two reasons that the velocity of spreading of negatively charged anionic species cannot be estimated by using Eq. (2). One reason is that the v_{gw} term it close to zero in the equation, the other reason is that anionic species do not sorb on clay minerals.

In order to obtain reliable information on the migration of negatively charged anionic species another option should be applied, namely to perform direct diffusion measurements.

4.3.2. Measurement of the diffusion coefficient of anionic species and HTO

To obtain approximately reliable information on the diffusion of radionuclides it is advised to use compact rock samples with their preserved consolidated structure. Various arrangements can be used. One option is to use a simple measuring arrangement, the break-through cells as mentioned in Section 2.3. The effective diffusion coefficients can be experimentally determined by using Eq (3).

It is also possible to collect information on the permeability and self diffusion of water itself by using a particular radiotracer, that is tritiated water, HTO.

Break-through measurements were carried out with HTO and with anionic long life-time fission product tracers ($^{99}TcO_4^-$ and ^{125}I) and additionally, with $H^{14}CO_3^-$. The compartments of c.a. 90 volume in the break-through cells were filled with ground water. The studied 8 mm thick borecore disc samples were obtained from the 570 m depth in the IB-4 borehole. Ground water in one side of the compartments was separately spiked with tracer amounts of the respective isotopes (10 MBq HTO, 6.2 MBq $^{99}TcO_4^-$ and 8 MBq $H^{14}CO_3^-$, respectively). These radionuclides emit β-radiation, the activities in small amounts of samples (10 microliter) taken regularly from the compartments were determined liquid scintillation detection. As an illustration, the measurements performed in a series with $H^{14}CO_3^-$ are shown in Figure 11 (next page).

The top part of the figure shows that after a minor decrease the activity (i.e. the concentration of $H^{14}CO_3^-$) is constant. The bottom part displays the break-through, the appearance and the increase of the concentration of the radionuclide in the ground water in the compartment on the other side of the sample disc. Note that only c.a. 2 % of the total amount of HCO_3^-

penetrates through the 8 mm thick disc within a year. The delay in the commencement of the break through indicates the interaction between the solute and solid phases. The intercept on the horizontal axis is c.a. 220 days, this is the value of α (retention coefficient) in Eq. (3).

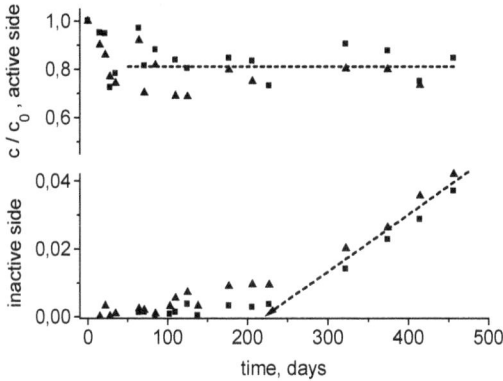

Figure 11. The relative activities of $^{14}CO_3^{2-}$ in the active (top) and originally inactive (bottom) compartments of the break-through cell at pH = 12 on two samples (marked with squares and triangles).

The delay in the break-through can probably be attributed to the exchange of radiocarbon to the carbonates in minerals of solid phase. Namely, the mesurements presented in Figure 11 were performed at pH=12. At this pH the 2 $HCO_3^- \leftrightarrow CO_3^{2-} + H_2O + CO_2$ equilibrium is shifted to the right, ie. CO_3^{2-} ions prevail over the hydrocarbonates, and some of the $^{14}CO_3^{2-}$ ions may be exchanged from the solution to the solid phase with carbonates in calcite (dolomite). The main data extracted from the measurements and evaluated in correspondence with Eq (3) are collected in Table 6. Further experimental details, description of samples and measurements performed with the β-emitting isotopes can be found in [28].

Migrating species	D_{eff} 10^{-12} m^2 s^{-1}	α
HTO	14	~ 0
$H^{14}CO_3^-$	~ 1	~ 0 [a]
	~ 1	1.5 [b]
$^{99}TcO_4^-$	~ 1	~ 0
$^{125}I^-$	~ 8 - 14 [c]	0.3 – 1.3 [c]

Remarks: measured at ([a]) pH = 8 and ([b]) pH= 12
([c]) values from different type measurements (see in [28])

Table 6. Effective diffusion coefficients and retention factors (α in Eq. 3) of neutral (HTO) and anionic migrating radionuclides

As for investigations related to the diffusion of iodide anion it should be mentioned that several more extended measurements have also been performed by using γ-emitting ^{125}I isotope (35.5 keV with half life-time only 60.2 days). The γ-radiation allows a more easy detection of the radiating isotope, for example presence and acitivity of ^{125}I in solid samples can directly be measured. The migration behaviour is expected to be the very same as the long half life time fission product [129]I (1.7 10⁷ years).

Two additional types of diffusion measurements were performed with [125]I. The in-diffusion was measured from spiked groundwater into borecore samples, and concentrations of iodine were determined in slices cut from the end of the solid borecore. In another arrangement the „in situ" conditions were modeled by placing the borecore samples under 100 bar pressure at 50 °C during the period of the measurements. These conditions correspond to the values of the ambient pressure and temperature in a rock located in 1000 m depth (characterising e.g. the former exploratory alpha drift). Two sets samples were used for these measurements, a few one was kept under these conditions for 118, another few for 340 days, the iodine contents were measured in the solid slices of sample borecore afterwards.

The results of the three types of measurements of diffusion coefficient of iodide (the break-through, the in-diffusion and the „in situ") were in good correspondence with similar values. Further experimental details and evaluation are described in [29].

As for the general evaluation of measurements with anions, data of Table 6 are in good correlation with the expectations. Namely, water and the anionic iodide exhibit high mobility, with similar D_{eff} values. This reflects the lack of interactions with clays. Hydrocarbonate and technetate anions move more slowly. In the first instance some interaction (exchange) may take place between carbonate ions in the solution and in the solid phase. Technetate may move slower due to the larger radius of the hydrated anion. Further on, it is worth comparing these values with the self diffusion coefficients of ions in strong aqueous electrolytes. For this end, the Bruggeman relation should be recalled which establishes connection between the measured effective diffusion coefficient (D_{eff}) and the self diffusion constant (D_s) through the porosity (ε) [30]:

$$D_{eff} = \sqrt{\varepsilon^3} \cdot D_s .$$ (4)

The self diffusion constant for ions in aqueous electrolytes is 1.7 10⁻⁹ m² s⁻¹ [31]. Taking into account the 1.5 – 3 % porosity, the measured D_{eff} values correlate well to the self-diffusion constant values characterising the electrolytes in aqueous soulutions. The similarity of the compared values reflects that there is no significant interaction between the anionic component and the minerals of the clay – they diffuse through the rock without any interaction with the pore walls (provided the α rock capacity factors are close to zero).

As an upper limit for the velocity of the migration of long life-time isotopes the obtained 10⁻¹² m² s⁻¹ can be considered. (The cations migrate more slowly since they interact and may became sorbed on the minerals.) As a very rough approximation the

$$x \approx \sqrt{D_{eff}\, t}$$ (5)

distance (radius) – time relation can be considered within which the radionuclides spread during their life-time. Inserting the long half life-times values into this approaching equation the order of magnitude of migration distance can be estimated. For illustration, for $^{99}TcO_4^-$ (half life-time 2.1 10^5 years) ca. 1.4 m is the distance, for the longer lifetime ^{129}I (1.7 10^7 years) ~ 30 m is the distance of spreading within one half lifetime period in Boda Claystone media. These distance values should be considered with precaution. First, only one half lifetime period was considered, the amount of radiation emitting isotopes will be only halved within this period. The second point is our starting precondition, namely the hydraulic flow was neglected. Thus, in real cases the distances of migration can be significanly longer.

Some properties of the other clayrock types considered for final disposal for nuclear wastes were also briefly mentioned in previous section 3.1 (Boom Clay, Opalinus Clay, and Callovo-Oxfordian argillite). Here, at the end of the present section dealing with diffusion constant measurements on Boda Claystone samples, a reference should also be given to an overview, which provides a detailed compilation and description of recent diffusion studies performed in the framework of the EC FUNMIG IP project on all the mentioned four types of clayrocks [32]. In general diffusion data obtained on Boda Claystone suit well within those collected on the other three types of rocks which had been extensively studied and characterized in the project.

4.3.3. Break-through studies with uranium

In the preceeding sections mostly sorption and diffusion properties of fission products were discussed on Boda Claystone samples. To consider the behaviour and interactions of the long life-time actinide components with minerals in Boda Claystone the major component of the high level nuclear wastes, uranium was also investigated in break-through experiments. Uranium is similar to neptunium in respect with the occurence of various oxidation states (as was shown for Np in Figure 2.) Depending on the Eh – pH conditions uranium may be stabilized either in cationic, or in anionic form or as stable neutral precipitate. The U-C-O-H diagram is shown – it can be applied for carbonate containing rocks and in correspondence hydrocarbonate contaning ground water, which is the case for Boda Claystone. Stability regions of iron are also shown and separated in black dashed lines (Figure 12 - based on [4]).

The life cycle of the disposal site is also shown in blue (similarly to Np in Fig 2). At the start uranium (or a part of it) may be dissolved in form uranyl ions (UO_2^{2+}) into the ground water. This species is a cation and prefers to ad- or chemisorb on the clay minerals. After having the site closed the potential decreases, the pH increases – neutral UO_2CO_3 may precipitate. In a subsequent stage with the further increase of pH negatively charged carbonates can be formed which tend to migrate in the aqueous phase. In the final stage with the decrease of potential the reduction prevails and stable UO_2 forms from the various migrating U(VI) uranyl carbonate species (shown as the orange encircled area in Figure 12). In the encircled region $Fe^{2+} \leftrightarrow Fe^{3+}$ redox processes may also proceed, maybe they can be coupled with the mentioned U(VI) → U(IV) reduction.

In the corresponding break-through experiments the ground water was saturated with uranyl acetate (using uranium with natural isotope composition, dominant ^{238}U with ca. 1% ^{235}U). The changes of activities of uranium were measured in the compartments of the break-through cells. Furthermore, the distribution of uranium along the path of migration in the bore core

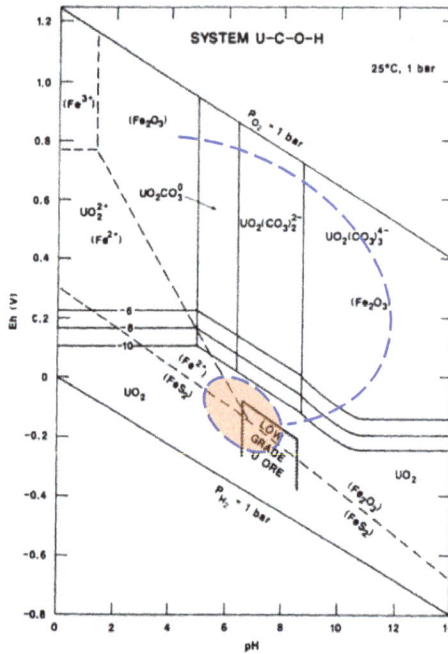

Figure 12. The Eh-pH Pourbaix-diagram of uranium. The dashed line represents the possible changes during the life cycle of the repository, the encircled region might be the final stage.

samples was also determined by high sensitivity laser ablation ion-coupled mass spectrometry (LA-ICP-MS) after finishing the experiments and dismounting the break-through cells. The change of concentration (activity) of uranyl ions in dependence of time is shown in Figure 13. Note the long time interval, the samples were kept in the cells for ca. five years.

Primarily the depletion of uranium is seen in the starting solution in Figure 13. In the very first period a sudden drop takes place, c.a. 20 % of uranium disappears from the solution, most probably due to precipitation of carbonates via the HCO_3^- $\leftrightarrow$ CO_3^{2-} equilibrium in the solution, where almost all the carbonates are consumed. After having this fast process finished the amount of uranium had been stabilised for c.a. one year at the 70 – 80 % level of the original concentration. Later on the starting solution is considerably depleted in uranium. Uranium is apparently trapped in the solid borecore disc, simultaneous measurements on liquid samples does not show any sign of through diffusion. Negligible increase of the amount of uranium is detected in samples taken from the opposite compartment of break-through cell (bottom curve in the Figure).

In correspondence, the LA-ICP-MS measurements directly prove the event of trapping. A typical distribution of concentration (intensity) of uranium is shown in Figure 14, which displays the distribution of uranium along the migration path, crossing the borecore dics. (Notice the logarithmic scale on the vertical intensity/concentration axis.)

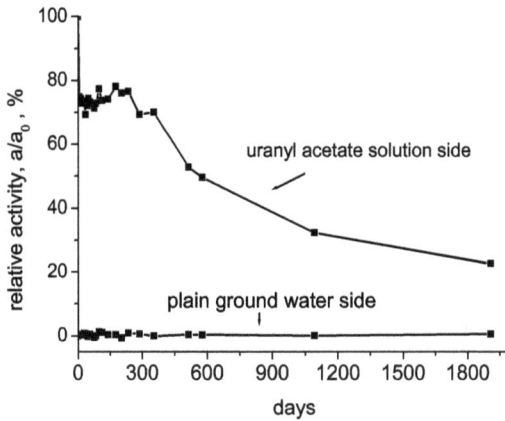

Figure 13. Changes of concentations of uranium in the two compartments of a break-through cell during ca. 5 years. Top curve is the original solution, the bottom curve is for the opposite compartment.

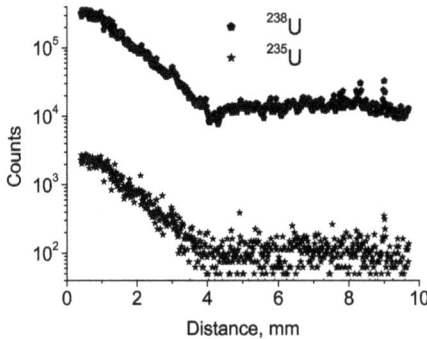

Figure 14. ^{238}U and ^{235}U contents measured in the sample after finishing the 5 year break-through experiment. The left side of sample contacted the uranyl solution. Note the logarithmic scale on the vertical axis.

The figure illustrates the sensitivity of the method, even the minor uranium component, ^{235}U (a few ppb in the sample, ~ 1 % in the natural uranium component), can convincingly be measured. It is seen that the characteristic longest migration distance was only c.a. 4 mm during the five years duration. (The horizontal sections between 4 and 10 mm distances refer to the ^{238}U and ^{235}U contents present originally in the sample before the start of measurements.) At the first approximation the drops of concentrations/intensities of ^{238}U and ^{235}U can be represented by straight lines in logarithmic scale, thus the decrease can be described some power dependence of distance covered by the migration. Further experimental details of the measurements are described in [33].

A rough estimation can be performed by using the square-root approximation mentioned with Eg (5). Inserting 4 mm and 1900 days into Eq (5), a nominal $D_{eff} \approx 10^{-13}$ m^2 s^{-1} can be obtained.

This value is less with two orders of magnitude than those obtained for the mobile iodine and HTO, and is less with one order of magnitude obtained for the less mobile hydrocarbonate and technetate (shown in Table 6). Thus this comparison with neutral and anionic species can also be considered as an indirect proof the measurements with uranium. The slow migration of uranium can probably be attributed to more than one single factor. Beside chemisorption of uranyl ions on clay minerals interaction with (hydro)carbonate anions may take place with forming partly immoblie carbonate species (see the Eh-pH diagram in Figure 12). Fe^{2+}/Fe^{3+} ratios were measured by Mössbauer spectroscopy in the surface layers of the borecore samples after having the uranium break-through measurements finished. They were the same as at the start. In consequence the $Fe^{2+} \rightarrow Fe^{3+}$ process coupled with $U(VI) \rightarrow UO_2$ reduction, as mentioned in the introductory part of this section, does not play role [33].

5. Summary

Certain claystones have been considered as appropriate media for long time isolation of high level nuclear waste depositories. Various interactions between the migrating anionic and/or cationic species of radionuclides with the constituents of minerals may take place. For example, the open, layered structure of clay minerals is advantageous for this purpose resulting in high ad- and chemisorption capacities to retard migration of long life-time radioisotopes. Characteristic related properties of different clayrock formations considered as possible hosts in Europe are briefly mentioned and compared. Boda Claystone is discussed in more detail, diagenesis, lithology and mineralogical properties are described. Results of various specific measurements on claystone samples are presented. Particular attention is devoted to sorption and diffusion measurements performed with anionic fission product and actinide radionuclides. From these results some characteristic distances which can be traversed by these radionuclides in the clay rock during the life time of the depository site are estimated.

Author details

Károly Lázár*
Centre for Energy Research, Institute of Isotopes, Hungarian Academy of Sciences, Budapest, Hungary

Zoltán Máthé
Mecsekérc Plc., Pécs, Hungary

Acknowledgement

The research leading to these results has received funding from the European Union's European Atomic Energy Community's (EURATOM) sixth and seventh Framework Programmes under grant agreements FP6-516514 (FUNMIG) and FP7-212287 (ReCosy), respectively. The authors are thankful for the fruitful informations and knowledge provided by the opportunity with the participation in these projects. Special thanks are due to the

* Coresponding Author

colleagues who participated the work in its different stages (J. Megyeri, P. Mell., T. Szarvas, and late L. Riess, at the Institute of Isotopes), and particular thanks are also due to colleagues who contributed to the characterisation of samples by analysing them with specific and unique methods (J.-C. Parneix and M. Perdicakis). The courtesy of PURAM for providing the samples for analysis is also appreciated.

6. References

[1] Ojovan M.I, Lee W.E, (2005) An Inroduction to Nuclear Waste Immobilisation, Elsevier, 310 p.
[2] Lieser K H (1997) Nuclear and Radiochemistry: Fundamentals and Applications, VCH, Wiley, Weinheim, 460 p.
[3] Extrapolation of Short Term Observations to Time Periods Relevant to the Isolation of LongLived Radioactive Waste, IAEA-TECDOC-1177, IAEA, Vienna, (2000) 104 p.
[4] Brookins D G, (1988), Eh-pH Diagrams for Geochemistry, Springer, Berlin 176 p.
[5] Brookins D G (1984) Geochemical Aspects of Radioactive Waste Disposal, Springer, Berlin. 347 p.
[6] Stability and Buffering Capacity of the Geosphere for the Long-term Isolation of Radioactive Waste (Application to Argillaceous Media), „Clay Club" Workshop Proceedings, Braunschweig, Germany, Dec. 2003, OECD 2004, NEA 5303, 244 p.
[7] Clay in Natural and Engineered Barriers for Radioactive Waste Confinement, Ed: Aranyossy J-F. Special issue of Physics and Chemistry of the Earth, Vol. 32 (2007) 965 p.
[8] The Use of Scientific and Technical Results from Underground Research Laboratory Investigations for the Geological Disposal of Radioactive Waste, IAEA-TECDOC-1243, IAEA, Vienna (2001) 76 p.
[9] Altmann S. (2009) RTD Component 3, in: Fundamental Processes of Radionuclide Migration, FUNMIG Workshop Proceedings, (Ed. Buckau G et al.) Forschungszentrum Karlsruhe, Wissenschaftliche Berichte FZKA 7461, 45 – 82.
[10] Dossier 2005 Argile, Safety Evaluation of a Geological Repository (2005) ANDRA 782 p
[11] Project Opalinus Clay, Safety Report, NAGRA Technical Report 02-05, (2002) 472 p.
[12] Haas J, Péró Cs, (2004), Mesosoic evolution of the Tisza Mega-unit. International Journal of Earth Sciences 93: 297-313.
[13] Konrád Gy, Sebe K, Halász A, Babinszki E, (2010), Sedimentology of a Permian playa lake: the Boda Claystone Formation, Hungary. Geologos 16: 27-41.
[14] Jámbor Á, (1964), Lower Permian formations of the Mecsek Mountains. Manuscript, Mecsek Ore Environment Company, Pécs (In Hungarian.)
[15] Barabás A, Barabás-Stuhl Á, (1998), Stratigraphy of the Permian formations in the Mecsek Mountains and its surroundings. In: Stratigraphy of Geological Formations of Hungary. Mol Plc. and Hung. Geol. Institute, Budapest, pp. 187-215. (In Hungarian.)
[16] Summary Report of the Site Characterisation Program of the Boda Siltstone Formation (1998), Vol 4. Ed: Máthé Z., Manuscript, Mecsek Ore Environment Co., Pécs. 76 p.
[17] Árkai P, Balogh K, Demény A, Fórizs I, Nagy G, Máthé Z, (2000), Composition, diagenetic and post-diagenetic alterations of a possible radioactive waste repository site: the Boda Albitic Claystone Formation, southern Hungary. Acta Geologica Hungarica, 43: 351-378.

[18] R-Varga A, Szakmány Gy, Raucsik B, Máthé Z, (2005), Chemical composition, provenance and early diagenetic processes of playa lake deposits from the Boda Siltstone Formation (Upper Permian), SW Hungary. Acta Geol. Hungarica 48: 49-68.

[19] R-Varga A, Raucsik B, Szakmány Gy, Máthé Z, (2006), Mineralogical, petrological and geochemical characteristics of the siliciclastic rock types of Boda Siltstone Formation. Bulletin of the Hungarian Geological Society, 136: 201-232.

[20] Digest on the Results of the Short-term Characterisation of the Boda Claystone Formation, Ed. Kovács L., Mecsekérc-Puram, Pécs-Paks, (1999) 61 p.

[21] Szűcs I., Csicsák J., Óvári Á., Kovács L., Nagy Z. (2004) Confinement performance of Boda Claystone Formation, Hungary, in: Stability and Buffering Capacity of the Geosphere for the Long-term Isolation of Radioactive Waste (Application to Argillaceous Media), „Clay Club" Workshop Proceedings, OECD 2004, NEA 5303. p. 209-224.

[22] Mösssbauer Mineral Handbook (1998) Eds: Stevens J.G. et al., Mössbauer Effect Data Center, 527 p.

[23] Lázár K, Máthé Z, Földvári M (2010) Various redox conditions in Boda Claystone as reflected in the change of Fe^{2+}/Fe^{3+} ratio in clay minerals, Journal of Physics: Conference series 217: 012053.

[24] Perdicakis M, Xu Y L, Lázár K., Máthé Z., Rouillard L., (2012) Voltammetric characterization of Boda Albitic Claystone: Comparison with Mössbauer spectroscopy data, Electroanalysis, submitted.

[25] Mell P, Megyeri J, Riess L, Máthé Z, Csicsák J, Lázár K (2006) Sorption of Co, Cs, Sr and I onto argillaceous rock as studied by radiotracers, Journal of Radioanalytical and Nuclear Chemistry, 268: 405-410

[26] Sipos P, Németh T, Máthé Z (2010) Preliminary results on the Co, Sr and Cs sorption properties of the analcime-containing rock type of the Boda Siltstone Formation, Central European Geology, 53: 67-78.

[27] Fedor F, Hámos G, Jobbik A, Máthé Z, Somodi G, Szűcs I (2008) Laboratory pressure pulse decay permeability measurement of Boda Claystone, Mecsek Mts., SW Hungary, Physics and Chemistry of the Earth 33: S45-S53.

[28] Lázár K, Megyeri J, Parneix J-C, Máthé Z, Szarvas T, (2009) Diffusion of anionic species (99TcO4-, H14CO3-) and HTO in Boda Claystone borecore samples, in: Fundamental Processes of Radionuclide Migration, FUNMIG Workshop Proceedings, (Ed. Buckau G et al.) Forschungszentrum Karlsruhe, Wissenschaftliche Berichte FZKA 7461, 199 – 204.

[29] Mell P, Megyeri J, Riess L, Máthé Z, Hámos G, Lázár K (2006) Diffusion of Sr, Cs, Co and I in argillaceous rock as studied by radiotracers, Journal of Radioanalytical and Nuclear Chemistry, 268: 411-417

[30] Rose D A, (1963) Water movement in porous materials, British Journal of Applied Physics, 14: 256-262

[31] Erdey Grúz T., Schay G, (1962) Theoretical Physical Chemistry, Tankönyvkiadó, Budapest, Vol 3. 447 p. (in Hungarian)

[32] Altmann S, Tournassat C, Goutelard F, Parneix J-C, Gimmi T, Maes N, (2012) Diffusion-driven transport in clayrock formations, 27: 463-478.

[33] Lázár K, Megyeri J, Mácsik Zs, Széles É, Máthé Z, (2011) Migration of uranyl ions in Boda Claystone samples, in: Redox Phenomena Controlling Systems, RECOSY 3rd Annual Workshop Proceedings, (Ed. Altmaier M, et al.) KIT Scientific Reports 7603, 91-97

Distribution and Origin of Clay Minerals During Hydrothermal Alteration of Ore Deposits

Miloš René

Additional information is available at the end of the chapter

1. Introduction

Hydrothermal alterations of host rocks (granites and metasediments) connected with origin Sn-W and U deposits are often accompanied by origin of chlorite, clay minerals and white mica (muscovite, hydromuscovite, phengite) [1, 2, 3, 4, 5, 6, 7]. Clay minerals originated usually in last stages of these alterations, when temperature of hydrothermal fluids is in range of 50–200 ℃. In area of Central European Variscan belt (Bohemian Massif) occur a few Sn-W- and U-ore deposits in which are evolved altered rocks with highly interested chlorite, clay minerals and white mica assemblages (Fig. 1). This paper is concentrated on description and discussion of chloritization and argillization, originated during alteration of host rocks series at selected Sn-W and U ore deposits in the area of the Bohemian Massif (Czech Republic).

2. Geological background

The Sn-W greisen deposits are connected with topaz-granite stocks in the Saxothuringian Zone of the Bohemian Massif (Krásno–Horní Slavkov ore district, Cínovec). The Krásno–Horní Slavkov ore district comprises topaz-bearing granite stocks evolved along the southeastern margin of the Krudum granite body in the Slavkovský les Mts. area (Fig. 2). The inner structure of these stocks (Hub, Schnöd and Vysoký Kámen) is well stratified, comprising partly greisenized topaz-albite granites, leucocratic topaz-albite granites and layers of alkali-feldspar syenites. In upper parts of the Hub and Schnöd stocks are evolved topaz-mica greisens, accompanied by partly greisenized topaz-albite granites and distinctly argillitized topaz-albite granites. The highly interested clay mineral assemblage occurs in Sn-W ore spots enclosed in greisens [8] and in argillitized topaz-bearing granites. The Cínovec granite stock is relatively small, elliptical, vertical stratified body occurred in the central part of the Altenberg-Teplice caldera. The borehole CS-1, located in the center of granite stock, transacted lepidolite-bearing granite at the top of the section (about 90 m thick), an intermediate zone of zinnwaldite-bearing granite (thickness about 640 m) and a lower zone of protolithionite-bearing granite to the depth 1596 m. In uppermost part of granite stocks occurs irregular topaz-mica greisen bodies together with flat Sn-W ore veins enriched also in zinnwaldite and quartz. Clay minerals occur usually as filling of small cavities in quartz and/or as filling of small fissures accompanied flat ore veins [9].

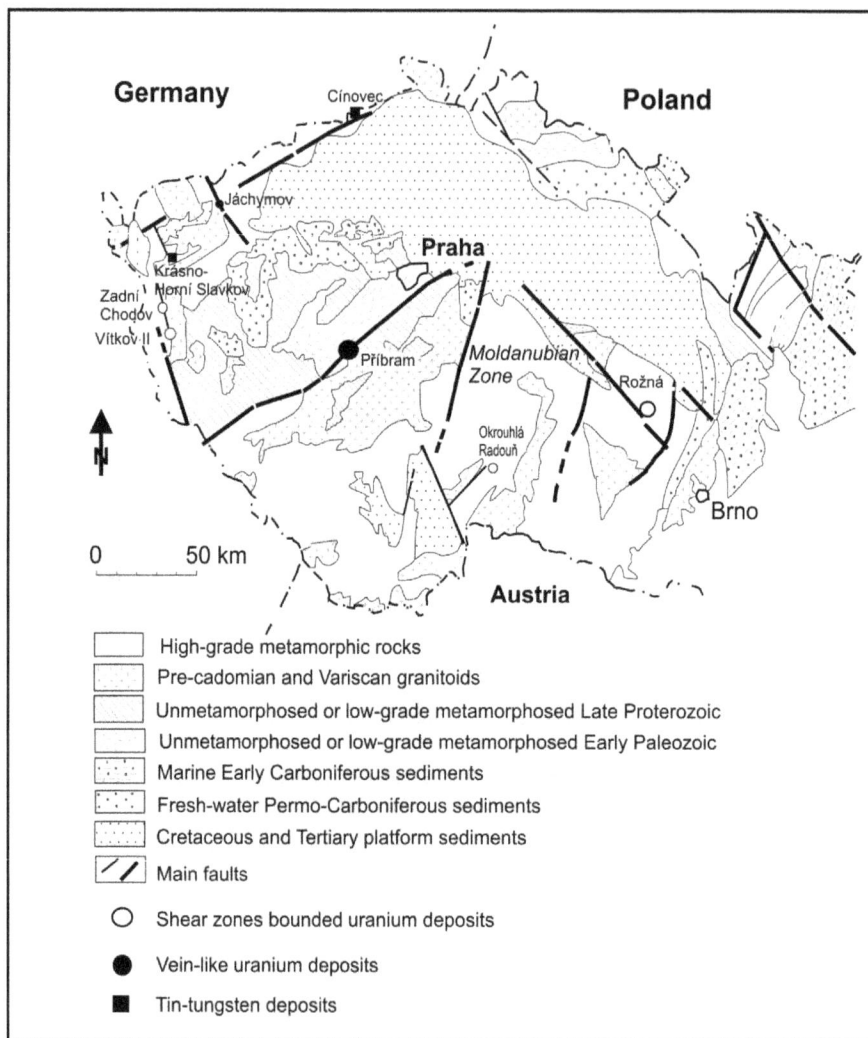

Figure 1. Geological sketch map of the Bohemian Massif.

Uranium ore deposits with a huge evolved argillization of country rocks occur in some shear zones of the Moldanubian Zone (Rožná–Olší ore district, Okrouhlá Radouň, Zadní Chodov, Vítkov II). The Moldanubian Zone represents a central, deeply eroded part of the Bohemian Massif. Therefore, in present-day section, it is composed dominantly of plutonic and high-grade metamorphic rocks (two-mica and biotite granites of the Moldanubian batholith, Třebíč pluton and Bor pluton).

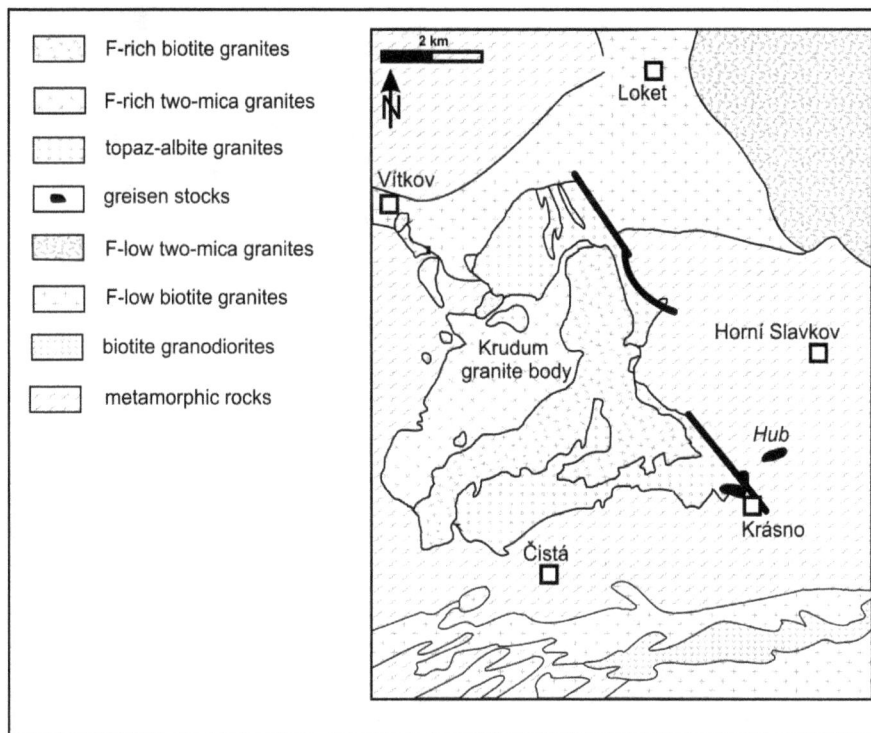

Figure 2. Geological sketch map of the Krásno–Horní Slavkov ore district.

The Rožná–Olší uranium district lies in the NE part of the Moldanubian Zone of the Bohemian Massif (Fig. 3). The high-grade metamorphic rocks of this zone were overthrust on its NE boundary by the Svratka Crystalline unit and on the easterly located Cadomian Brunovistulian foreland. The high-grade paragneisses of the Moldanubian Zone are subdivided into Monotonous, Varied and Gföhl Unit. The Rožná–Olší ore district is located in the uppermost Gföhl unit. The host rocks of the Rožná U-deposit consist mainly of biotite paragneisses and amphibolites with small bodies of calc-silicate rock, marble, serpentinite and pyroxenite. The subsequent exhumation of these rocks series to middle crustal levels was associated with kilometer-scale isoclinal folding. Longitudinal N–S to NNW–SSE striking ductile shear zones (Rožná and Olší shear zones) dip WSW at an angle of 70–90º

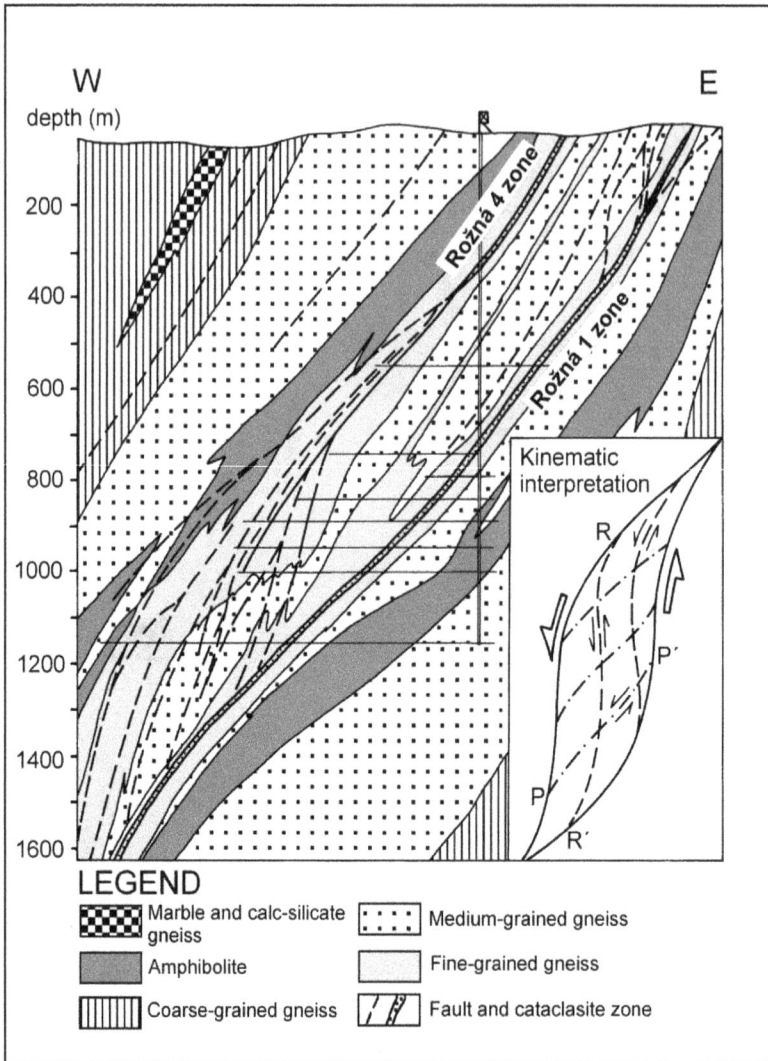

Figure 3. Cross section of the Rožná uranium deposit.

and strike parallel to the tectonic contact between the Gföhl unit and the Svratka Crystalline Unit. The main longitudinal faults of the Rožná shear zone are designated as Rožná 1 (R1) and Rožná 4 (R4) and host the main part of the disseminated uranium mineralization. The less strongly mineralized Rožná 2 (R2) and Rožná 3 (R3) fault zones host numerous separate pinnate carbonate veins. Longitudinal fault structures are crosscut and segment by steep, ductile to brittle NW–SE and SW–NE-striking fault zones that host post-uranium carbonate-quartz-sulfide mineralization. Uranium mineralization forms (i) disseminated coffinite>uraninite>U-Zr-silicate ore in chloritized, pyritized, carbonatized, and graphite-enriched cataclastites of longitudinal faults, (ii) uraninite>coffinite ore in carbonate veins, (iii) disseminated coffinite>uraninite in desilicified, albitized, and hematitized gneiss (episyenite) adjacent to longitudinal faults and (iiii) mostly coffinite ore bound to the intersection of the longitudinal structures. Total mine production of the Rožná-Olší ore district was 23,000 tons U with average grade of 0.24 % U. The Rožná uranium deposits is the last recently mined uranium deposit in Central Europe with annual production about 300 t U [10].

Okrouhlá Radouň uranium deposit is formed by NNW–SSE-striking shear zone occurred on the northeastern margin of the Klenov two-mica granite body in the southern part of the Bohemian Massif. Host-rock series of this ore deposit are formed by high-grade metasediments of the Moldanubian Varied group and peraluminous two-mica granites of the Moldanubian batholith. The shear zone is filled with cataclasites formed by host rocks, altered to clay minerals-rich and chlorite-rich assemblages with uranium mineralization enriched in coffinite, partly also in pitchblende [11, 12, 13]. Uranium ore deposits in the Bor pluton (Zadní Chodov, Vítkov II) are located in N-S to NW-SE shear zones evolved in biotite monzogranites of I/S-type [14, 15]. The hydrothermal alterations associated with uranium mineralization are represented particularly by the removal of quartz, chloritization, albitization, hematitization and origin of younger generations of chlorite and white mica (muscovite, phengite). Shear zones evolved on the west margin of the Bor pluton (Zadní Chodov), on the boundary between granites of the Bor pluton and metasediments of the Moldanubian Zone are distinctly enriched in more generation of chlorite accompanied by various clay minerals (smectite) [16].

3. Analytical methods

Clay minerals and chlorites were analyzed in polished thin sections using a CAMECA SX 100 electron microprobe working in WDX mode employing the PAP matrix correction program [17] at the Institute of Geology of the Academy of Sciences of the Czech Republic. The operating conditions were 15 kV acceleration voltage, 15 nA beam current, and 1–2 μm beam diameter. Counting times on the peaks were 10–30 seconds depending on the element. Background counts were measured in each case in half the time for peak measurement on both sides of the peak. Calibrations were done using standard sets from SPI. Standards included fluorite (F), jadeite (Na, Al), diopside (Ca), leucite (K), magnetite (Fe), quartz (Si), periclase (Mg), rhodonite (Mn), rutile (Ti), spinel (Cr) and tugtupite (Cl). Detection limits for these elements are as follows: F 0.09–0.15 wt%, other elements 0.03–0.20 wt%. Formulae of chlorite were calculated in relation to 36 (O, OH) atoms per formula unit (apfu), formulae of white mica and illite were calculated in relation to 24 (O, OH) apfu and formulae of clay minerals (kaolinite, smectite) were calculated on the basis of 18 (O, OH). For these calculations was used MINPET software. For calculation of chlorite thermometry was used a six-component chlorite solid solution model according Walshe (1986) [18].

Major elements in whole rock samples were determined by X-ray fluorescence spectrometry using the PANanalytical Axios Advanced spectrometer at Activation Laboratories Ltd., Ancaster, Canada. Trace elements were determined by ICP MS (a Perkin Elmer Sciex ELAN 6100 ICP mass spectrometer) at Activation Laboratories Ltd., Ancaster, Canada (Table 1). Whole rock samples enriched in clay minerals were used for sampling of clay mineral fractions. The size fraction of clay minerals in size below 4 μm was prepared by conventional sedimentation method. X-ray diffraction (XRD) analysis of clay minerals in clay-size fractions were obtained on untreated, ethylene-glycol solvated and heated samples using a Philips PW 7310 diffractometer with CuK$_\alpha$ radiation (40 kV, 40 mA) and Ni filter standard set.

4. Results

4.1. Petrology

The investigation of clay minerals assemblage occurred in Sn-W ore deposits of the Saxothuringian Zone (Bohemian Massif, Czech Republic) was concentrated on assemblages occurring in Sn-W ore pockets and highly argillized topaz-albite granites of the Hub stock (Krásno–Horní Slavkov ore district). The ore pockets are globular or even irregular bodies tens of centimetres in size, with a very high proportion of cassiterite, which are enclosed in topaz-mica greisens. Quartz, Li-micas, fine flakes of white mica (muscovite) and clay minerals (dickite, kaolinite, very rare cookeite) are the accompanying minerals of these ore pockets. Clay minerals matter filling small cavities between a bigger cassiterite grains has a grey appearance being a mixture of dickite, Li-, Al-chlorite (cookeite) with dickite and white mica. The identification of cookeite and illite was performed by X-ray diffraction method.

In highly argillized topaz-albite granites a complex clay minerals assemblage was identified using X-ray diffraction method and microprobe analysis. This very fine-grained assemblage enclosed between bigger grains of quartz, topaz and tables of Li-mica is formed of smectite, illite, kaolinite, dickite, chlorite and corrensite. Corrensite was identified on the basis of 29.2 Å XRD reflection on natural, oriented sample and 31.1 Å reflection after ethylene glycol treatment. These XRD reflections are significant for identification of corrensite [31]. In greisenized topaz-albite granites occurs sometimes also white mica (muscovite, hydromuscovite).

Disseminated uranium mineralization occurred in shear zones of the Rožná, Okrouhlá Radouň, Zadní Chodov and Vítkov uranium deposits comprises usually three stages (pre-ore, ore and post-ore stages). The first two stages are of the late-Variscan age; the last stage is very probably of the post-Variscan age. The pre-ore stage is characterized by a huge occurrence of inherited chlorite originated by chloritization of biotite (chlorite I). In syn-ore stage originated authigenic chlorite (chlorite II), together with authigenic Mg-Fe chlorite (chlorite III) occurred often as filling of small cavities in intensively altered paragneisses (Rožná, Okrouhlá Radouň, Zadní Chodov) and/or in altered granites (Okrouhlá Radouň, Vítkov II). During pre-ore stage originated also as relatively rare mineral white mica

wt%	Ko-55	R-1	R-2	Re-503	Re-510	OR-99	ZCH-6
SiO$_2$	75.52	51.85	45.76	57.42	53.81	49.42	44.30
TiO$_2$	0.03	1.30	0.98	0.62	0.08	0.30	0.33
Al$_2$O$_3$	13.23	18.36	15.37	18.76	18.35	15.47	9.45
Fe$_2$O$_3$ tot.	1.75	9.24	6.30	4.73	1.23	6.25	12.44
MnO	0.11	0.14	0.15	0.14	0.14	0.07	0.24
MgO	0.20	2.85	2.25	1.50	0.55	2.20	14.97
CaO	1.24	3.69	11.10	8.27	8.39	6.64	1.54
Na$_2$O	0.20	5.95	6.25	7.99	6.24	0.46	0.23
K$_2$O	1.65	1.89	0.97	0.55	1.10	1.38	0.01
P$_2$O$_5$	0.19	0.30	0.21	0.38	0.20	0.29	0.49
L.O.I.	5.40	4.41	9.52	0.33	9.32	17.55	15.81
Total	99.52	99.98	98.86	100.69	99.41	100.03	99.81
ppm							
U	16	877	833	232	353	436	4553
Th	3	4	6	20	7	8	21
Y	4	33	39	23	11	28	69
Zr	25	243	261	186	51	169	118
Ba	8	456	497	366	388	3	35
Rb	886	88	40	15	55	81	1
Sr	17	230	466	176	296	197	46
La	1.55	12.00	21.00	42.90	11.00	19.10	12.80
Ce	2.13	27.70	44.80	82.70	19.51	40.10	35.10
Pr	0.25	3.85	5.50	9.44	1.94	4.40	5.69
Nd	1.29	17.20	22.70	35.00	7.13	17.10	30.60
Sm	0.40	5.61	5.46	6.53	1.83	4.40	14.40
Eu	0.007	0.65	1.20	1.14	0.066	2.40	7.14
Gd	0.35	6.14	6.25	6.24	1.89	4.40	13.50
Tb	0.09	1.34	1.22	0.83	0.34	0.82	2.72
Dy	0.61	9.13	7.26	4.85	2.45	5.40	15.30
Ho	0.09	1.86	1.47	0.86	0.48	1.10	2.74
Er	0.27	6.10	4.06	2.54	1.45	2.40	7.45
Tm	0.06	0.99	0.67	0.38	0.20	0.44	1.11
Yb	0.36	6.68	4.22	2.25	1.49	3.20	6.90
Lu	0.04	1.03	0.60	0.31	0.22	0.46	0.86
ΣREE	7.50	100.29	126.41	195.97	50.00	105.72	156.31
La$_N$/Yb$_N$	2.91	1.21	..3.36	12.88	4.96	4.03	1.25
Eu/Eu*	0.06	0.34	0.63	0.55	0.11	1.67	1.56
Th/U	0.188	0.005	0.007	0.087	0.020	0.018	0.005

Table 1. Representative analyses of altered rocks from the Sn-W and U ore deposits of the Bohemian Massif.
Ko-55 – argillized topaz-albite granite, Hub stock, Krásno–Horní Slavkov ore district, R-1, R-2 – altered biotite gneiss, Rožná uranium deposit, Re-503 – altered biotite gneiss, Okrouhlá Radouň uranium deposit, Re-510 – altered two-mica granite, Okrouhlá Radouň uranium deposit, OR-99 – altered biotite

gneiss, Okrouhlá Radouň uranium deposit, ZCH-6 altered biotite gneiss, Zadní Chodov uranium deposit. REE – rare earth elements, La_N/Yb_N = LREE/HREE (light rare earth elements/heavy rare earth elements, $Eu/Eu^* = Eu_N/\sqrt{(Sm_N \times Gd_N)}$. Normalising values of chondrites are from Taylor and McLennan [29].

(muscovite, hydromuscovite, phengite). For syn-ore stage is origin of various clay minerals (Fe-illite, smectite, kaolinite) significant. Compared with voluminous pre-ore and syn-ore stage alteration, post-ore stage alteration is usually restricted to origin of small authigenic chlorite-carbonate veins and/or veilets and disseminations of chlorite in carbonatized host rocks (chlorite IV). The origin of the youngest chlorite is sometimes accompanied by origin of clay minerals (illite, kaolinite).

5. Geochemistry of altered rocks

The chemical composition of arigillized topaz-albite granites connected with Sn-W mineralization was investigated in area of the Hub stock (Krásno–Horní Slavkov ore district). Argillitized granites are in comparison with original topaz-albite granite enriched in CaO and MgO and depleted in alkalies (Fig. 4). During argillization of topaz-albite granites were also accessory minerals (monazite and zircon) partly dissoluted and argillized granites were depleted in REE (Fig. 5).

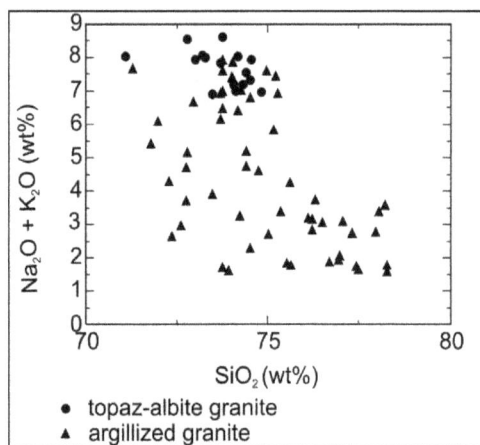

Figure 4. Plot of Na_2O + K_2O (wt%) vs. SiO_2 (wt%) for arigillized granites from the Krásno–Horní Slavkov ore district.

Geochemistry of altered high-grade metamorphic rocks was studied in the area of the Rožná, Okrouhlá Radouň and Zadní Chodov uranium ore deposits. The distribution of REE in barren, pre-ore altered (desilicified, albitized, hematitized and chloritized) biotite paragneisses of all three examined uranium deposits usually display patterns similar to those of the parent paragneisses. Barren, syn-ore argillized, chloritized and in the Okrouhlá Radouň uranium deposit also strongly carbonatized rocks show significantly lower bulk contents of REE (49–232 ppm) and relatively high LREE/HREE ratios (8.7–17.6) in comparison with hydrothermally unaffected gneisses. Higher LREE/HREE ratio (12.8–16.1), i.e. high depletion on HREE was found in graphitised cataclastites from the Rožná uranium deposit, which are characterized by the lowest bulk content of REE (49–98 ppm) (Fig. 6).

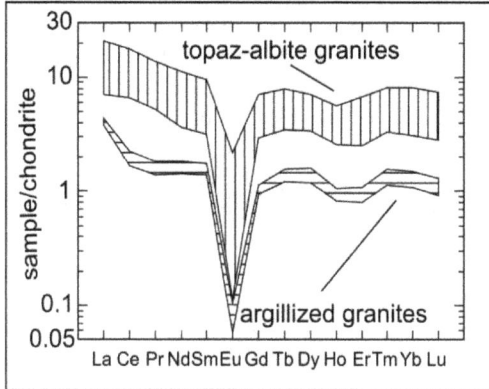

Figure 5. Chondrite normalized REE patterns for topaz-albite granites and argillized granites from the Krásno–Horní Slavkov ore district. Normalising values are from Taylor and McLennan [29].

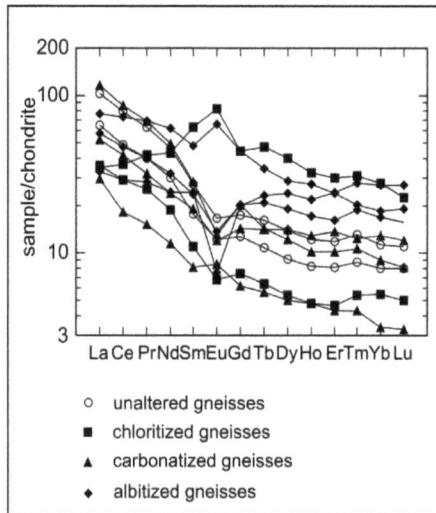

Figure 6. Chondrite normalized REE patters for hydrothermally altered rocks from the Rožná and Okrouhlá Radouň uranium deposits. Normalising values are from Taylor and McLennan [29].

The geochemistry of altered granites connected with uranium mineralization was studied in the Okrouhlá Radouň and Vítkov II uranium deposits. Hydrothermal alteration of two-mica granites from the southern part of the Okrouhlá Radouň uranium deposit is characterized by a higher Fe_2O_3/FeO ratio and by a significant depletion in SiO_2 contents. Highly altered granites typically show high contents of Na$_2$O and usually low contents of K$_2$O. The content of K$_2$O is higher only at the presence of higher amounts of newly originated white mica (muscovite) and/or illite. The later carbonatization of quartz-depleted altered granites is characterized by high contents of CaO and CO$_2$. The hydrothermal alteration of granites is, due to the dissolution of K-feldspar, connected with a depletion in Rb and sometimes also with the evolution of a prominent negative europium anomaly (Eu/Eu* = 0.11). The higher content of HREE in altered granites is connected with the origin of uranium mineralization and a higher concentration of HREE in coffinite. Chloritization of porphyric biotite granite from the Bor pluton at the Vítkov II uranium deposit are accompanied by the silica removal, which continued during the argillization. A moderate increase in TiO$_2$ and P$_2$O$_5$ contents occurred in the course of the hydrothermal alterations. Contents of Al$_2$O$_3$ and Fe$_2$O$_3$ increased during the chloritization and argillization. In the course of chloritization, the content of FeO and MgO increased considerably, which is reflected by the formation of chlorites I richer in iron. The MgO content also increased sizeably during argillization. The content of CaO increases in the granites affected by carbonatization and it decreases in the rocks affected by chloritization and in majority of argillized granites. The content of Na$_2$O increases in the rocks affected by albitization, it is considerably lower in the argillized granites.

6. Composition of chlorite

In altered paragneisses and granites of above-mentioned uranium ore deposits chlorite occurs usually in four generations (chlorite I, II, III, IV). The main portions of chlorite are formed by inherited chlorite I and inherited to authigenic chlorite II, which occurs in shear zones. These shear zones host the main part of the disseminated uranium mineralization. The chlorite I often preserves the morphology of original biotite, whereas chlorite II forms fluidal aggregates in cataclastites, represented the predominant filling of shear zones. The both later chlorite generations (chlorite III and IV) crystallized in free voids of rocks originated due to quartz dissolution. The composition of these four chlorite generations in individual uranium deposits is quite different. The inherited chlorite of pre-ore and syn-ore stage is pycnochlorite to brunsvigite (Fig. 7, 8, 9).

In the Rožná uranium deposit occurs also Mg-rich inherited to authigenic chlorite (diabantite), which forms fluidal aggregates in shear zones (matrix chlorite). For inherited chlorite from the Rožná uranium deposit is significant relatively high content of Si (up to 7.21 apfu). The content of Si in inherited chlorite from pre-ore stage of the Okrouhlá Radouň, Zadní Chodov and Vítkov II uranium deposits is distinctly lower (4.38–6.44) (Fig. 8, 9). Likewise occur differences in Mg/(Mg + Fe) ratio. Distinctly Fe-enriched inherited chlorites (brunsvigite) occur in shear zones of the Rožná and Okrouhlá Radouň uranium deposits with Mg/(Mg + Fe) ratio from 0.13 to 0.43. However, the inherited to authigenic chlorite from the Rožná uranium deposit is Mg-enriched chlorite (diabantite) with Mg/(Mg + Fe) ratio from 0.48 to 0.69 (Fig. 7).

The authigenic chlorites from all these uranium deposits are in comparison with inherited chlorites enriched in Mg. The lower enrichment in Mg displays the authigenic chlorites from the Rožná and Okrouhlá Radouň uranium deposits, whereas high enrichment in Mg

wt.%	Ro-5	Ro-16	Ro-27	Ra-13	Ra-54	ZCH-1	Vi-18
SiO_2	29.81	30.36	31.91	27.47	25.09	27.28	26.03
TiO_2	0.36	0.00	1.74	0.04	0.12	0.00	0.11
Al_2O_3	14.34	16.32	13.97	20.85	20.60	20.40	20.41
Cr_2O_3	0.02	0.00	0.00	0.01	0.02	0.05	0.00
FeO	33.66	26.75	17.70	22.49	32.95	20.73	25.99
MnO	0.42	0.27	0.31	0.19	0.21	0.15	0.37
MgO	8.77	13.92	19.54	16.40	9.48	19.09	13.95
CaO	0.28	0.27	0.00	0.14	0.04	0.00	0.07
Na_2O	0.62	0.00	0.60	0.06	0.00	0.06	0.20
K_2O	0.10	0.00	1.31	0.07	0.02	0.04	0.09
F	0.00	0.00	0.00	0.11	0.00	0.00	0.00
Cl	0.00	0.00	0.00	0.00	0.01	0.00	0.00
H_2O calc.	10.97	11.41	11.72	11.56	11.03	11.71	11.26
O=(F, Cl)	0.00	0.00	0.00	0.05	0.00	0.00	0.00
Total	99.35	99.32	98.80	99.34	99.57	99.51	98.48
apfu							
Si^{4+}	6.52	6.38	6.53	5.67	5.46	5.59	5.54
Al^{IV}	1.48	1.62	1.47	2.33	2.54	2.41	2.46
Al^{VI}	2.21	2.42	1.90	2.75	2.73	2.51	2.26
Ti^{4+}	0.06	0.00	0.27	0.01	0.02	0.00	0.02
Fe^{2+}	6.15	4.70	3.03	3.89	5.99	3.55	4.63
Cr^{2+}	0.00	0.00	0.00	0.00	0.00	0.01	0.00
Mn^{2+}	0.08	0.05	0.05	0.03	0.04	0.03	0.07
Mg^{2+}	2.86	4.36	5.96	5.05	3.07	5.83	4.43
Ca^{2+}	0.07	0.06	0.00	0.03	0.01	0.00	0.02
Na^{1+}	0.26	0.00	0.24	0.02	0.00	0.02	0.08
K^{1+}	0.03	0.00	0.34	0.02	0.01	0.01	0.02
F^{1-}	0.00	0.00	0.00	0.14	0.00	0.00	0.00
Cl^{1-}	0.00	0.00	0.00	0.00	0.01	0.00	0.00
OH^{1-}	16	16	16	15.92	16	16	16
O	36	36	36	36	36	36	36
Mg/(Mg + Fe)	0.32	0.48	0.66	0.57	0.34	0.62	0.49

Table 2. Representative analyses of chlorites from the Rožná, Okrouhlá Radouň, Zadní Chodov and Vítkov II uranium deposits.
Ro-5 – inherited chlorite, ore stage, Rožná, Ro-16 –authigenic chlorite, post-ore stage, Rožná,
Ro-27 – inherited to authigenic chlorite, pre-ore stage, Rožná, Ra-13 – authigenic chlorite, ore stage, Okrouhlá Radouň, Ra-54 –inherited chlorite, pre-ore stage, Okrouhlá Radouň, ZCH-1 – inherited chlorite, pre-ore stage, Zadní Chodov, Vi-18 –inherited chlorite, pre-ore stage, Vítkov II.

displays chlorites from the West Bohemian uranium deposits (Zadní Chodov, Vítkov II) (Fig. 7, 8, 9). The Mg/(Mg + Fe) ratio for authigenic chlorite from syn-ore and post-ore stage of the Rožná uranium deposit is quite similar (0.45–0.55), whereas the values of this ratio for authigenic chlorite from the Okrouhlá Radouň uranium deposit are partly lower (0.30–0.57). However, the Mg/(Mg + Fe) ratio in authigenic chlorites from the Zadní Chodov and Vítkov II uranium deposits is distinctly higher (0.74–0.88).

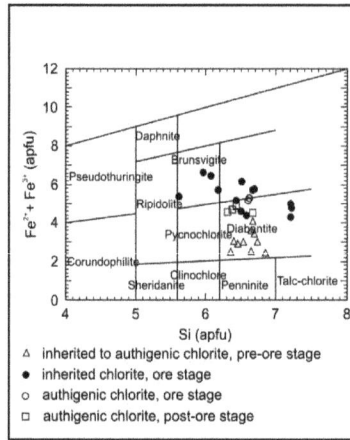

Figure 7. Classification diagram for chlorite from the Rožná uranium deposit according Hey [30].

From chemical composition of chlorites can be estimated temperatures of hydrothermal alterations in these ore deposits. The used chlorite thermometer for chlorites from pre-ore stage in the Rožná uranium deposit yielded temperatures from 219 ºC to 310 ºC. Authigenic syn-ore chlorites from all four uranium deposits indicate a temperature range from 145 ºC to 210 ºC. Authigenic post-ore chlorites are relatively rare and yielded temperatures for 150 ºC to 170 ºC.

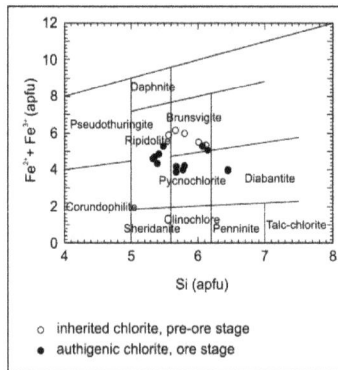

Figure 8. Classification diagram for chlorite from the Okrouhlá Radouň uranium deposit according Hey [30].

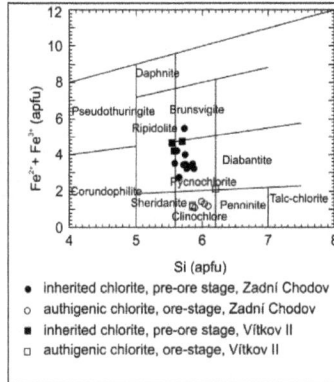

Figure 9. Classification diagram for chlorite from the Zadní Chodov and Vítkov II uranium deposits according Hey [30].

7. Composition of white mica

The white mica occurs in altered rocks of the Sn-W and U ore deposits as relatively rare mineral. The majority of white mica has composition of muscovite with variable content of water (hydromuscovite), Fe, Mg, Na and Ca (Table 3, Fig. 10).

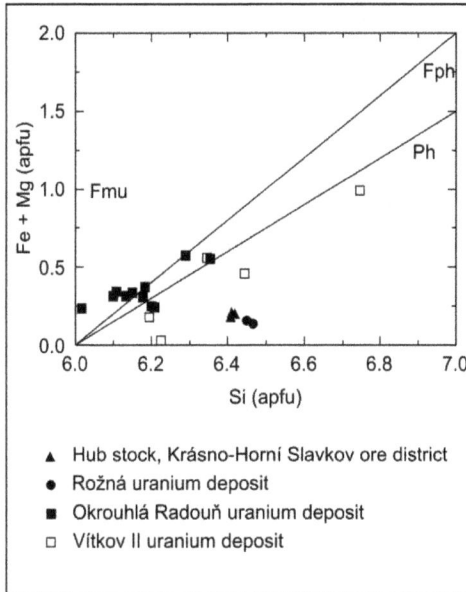

Figure 10. Plot of Fe + Mg (apfu) vs. Si (apfu) for white mica from the Krásno–Horní Slavkov ore district, Rožná, Okrouhlá Radouň and Vítkov II uranium deposits.

wt.%	Hub-27	Hub-30	Ro-12	Ra-1	Ra-2	Vi-1	Vi-3	Hub-18	Hub-19	SPO-2
SiO_2	48.30	48.87	48.82	46.63	45.45	45.96	47.98	50.90	50.94	48.52
TiO_2	0.06	0.05	0.00	0.12	0.02	0.32	0.40	0.03	0.01	0.08
Al_2O_3	34.38	34.69	34.37	35.52	37.23	34.07	33.25	34.83	35.52	26.03
Cr_2O_3	0.01	0.01	0.00	0.05	0.02	0.00	0.00	0.03	0.00	0.00
FeO	1.41	1.28	0.24	1.48	1.06	0.64	1.05	0.98	1.42	6.82
MnO	0.04	0.06	0.00	0.02	0.08	0.02	0.00	0.19	0.16	0.04
MgO	0.27	0.30	0.56	0.73	0.62	0.54	2.25	0.19	0.28	1.78
CaO	0.05	0.05	0.00	0.01	0.00	2.62	0.00	0.29	0.46	0.15
Na_2O	0.16	0.19	0.40	0.69	0.67	0.02	0.26	0.12	0.15	0.06
K_2O	10.10	10.21	9.69	10.55	10.78	10.07	10.21	8.32	7.13	9.31
F	0.91	0.90	0.00	0.08	0.03	0.00	0.00	0.31	0.15	0.00
Cl	0.01	0.02	0.00	0.01	0.00	0.00	0.00	0.00	0.03	0.00
H_2O calc.	4.08	4.13	4.53	4.49	4.52	4.45	4.53	4.50	4.60	4.30
Total	99.78	100.76	98.61	100.38	100.48	98.71	99.93	100.69	100.85	97.09
apfu										
Si^{4+}	6.41	6.42	6.47	6.18	6.02	6.19	6.35	6.57	6.53	6.77
Al^{IV}	1.59	1.58	1.53	1.82	1.98	1.81	1.66	1.43	1.47	1.23
Al^{VI}	3.78	3.79	3.83	3.72	3.82	3.60	3.52	3.86	3.89	3.04
Ti^{4+}	0.01	0.01	0.00	0.01	0.00	0.03	0.04	0.00	0.00	0.01
Fe^{2+}	0.16	0.14	0.03	0.16	0.12	0.07	0.12	0.11	0.15	0.80
Cr^{2+}	0.00	0.00	0.00	0.01	0.00	0.00	0.00	0.00	0.00	0.00
Mn^{2+}	0.00	0.01	0.00	0.00	0.01	0.00	0.00	0.02	0.02	0.01
Mg^{2+}	0.05	0.06	0.11	0.14	0.12	0.11	0.44	0.04	0.05	0.37
Ca^{2+}	0.01	0.01	0.00	0.00	0.00	0.38	0.00	0.04	0.06	0.02
Na^{1+}	0.04	0.05	0.10	0.18	0.17	0.01	0.07	0.03	0.04	0.02
K^{1+}	1.71	1.71	1.64	1.78	1.82	1.73	1.72	1.37	1.17	1.66
OH^{1-}	4.00	4.00	4.00	4.00	4.00	4.00	4.00	4.00	4.00	4.00
O	24.00	24.00	24.00	24.00	24.00	24.00	24.00	24.00	24.00	24.00
Mg/(Mg + Fe)	0.25	0.30	0.80	0.47	0.51	0.60	0.79	0.26	0.26	0.32

Table 3. Representative analyses of white mica and illite from the Krásno—Horní Slavkov ore district, Rožná, Okrouhlá Radouň and Vítkov II uranium deposits.
Hub-27, 30 – muscovite, argillized granite, Krásno–Horní Slavkov ore district, Ro-12 – muscovite, altered gneiss, Rožná uranium deposit, Ra-1, 2 – muscovite, altered two-mica granite, Okrouhlá Radouň uranium deposit, Vi-1, 3 – muscovite, altered biotite granite, Vítkov II uranium deposit, Hub-18, 19 – illite, argillized granite, Krásno–Horní Slavkov ore district, SPO-2 – illite, altered gneiss, Rožná uranium deposit.

The lowest content of Fe (0.01–0.16 apfu), Mg (0.00–0.06 apfu), Na (0.01–0.05 apfu) and Ca (0.00–0.01 apfu) displays muscovite from greisenized topaz-albite granites of the Hub stock (Krásno–Horní Slavkov ore district). Similar low contents of these elements were found in white mica from the Rožná uranium deposit. The white mica from the Okrouhlá Radouň uranium deposit is partly enriched in Fe (up to 0.43 apfu) and Mg (0.11–0.39 apfu). The white mica from this uranium deposit displays also enrichment in Na (0.12–0.22 apfu). The white mica from the Vítkov II uranium deposit has highly variable content of Mg (0.0–0.73 apfu) and analysis with the highest content of Mg can be classified as phengite (Fig. 10).

8. Composition of clay minerals

Chemical composition of clay minerals (illite, kaolinite, smectite, corrensite) was determined only for illite, kaolinite and smectite. Corrensite is distinctly rare mineral, which was not found in thin sections that are analysed by microprobe. Kaolinite and illite were identified by XRD method in all examined deposits. Chemical analyses of these clay minerals were plotted in the $M^+R^3 - 2R^3 - 3R^2$ diagram modified by Velde [19] (Table 4, Fig. 11). However, they are analysed only in argillized granites from the Krásno–Horní Slavkov ore district and in altered rocks from the Rožná and Okrouhlá Radouň uranium deposits. In kaolinite from argillized granites of the Hub stock (Krásno–Horní Slavkov ore district) content of Si is lower (3.45–4.08 apfu) than its content in kaolinite from altered rocks in the Okrouhlá Radouň uranium deposit (4.03–4.17 apfu). Some differences in chemical composition of kaolinite from both ore deposits display also contents of Fe, Mg, Ca and K. In kaolinite from the Krásno–Horní Slavkov Sn-W ore district occur enrichment in Fe (0.02–0.26 apfu), Ca (0.00–0.07 apfu) and K (0.00–0.10 apfu). Contents all these elements in kaolinite from the Okrouhlá Radouň uranium deposit are lower – Fe (0.01–0.08 apfu), Ca (0.01–0.06 apfu) and

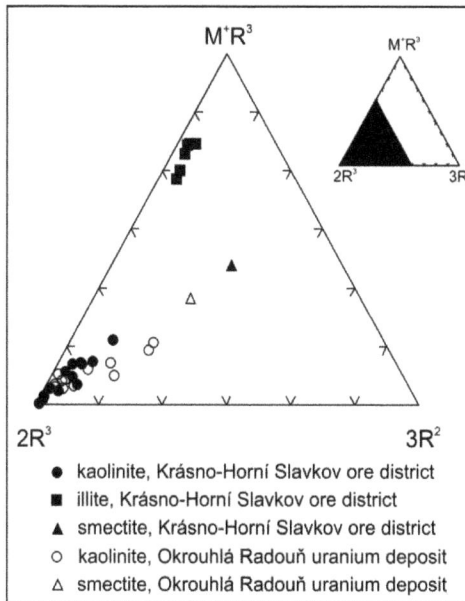

Figure 11. Plot of $M^+R^3 - 2R^3 - 3R^2$ according Velde [19] for clay minerals from the Krásno–Horní Slavkov ore district and Okrouhlá Radouň uranium deposit.

wt.%	Hub-13	Hub-30	Hub-36	Ra-12	Ra-15	Ra-16	Ra-13
SiO$_2$	46.80	46.92	39.52	48.38	50.57	47.27	52.05
TiO$_2$	0.00	0.02	0.00	0.02	0.00	0.00	0.00
Al$_2$O$_3$	38.61	38.01	14.06	36.66	30.43	37.32	27.18
Cr$_2$O$_3$	0.01	0.00	0.00	0.00	0.00	0.01	0.02
FeO	0.30	0.23	1.43	0.31	1.10	0.11	1.49
MnO	0.00	0.06	0.11	0.00	0.02	0.06	0.00
MgO	0.07	0.08	0.06	0.46	3.06	0.28	3.49
CaO	0.15	0.08	0.08	0.21	0.58	0.46	0.64
Na$_2$O	0.04	0.05	0.02	0.02	0.01	0.01	0.06
K$_2$O	0.20	0.06	0.36	0.03	0.47	0.01	1.85
F	0.09	0.08	0.00	0.00	0.00	0.00	0.00
Cl	0.00	0.01	0.00	0.01	0.03	0.00	0.05
H$_2$O calc.	13.88	13.80	13.74	13.96	13.82	13.85	13.74
O=(F, Cl)	0.04	0.04	0.00	0.00	0.01	0.00	0.01
Total	100.11	99.36	100.38	100.36	100.08	99.49	100.56
apfu							
Si^{4+}	4.03	4.06	3.45	4.16	4.39	4.09	4.54
Al^{IV}	0.00	0.00	0.55	0.00	0.00	0.00	0.00
Al^{VI}	3.92	3.88	4.08	3.71	3.11	3.81	2.79
Ti^{4+}	0.00	0.00	0.00	0.00	0.00	0.00	0.00
Fe^{2+}	0.02	0.02	0.10	0.02	0.08	0.01	0.11
Cr^{2+}	0.00	0.00	0.00	0.00	0.00	0.00	0.00
Mn^{2+}	0.00	0.00	0.01	0.00	0.00	0.00	0.00
Mg^{2+}	0.01	0.01	0.01	0.06	0.40	0.04	0.45
Ca^{2+}	0.01	0.01	0.01	0.02	0.05	0.04	0.06
Na^{1+}	0.01	0.01	0.00	0.00	0.00	0.00	0.01
K^{1+}	0.02	0.01	0.04	0.04	0.05	0.01	0.21
F^{1-}	0.05	0.04	0.00	0.00	0.00	0.00	0.00
Cl^{1-}	0.00	0.00	0.00	0.00	0.01	0.00	0.02
OH^{1-}	7.98	7.98	8.00	8.00	8.00	8.00	7.99
O	18.00	18.00	18.00	18.00	18.00	18.00	18.00
Mg/(Mg + Fe)	0.29	0.37	0.07	0.73	0.83	0.82	0.81

Table 4. Representative analyses of clay minerals from the Krásno–Horní Slavkov ore district and Okrouhlá Radouň uranium deposit.
Hub-13, 30, 36 – kaolinite, argillized granite, Krásno–Horní Slavkov ore district, Ra-12, 15, 16 – kaolinite, argillized gneiss, Okrouhlá Radouň uranium deposit, Ra-13 – smectite, argillized gneiss, Okrouhlá Radouň uranium deposit.

K (0.00–0.08 apfu). However, the content of Mg (0.02–0.45 apfu) is partly higher in kaolinite from the Okrouhlá Radouň ore deposit, whereas kaolinite from the Krásno–Horní Slavkov ore district is in Mg depleted (0.00–0.07 apfu). Illite from both type deposits is partly enriched in Fe (0.07–0.11 apfu). Smectite from the Krásno–Horní Slavkov ore district is enriched in Fe (0.25 apfu) and Ca (0.15 apfu), whereas smectite from the Okrouhlá Radouň ore deposit is enriched in Mg (0.45 apfu).

9. Discussion

The associations of chlorites and clay minerals and their chemistry were studied in unconformity-type uranium deposits occurred in Canada [6] and Australia [1, 2]. In these deposits were studied layered silicates in mineralized shear zones, which are evolved in various altered metasediments. In both areas are clay minerals represented predominantly by chlorites and smectites. Authigenic chlorites from unconformity-type uranium deposits occurred in Australia are enriched in Mg [1,2]. The clay minerals assemblage is usually enriched in illite-smectite mixed layer minerals. Illite-smectite mixed layer minerals together with kaolinite and chlorite are characteristic for argillized granitic rocks occurred in uranium ore deposits from the French Massif Central [3]. The temperatures, which were determined by chlorite thermometry in the Rožná, Okrouhlá Radouň and Vítkov II uranium deposits can be well correlated with temperatures obtained by study of fluid inclusions [10, 20].

The greisenized and argillized granites connected with Sn-W mineralization are usually enriched in kaolinite and illite [4, 5, 7]. In greisens from Sn-W ore deposits evolved in the Saxothuringian Zone of the Bohemian Massif was also found dickite [9], together with smectite and illite-smectite mixed layer minerals [21]. The cookeite, which was found in cassiterite–enriched pockets in the Krásno–Horní Slavkov ore district [8] occurs also in other tin ore deposits [22]. Temperature of the cookeite origin can be estimated from data on cookeite stability [23] at 270–350 °C. This temperature can be well correlated with temperatures of clay minerals origin in ore pockets obtained by fluid inclusion study [24]. Occurrences of corrensite were recorded usually from hydrothermally altered intermediate to basic volcanic rocks [25, 26, 27]. Similar association of corrensite with kaolinite, illite and mixed-layer illite/smectite was found in granitic cupola of the Montebras, France [28]. According to Velde [19], the thermal stability of corrensite ranges from 180–200 °C to 280 °C. This temperature is in agree with the homogenization temperatures, which were found in quartz-fluorite veinlets occurred in arigilized greisens of the Krásno–Horní Slavkov ore district [24].

Author details

Miloš René
Institute of Rock Structure and Mechanics, v.v.i., Academy of Sciences of the Czech Republic, Prague, Czech Republic

Acknowledgement

This study was financially supported by Grant Agency of the Czech Republic (Project No. 205/09/540) and by Ministry of Education, Youth and Sports (Project No. ME10083). The detailed comment by editor (M. Valaskova) helped to improve the initial draft of the manuscript.

10. References

[1] Beaufort D, Patrier P, Laverret F, Bruneton P, Mondy J, (2005) Clay-alteration associated with Proterozoic unconformity-type uranium deposits in the East Alligator rivers uranium field, Northern Territory, Australia. Econ Geol 100: 515–536.

[2] Nutt C J (1989) Chloritization and associated alteration at the Jabiluka unconformity-type uranium deposit, Northern Territory, Australia. Canad Mineral 27: 41–58.

[3] Patrier P, Beaufort D, Bril H, Bonhommé M, Fouillac A M, Aumáitre R (1997) Alteration-mineralization at the Bernardan U deposit (Western Marche, France). The contribution of alteration petrology and crystal chemistry of secondary phases to a new genetic model. Econ Geol 92: 448–467.

[4] Pouliot G, Barondeau B, Sauve P, Davis M (1978) Distribution of alteration minerals and metals in the Fire Tower zone at Brunswick tin mines Ltd., Mount Pleasant area, New Brunswick. Canad Mineral 16: 223–237.

[5] Psyrillos A, Mannig D A C, Burley S D (2001) The nature and significance of illite associated with quartz-hematite hydrothermal veins in the St. Austell pluton, Cornwall, England. Clay Minerals 36: 585–597.

[6] Rimsaite J (1978) Layer silicates and clays in the Rabbit lake uranium deposit, Saskatchewan. Pap Geol Surv Canada 78-1A: 303–315.

[7] Sainsbury C L (1960) Metallization and post-mineral hypogene argillization, Lost river tin mine, Alaska. Econ Geol 55: 1478–1506.

[8] Melka K, Košatka M, Zoubková J (1991) The occurrence of dioctahedral chlorite in greisen. In: Proc 7th Euroclay Conf. Dresden. pp. 757–760.

[9] Melka K, Štemprok M (1961) The determination of dickite from Cínovec (Zinnwald), Czechoslovakia. Acta Univ Carol Geol Suppl 1: 307–317.

[10] Kříbek B, Žák K, Dobeš P, Leichmann J, Pudilová M, René M, Scharm B, Scharmová M, Hájek A, Holeczy D, Hein U F, Lehmann B (2009) The Rožná uranium deposit (Bohemian Massif, Czech Republic): shear zone-hosted, late Variscan and post-Variscan hydrothermal mineralization. Mineral Deposita 44: 99-128.

[11] René M, Matějka D, Klečka M (1999) Petrogenesis of granites of the Klenov massif. Acta Montana Ser AB 7: 107-134.

[12] René M, Matějka D, Nosek T (2003) Geochemical constraints on the origin of a distinct type of two-mica granites (Deštná – Lásenice type) in the Moldanubian batholith (Czech Republic). Acta Montana Ser A 23: 59-76.

[13] René M (2005) Geochemical constraints of hydrothermal alteration of two-mica granites of the Moldanubian batholith at the Okrouhlá Radouň uranium deposit. Acta Geodyn Geomater 2: 1–17.

[14] René M (2000) Petrogenesis of the Variscan granites in the western part of the Bohemian Massif. Acta Montana Ser A 15: 67-83.

[15] Siebel W, Breiter K, Wendt I, Höhndorf A, Henjes-Kunst F, René M (1999) Petrogenesis of contrasting granitoid pluton in western Bohemia (Czech Republic). Mineral Petrol 65: 207-235.

[16] Bareš M, Fiala V (1982) Hydrothermogenous clay minerals of some localities in Moldanubian rocks. In: 9th Conf. On Clay Mineralogy and Petrology. Zvolen. pp. 233–240.

[17] Pouchou JL, Pichoir F (1985) "PAP" (φ-ρ-Z) procedure for improved quantitative microanalysis. In: Armstrong J T, editor Microbeam analysis. San Francisco. San Francisco Press. pp. 104–106

[18] Walshe J L (1986) A six-component chlorite solid solution model and the conditions of chlorite formation in hydrothermal and geothermal systems. Econ Geol 81: 681–703.

[19] Velde (1977) Clays and clay minerals in natural and synthetic systems. Amsterdam. Elsevier. 325 p.

[20] René M (1997) Fluid system of the Vítkov II uranium deposit (Bor pluton, western part of Bohemian massif, Czech republic). Mitt Österr miner Ges 142: 19–20.

[21] Mach Z (1979) Association of clay minerals with tin-tungsten mineralization in the surroundings of Krásno near Horní Slavkov. Acta Univ Carol Geol 1979: 105–132 (in Czech)

[22] Ren Shuang K, Eggleton R A, Walshe L (1988) The formation of hydrothermal cookeite in the breccia pipes of the Ardlethan tin field, New South Wales, Australia. Canad Mineral 26: 407–412.

[23] Vidal O, Goffé B (1991) Cookeite $LiAl_4(Si_3Al)O_{10}(OH)_8$: Experimental study and the thermodynamical analysis of its compatibility relations in the $Li_2O–Al_2O_3–SiO_2–H_2O$ system. Contrib Mineral Petrol 108: 72-81.

[24] Dolníček Z, René M, Prochaska W, Kovář M (2012): Fluid evolution of the Hub stock, Horní Slavkov–Krásno Sn–W ore district, Bohemian Massif, Czech Republic. Mineral. Deposita DOI: 10.1007/s00126-012-0400-0.

[25] Garvie L A J, Metcalfe R (1997) A vein occurrence of co-existing talc, saponite, and corrensite, Builth Wells, Wales. Clay Minerals 32: 223–240.

[26] Dekayir A, Amouric M, Olives J (2005) Clay minerals in hydrothermally altered basalts from Middle Atlas, Marocco. Clay Minerals 40: 67–77.

[27] Jiménez-Millán J, Abad I, Nieto F (2008) Contrasting alteration processes in hydrothermally altered dolerites from the Betic Cordillera, Spain. Clay Minerals 43: 267–280.

[28] Dudoignon P, Beaufort D, Meunier A (1988) Hydrothermal and supergenne alterations in the granitic cupola of Montebras, Creuse, France. Clays Clay Minerals 36: 505–520.

[29] Taylor S R, McLennan S M (1985) The continental crust: Its composition and evolution. Oxford. Blackwell. 312 p.

[30] Hey M H (1954) A new review of the chlorites. Mineral Mag 30: 277–292.

[31] Bailey S W (1982) Nomenclature for regular interstratifications. Amer Mineral 67: 394–398.

Chapter 5

Kolsuz-Ulukisla-Nigde Clays, Central Anatolian Region – Turkey and Petroleum Exploration

Burhan Davarcioglu

Additional information is available at the end of the chapter

1. Introduction

Clays are naturally occurring, fine-grained minerals under surface conditions mostly as alteration products with distinct crystal structures. They may show plastic behavior when mixed with sufficient water and become stiff when dried or cooked. They, having different physical properties, occur in three modes: (I) surface clays that may be old or very young sedimentary formations, as suggested by the name, they occur near surface, (II) shales that have been subjected to pressures to become rigid and layered due to various reasons (e.g., tectonic or subsidence-related), and (III) fire clays that are mined out from deeper sites comparing to the other two and they behave refractory and include less impurities thus they are physically and chemically more uniform.

Some clay minerals having large surface area, high ion exchange capacity and molecular grid properties have been pioneered for the development of many new products (Falaras et al., 2000). Clay-organic complex structures have been a research subject particularly since 1930's. Results of these researches made benefical contribution to the process of expanding their use in diverse areas (Smith, 1934). Besides, progresses in analytical techniques that made possible to determine mineralogical and chemical compositions of the clays appreciably expedited this process.

Clay minerals are formed as a result of changes in temperature, pressure, geochemical, and physical conditions (Murray, 1999). Although clay minerals could be resulted from weathering, sedimentation, burial, diagenesis and hydrothermal alteration processes in general, occurrence of monomineralic clay deposits is scant. Even in an ordinary clay sample, several clay species could occur together. Clay minerals can be categorized in 4 subgroups in natural environments: (I) kaolinite group, (II) smectite group, (III) illite group, and (IV) chlorite group (Murray, 1991).

The relationship between heat and pressure with the formation of clay minerals and its consequence on the formation of petroleum has been an imperative subject for research past 40 years (Perry and Hower, 1970; Dypvik, 1983). In the recent years one of the methods in the petroleum explorations is organic maturation and the other is clay mineral diagenesis. In both the clay mineral diagenesis and the organic maturation, clay minerals show structural changes with the changing temperature, which reveals the degree of metamorphism. The principal factors including pressure, temperature, depth and burial that are efficient during the conversion of the clay minerals and hydrocarbon formation and primary migration can be explained through clay mineral diagenesis and organic maturation (Dunoyer de Segonzag, 1970; Dypvik, 1983).

Clay mineral characterization could be carried out employing spectroscopic methods for various purposes in the geological sciences (Heroux et al., 1979). In the literature, there is a voluminous research on determination of clay mineral chemistry using diverse techniques. Today, one of the most preferred methods is the FTIR (Fourier Transform Infrared) Spectroscopy. There is a significant increase in number of studies using this method dealing with the clay characterization in Turkey (Akyuz and Akyuz, 2003; Davarcioglu et al., 2005; Davarcioglu and Kayali, 2007; Davarcioglu et al., 2007; Davarcioglu and Ciftci, 2009). One of such studies is on the quantitative and qualitative characterization of Central Anatolian clay deposits and diatomites by employing the spectroscopic methods (Kayali et al., 2005; Davarcioglu et al., 2008; Davarcioglu, 2009). The Central Anatolia is one of the richest in occurrence of clay deposits in the world. Therefore, investigation of these deposits, their quantitative and qualitative characterization is highly important.

In the XRD measurements, characteristic peak of glycol-saturated montmorillonite is 17 A° peak (Cradwick and Wilson, 1972). As it disappears, mineralogical conversion of montmorillonite to illite becomes evident. As a consequence of chemical reactions depending on increasing depth and temperature, increase of Al and decrease of Si in montmorillonite's tetrahedral sheet results in changes in structure and by taking-up K ions available in the environment due to feldspar alteration, montmorillonite converts to illite (Weaver, 1960; Suchy et al., 2007). Expelled water as result of this change results in an increase in salinity of connate water already present in shale. Mg, Si and Ca ions, products of such conversion, could form minerals like calcite and dolomite facilitated by increasing temperature and these new minerals deposit within shale. This in turn results in a decrease in porosity and permeability and an increase in density (Bishop et al., 2002; Dunoyer de Segonzag, 1970).

Montmorillonite, being very sensitive clay mineral to temperature and depth, play an important role in both oil formation and its migration. As a result of reactions occurring in association with increasing temperature and depth, montmorillonite converts to illite losing its structural water (smectite converts to illite in alkaline environment). This conversion occurs at about 2500-4500 m burial depth and at 100 °C. In this process, K contents of pore waters show increase. Organic matter requires H ions to become hydrocarbon. H ions facilitate structural break-down (so-called cracking) of hydrocarbons with large organic molecules to smaller ones. During the loss of water from montmorillonite, H ion concentration

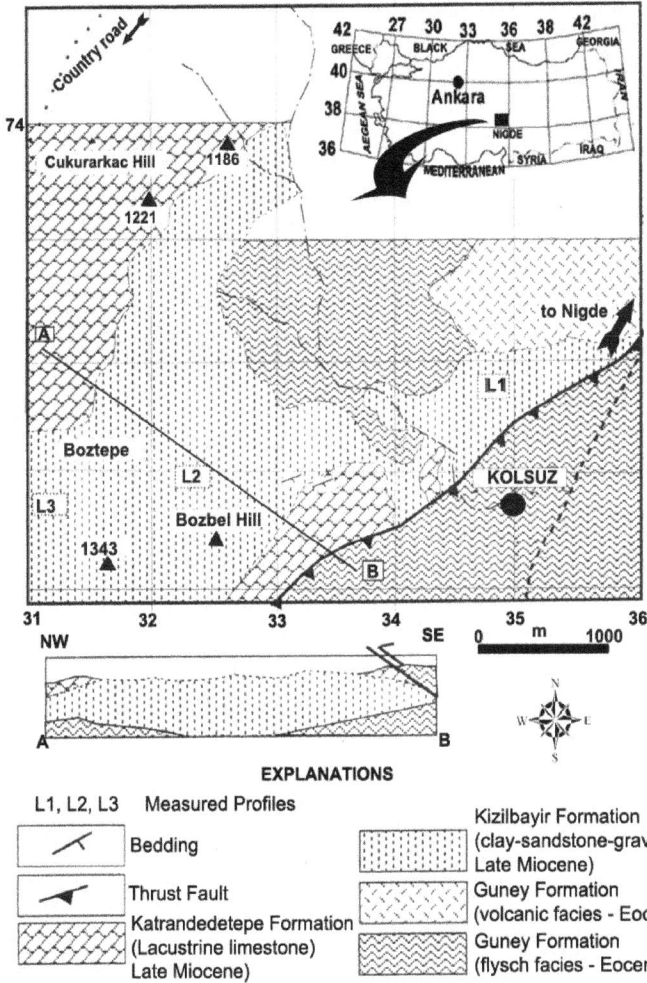

Figure 1. Generalized geological map of the Kolsuz-Ulukisla region and its around (adapted from Kayali et al., 2005).

of the environment raises. Water expulsion takes place at fairly steady and regular temperature. Some of the layers start losing water at 50-60 °C through 300 °C. About 70-80% of the water is expelled between 120-160 °C that corresponds to the formation of petroleum. This study suggests that results acquired from the organic maturation can be obtained through spectral analysis of structural changes occurring in clay minerals.

One of the geological studies in the Kolsuz area (Ulukisla-Nigde, Central Anatolian) deals with the clay profiles. In this study, using columnar sections and local observations, distribution of clay horizons and their lithostratigraphic relations were investigated (Oktay, 1982). On the other hand, sedimentological, mineralogical and chemical properties of the clays of the same area were studied in depth later by Gurel (1999). The Kizilbayir Formation outcropping in the north of Ulukisla basin is the key formation due to its association with the clay occurrences. The formation starts with gravel-bearing red-green clays at the base, progressing upward, large-scale cross-bedded conglomerate, sandstone with clay intercalations occur and conglomerate with mudstone interbeds dominate towards the top (Demirtasli et al., 1986). Thickness of the formation in the area ranged from 10 to 250 m.

However, no comprehensive study on the qualitative and the quantitative characteristics of the clays occurring in this area was available to date. Thus, this study aims to investigate clays of the area using the spectroscopic method and observed structural changes in the clay structures were interpreted in terms of petroleum formation and their possible use in petroleum exploration in the area.

2. Experimental

2.1. Preparation of samples

A combined profile representing the constructed profiles chosen for this and nearby area was shown in Figure 1. The clay samples were taken from three different levels shown in the litostratigraphic columnar section of a selected locality in the Kolsuz-Ulukisla region (Figure 2). The samples taken from lower level, middle level and upper level labeled as (Kk1), (Kk2), and (Kk3), respectively. Initially about 1 kg of samples were collected, and 20-40 g splits were prepared for further analyses. Samples were heated at 110 °C for 24 hours and crushed to powder and screened using an 80 mesh sieve. Clay fractions were prepared following the procedure including removal of carbonates, sulfates and organic matter by dissolving, through washing to acquire stable suspension, and siphoning (to acquire the clay fraction - <2 micron) (Brown, 1961; Gundogdu, 1982).

In general, the clay minerals contain significant amount of water. Conversely, they contain less alkaline and alkaline earth elements. Absorption bands due to the water molecules occupy large spectral fields, the ones critical for identification of clay minerals. Thus in order to minimize this undesired overlap, samples for the FTIR measurements were prepared through clay concentration without employing a centrifuge. However for the chemical analyses, samples were analyzed as bulk sample without concentrating for clay fraction. Organic matter was removed through boiling in H_2O_2 solution and then samples were dried in an oven at 110 °C for 24 hours.

2.2. FTIR measurements

Samples collected from the study area were prepared applying the disc technique (mixing ~1 mg clay sample with ~200 mg KBr) and put in molds. These intimate mixtures were then pressed at very high pressure (10 tons per cm^2) to obtain the transparent discs, which were then placed in the sample compartment. Bruken Equinox 55 Fourier transform FTIR

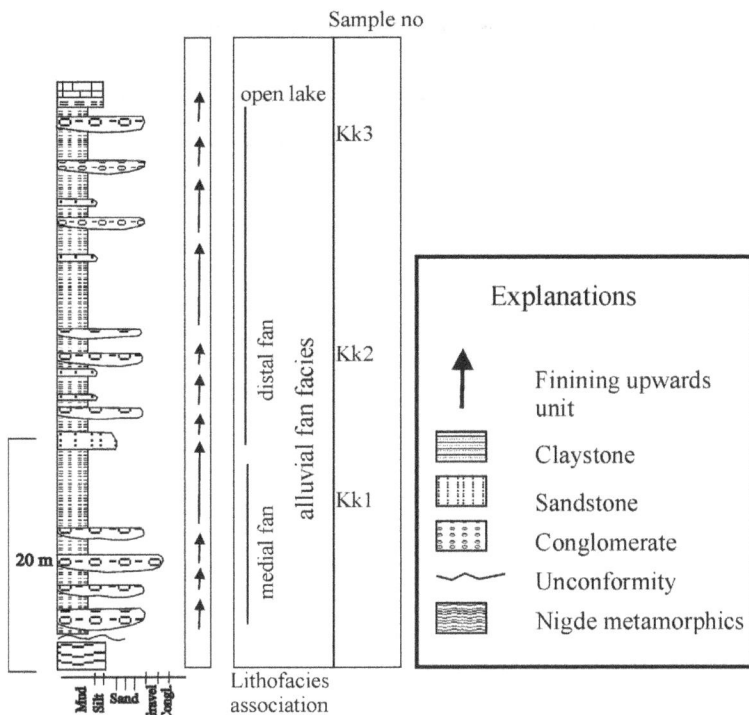

Figure 2. Generalized litostratigraphic columnar section of the Kolsuz-Ulukisla region (adapted from Kayali et al., 2005).

spectrophotometer (Department of Physics, METU, Ankara-Turkey) was used for the IR spectral measurements of these samples with standard natural clay and the spectra were recorded over the range of 5000-370 cm^{-1} (% transmission versus cm^{-1}). Before taking the spectra measurements of the samples, spectrophotometer was calibrated with polystryrenes and silicate oxide of thickness 0.05 nm.

The infrared spectra of the illite (IMt-1; Silver Hill, Montana, USA), illite-smectite mixed layer (ISMt-1; Mancos Shale, Ord.), montmorillonite (SCa-3; Otay, San Diego Country California, USA), Ca-montmorillonite (STx; Gonzales Country, Texas, USA), Na-montmorillonite (SWy-1; Crook Country, Wyoming, USA), kaolinite (KGa-1; Washington Country, Georgia, USA), chlorite (ripidolite, CCa-1; Flagstaff Hill, El Dorato Country, California, USA), and palygorskite (PFI; Gadsden Country, Florida, USA) known as standard natural clay minerals ("The World Source Clay Minerals") were taken (Table 1), and then the spectra of anhydrite, gypsum, and mixtures of the illite+quartz+feldspar, quartz+feldspar have been taken for the analyses of subject samples.

Along with XRD (X-ray powder diffraction) analysis, the functional groups in the clay minerals structures could only be determined through the FTIR spectra. Therefore, qualitative and quantitative analysis of the minerals by employing the FTIR spectroscopy is very important and promising.

Assignment	I	ISmML	Na-mont	C	P
			Wavenumber (cm⁻¹)		
v(OH) stretching	-	3685 (shoulder)	-	-	-
Inner-layer OH, (Al-O...H) stretching	-	-	3680	-	-
v(OH) stretching	-	-	-	3662	-
v(OH) stretching	-	-	3627 (shoulder)	-	3627
v(OH) stretching	3622	3622	3622	-	-
v(OH) stretching	-	-	-	3565	3546
v(OH) stretching	-	-	-	3434	-
δ(water-OH) stretching	-	-	-	-	3408, 3266
δ(water-OH) scissoring	-	-	-	-	1731
δ(water-OH) scissoring	-	-	-	-	1673, 1640
v(Si-O) stretching	-	-	-	-	1163, 1114
v(Si-O) normal to the plane stretching	1090	1090	-	-	-
v(Si-O) planar stretching	1031	1031	-	-	-
v(Si-O) stretching	-	-	-	-	1020
v(Si-O) stretching	-	-	-	988	-
(Al-Al-OH) deformation	916	916	920	-	-
OH deformation	-	-	-	-	905
(Al-Fe-OH) deformation	-	-	890	-	-
(Al-Mg-OH) deformation	832	810	875	-	-
M-OH stretching	-	-	805	819	-
Si-O deformation	-	-	-	766	-
(Al-O-Si) inner surface vibration	756	-	-	-	-
(Al-O-Si) inner layer vibration	-	750	-	-	-
OH deformation	688, 622	622	620	667	-
(Si-O-Al) deformation	-	-	-	543	-
(O-Si-O) bending	525	525	520	-	528
(O-Si-O) bending	468	468	468	-	469
(Si-O-Mg) deformation	-	-	-	441	-
(O-Si-O) bending	-	-	-	-	426

Table 1. Fundamental vibration frequencies of standard natural clays (I=illite, ISmML=illite-smectite mixed layer, Na-mont=Na-montmorillonite, C=chlorite, P=palygorskite).

A second treatment was employed only to the clay sample taken from the lower level (Kk1) to see whether there is a change in the structure of the samples or not due to FTIR spectrum measurements. For this procedure, HCl, bicarbonate (Na₂CO₃), and sodiumdithionit (Na₂S₂O₄), and sodiumstrate (Na₃C₆H₅O₇) liquids were added to the sample to remove carbonates (mainly calcite and dolomite), amorphous materials and manganese oxides, which were expected to be present. This mixture was treated in an oven at 120 °C for 24 hours and washed using ethyl alcohol until complete removal of those unwanted components was achieved. The FTIR spectrum of the precipitate was then taken.

3. Chemical and minerological analyses

The chemical analyses of the Kolsuz-Ulukisla clay samples (dried in an oven at 110 °C for 24 hours) were carried out at the ACME-Canada laboratories by means of XRF-ICP (X-ray fluorescence spectrometry-Inductively Coupled Plasma) technique. Major oxide composition of the samples representing the lower, middle and upper parts of the profile was given in Table 2. These data suggest that the clays of the region are essentially rich in

SiO_2, Al_2O_3, and CaO. Main cause of these enrichments was due to ascending briny and carbonated waters through capillary actions and precipitation due to transpiration in arid and semi-arid regions. On the other hand, aluminum enrichment could be due to presence of either other aluminum silicates (such as K-feldspars) or Al^{3+} being in the clay structures.

	Lower part (Kk1)	Middle part (Kk2)	Upper part (Kk3)
SiO_2	46.50	48.57	49.12
TiO_2	0.66	0.76	0.47
Al_2O_3	12.74	13.97	9.46
Fe_2O_3	7.28	8.10	4.24
MnO	0.14	0.14	0.15
MgO	4.68	5.09	2.90
CaO	9.37	6.94	15.04
Na_2O	0.96	1.10	1.72
K_2O	2.31	2.64	1.65
Cr_2O_3	0.028	0.027	0.029
P_2O_5	0.06	0.15	0.10

Table 2. Major oxide composition of the studied profile (in %).

XRD measurements were employed to determine the mineral phases included in the same samples (Siemens D-5000 Diffract AT V 3.1 diffractometer, CuKα radiation λ=1.54056 A° and 0.03 steps; General Directorate of Mineral Research and Exploration laboratories-MTA, Ankara-Turkey). According to the XRD measurements (Figures 3 and 4), subject clays are found to be composed of abundant chlorite (45%), illite (32%), quartz (20%), smectite (3%), feldspar, calcite, and trace quantities of palygorskite and Fe-oxide minerals. In the Kk1 lower part of the clay profile (Figure 2), amount of quartz tend to decrease while smectite, illite and chlorite show significant increase. Whereas at the top of the Kk1, just opposite of this abundance trend was observed and palygorskite was totally absent.

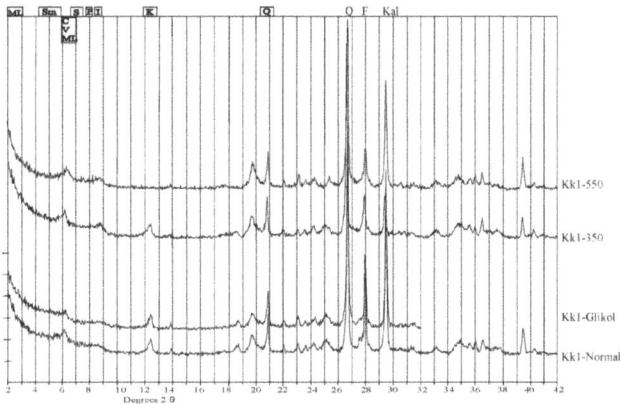

Figure 3. XRD pattern of the clay samples belonging to the lower level (Kk1) of Kolsuz-Ulukisla clays (Q=quartz, F=feldspar, I=illite, ML=mixed layer clay, C=chlorite, S=smectite).

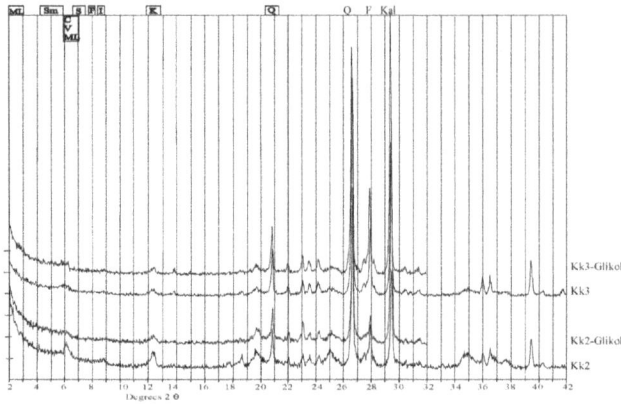

Figure 4. XRD patterns of the clay samples from the lower level (Kk2) and the upper level (Kk3) of Kolsuz-Ulukisla clays (Q=quartz, F=feldspar, I=illite, ML=mixed layer clay, C=chlorite, S=smectite).

DTA (Differantial Thermal Analysis) and TGA (Thermogravimetric Analysis) measurements have been carried out for the determinations of the thermal behaviour of the clay samples (Figure 5). Measurement were carried out in the MTA Labs (Ankara-Turkey) using a Rigaku Thermal Analyzer Ver. 2.22EZ (SN#39421). Here smectite peak falls in the same field with the one of chlorite. Minute endothermic peak of smectite occurs in between ~100-250 °C, and second endothermic peak appears at ~700 °C and shallow endothermic/exothermic peak is observed at ~800-900 °C. Best observed endothermic peak of chlorite in the DTA-grams is the one observed between 500-600 °C. This peak may shift toward ~700 °C due to the iron content. Subsequently this peak may fall in the same interval with smectite's peak at ~700 °C. Chlorite's exothermic peak occurs at 750 °C (Kok, 2006; Kok and Smykatz-Kloss, 2009; Yener et al., 2007).

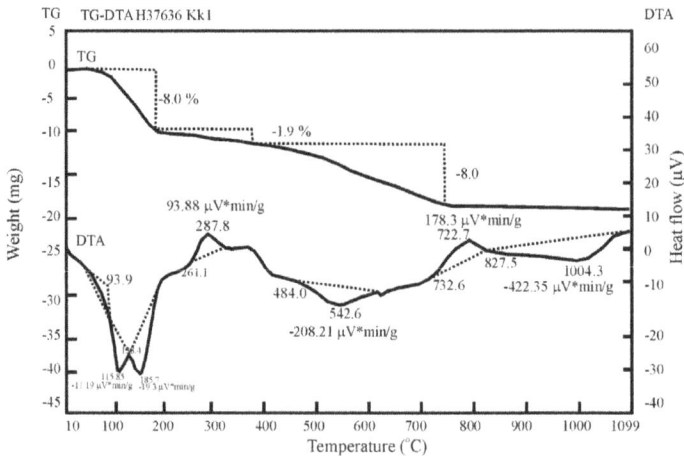

Figure 5. DTA-TGA measurements of the Kolsuz-Ulukisla clays.

4. Petroleum formation

In the recent years, one of the methods used in petroleum explorations is organic maturation and the other diagenesis of clay minerals. Results acquired through the first approach could be attained using the second method. Fundamental nature of the second approach can be explained as follow: clays that immature or recently deposited may contain smectite (montmorillonite), illite and kaolinite depending on the source area. With subsidence, these minerals lose their water content and are subjected to mineralogical transformations. Smectite converts to illite within the range of oil formation temperature (60-150 °C) (Weaver, 1960; Dunoyer de Segonzag, 1970). When the upper limit of this range was approached, kaolinite and illite convert to mica, if ferromagnesian minerals are available in the environment; these minerals transform to chlorite instead (Figure 6). Consequently, clays can give an idea about be the degree of maturation of a sample of interest. For the fields where oil explorations are carried out, clays can be used to answer following questions: (I) which layers has source rock potential, their regional coverage and relationship with the paleogeography, (II) source rocks occurring what part and depth of the basin and which time interval have enough maturation, (III) when and how the oil migration occurred, (IV) the relationship with oil formation and oil migration (Hunt, 1995). However, time and duration plays an important role in this process accompanying the mineralogical transformations.

Mineralogical changes of the clay minerals are closely related to the temperature and water chemistry of the environment. Thus, they reflect better and more precise transformation temperature (paleotemperature) of the clay minerals. Reflectance degree of vitrinite, a major organic component of coals, shows increase with increasing degree of metamorphism (Teichmuller, 1987). In diagenesis stage, reflectance degree of vitrinite is 5%. In this stage, organic matter is not mature enough to produce oil and the second stage is catagenesis (boundary is 2%). While oil formation takes place between 0.7% and 1.3%, wet gas occurs between 1.3% and 2%. Lower boundary of metagenesis is 4%, under which metamorphism starts (Bozkaya and Yalcin, 1996).

During diagenesis and metamorphism, mineralogical changes occurring in the clay structures give extent of such events of the sedimentary rocks. Structural characteristics and parameters of the clay minerals are not unique for all depositional environments. However, depositional environments are characteristic to formation of certain clay minerals. Major parameters of sedimentary environment including pressure, temperature, subsidence, time, proton-electron concentration and metal-ion concentration greatly affect the clay mineral structures (Bozkaya and Yalcin, 1996). Most of the petroleum source rocks contain various clay minerals. Source rock properties like porosity and permeability vary depending on variety, abundance and distribution of the clay minerals (Bayar et al., 1987).

In the study area, the Kolsuz clays are loosely cemented with calcite, silica and Fe-oxides. Gravels are of various origins including sandstone, greywacke, claystone, limestone, marl, volcanics, granodiorite, gabbro, quartzite, chert, and serpentinite (Kayali et al., 2005). Petrographic investigations on thin-sections of 12 samples from the Kizilbayir Formation

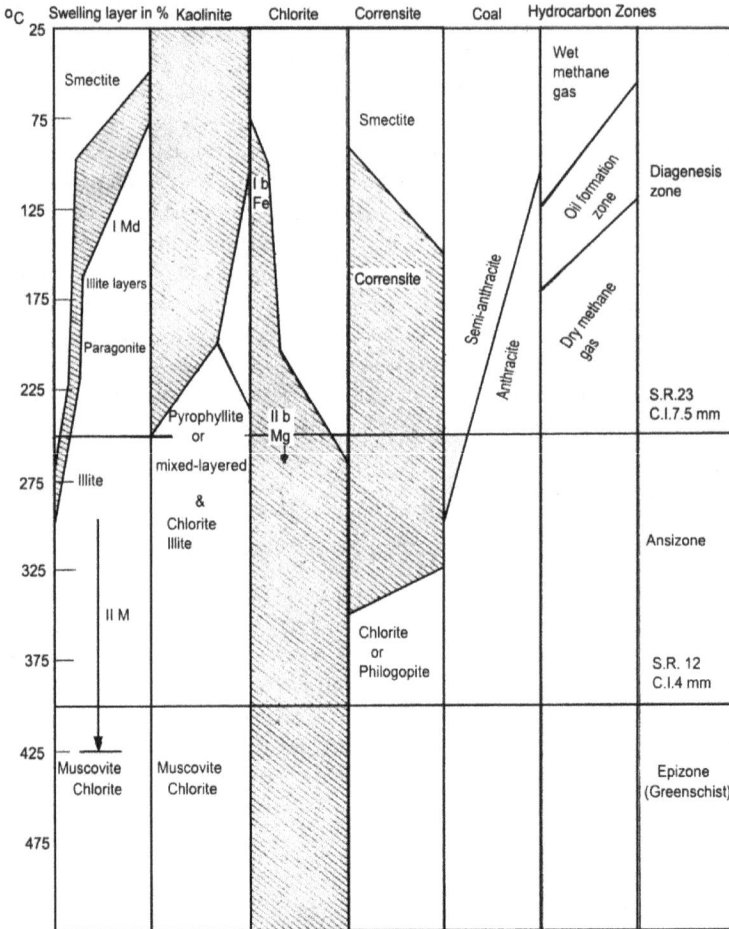

Figure 6. Diagenesis of clay minerals and oil formation zones in terms of temperature (adapted Weaver, 1960; Suchy et al., 2007).

indicated prevalent presence of feldspars, quartz, calcite, lithic fragments of volcanic, magmatic and metamorphic rocks, of carbonates (e.g., limestone and marl) and clastic sedimentary rocks (e.g., sandstone, claystone and greywacke) cemented chiefly by calcium carbonate (Gurel 1999; Kayali et al., 2005).

5. Results and discussions

The clay samples were taken from three different levels shown in the litostratigraphic columnar section of a selected locality in the Kolsuz-Ulukisla region (Figure 2). The FTIR spectra of the samples taken from lower level (Kk1), middle level (Kk2), and upper level (Kk3) of Kolsuz-Ulukisla (Turkey) clay profile are given in Figures 7 through 9, respectively. Interpretation of the observed vibrational bands in these samples were carried out by comparing with those found in the world clay standards with known fundamental vibration frequencies for illite, illite-smectite mixed layer, Na-montmorillonite, chlorite (ripidolite), palygorskite (Table 1) and the other clay standards. Results are listed in Table 3.

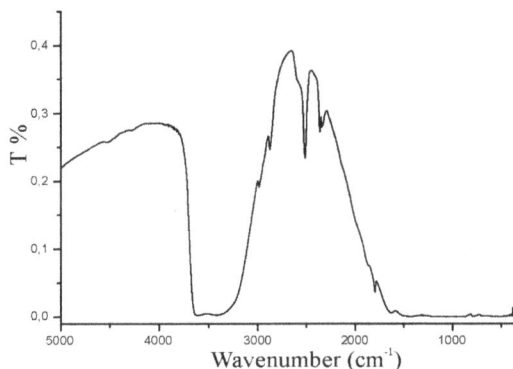

Figure 7. FTIR spectrum of the clay sample taken from the lower level (Kk1) of Kolsuz-Ulukisla.

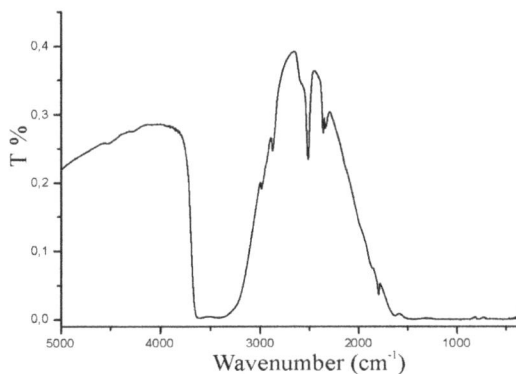

Figure 8. FTIR spectrum of the clay sample taken from the middle level (Kk2) of Kolsuz-Ulukisla.

As seen from Table 3, all of the samples belonging to the lower, middle, and upper levels of Kolsuz-Ulukisla (Turkey) clay profiles Kk1, Kk2, and Kk3, respectively, include illite, illite-smectite mixed layer, Na-montmorillonite, chlorite, palygorskite, calcite, feldspar and quartz minerals.

The FTIR spectrum of the sample representing (as summarized in FTIR measurements) lower level (Kk1) of Kolsuz-Ulukisla profile has been taken to see effects of the heat treatments on the structure of the subject samples (Figure 10). The assignments of the vibration frequencies of this spectrum were carried out following the same procedure

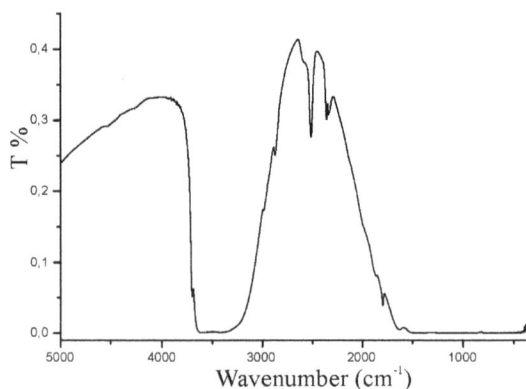

Figure 9. FTIR spectrum of the clay sample taken from the upper level (Kk3) of Kolsuz-Ulukisla.

Assignment	Kk1	Kk2 Wavenumber (cm⁻¹)	Kk3	Mineral
ν(OH) stretching "shoulder"	-	-	3680	Illite
Inner-layer OH, (Al-O...H) stretching	-	-	3680	Na-Montmorillonite
ν(OH) stretching	-	-	3660	Chlorite
ν(OH) stretching	3622	3622	3622	Illite/Na-Montmorillonite
ν(OH) stretching	2515	2515	2515	Calcite
δ(water-OH) scissoring	1731	1731	1731	Palygorskite
δ(water-OH) scissoring	1673	1673	1673	Palygorskite
ν(Si-O) normal to the plane stretching	1090	1090	1090	Illite-smectite mixed layer
ν(Si-O) planar stretching	1031	1031	1031	Illite-smectite mixed layer
ν(Si-O) stretching	988	988	988	Chlorite
(Al-Al-OH) deformation	920	920	917	Na-Montmorillonite
OH deformation	905	905	905	Palygorskite
M-OH stretching	819	819	817	Chlorite
M-OH leaning	805	805	805	Na-Montmorillonite
ν(OH) stretching	798	798	798	Quartz/Amorphous Silica/Feldspar
ν(OH) stretching	788	788	788	Quartz
ν(OH) stretching	697	697	697	Quartz
OH deformation	620	620	620	Na-Montmorillonite
(Si-O-Al) deformation	543	543	543	Chlorite
(O-Si-O) bending	520	520	520	Na-Montmorillonite
(O-Si-O) bending	468	468	468	Na-Montmorillonite/Palygorskite/Illite
(Si-O-Mg) deformation	441	441	441	Chlorite

Table 3. FTIR spectrum analysis results of Kolsuz-Ulukisla clay samples Kk1, Kk2, and Kk3 belonging to lower, middle, and upper levels, respectively.

applied to the spectra of the other samples and the results are given in Table 4. When results listed in Table 4 interpreted with the one in Table 3, it can be seen that the sample, Kk1, is composed of illite, calcite, chlorite, feldspar and quartz. Band assigned as vibrational frequencies for Na-montmorillonite and palygorskite were not observed after the thermal treatment. However, for illite at 916 and 833 cm⁻¹ (Al-Mg-OH) deformation and at 525 and 470 cm⁻¹ (O-Si-O) bending frequency bands were observed after the thermal treatment. This bands are resulted due to re-arrangement of Al and Mg atoms within the crystal structure during the thermal treatment (Bishop et al., 2002).

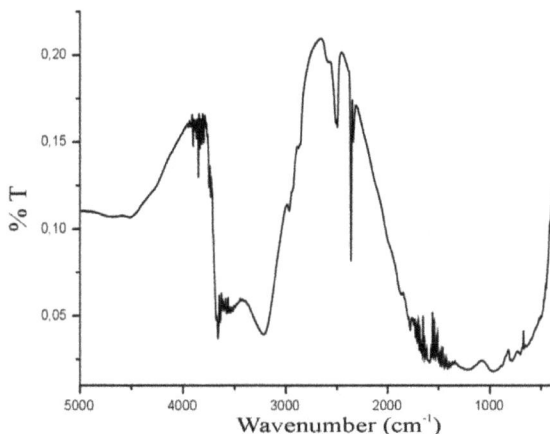

Figure 10. FTIR spectrum of the clay sample Kk1 after thermal treatment.

Assignment	Kk1 Wavenumber (cm⁻¹)	Mineral
v(OH) stretching "shoulder"	3680	Illite
v(OH) stretching	3662	Chlorite
v(OH) stretching	3622	Illite
v(OH) stretching	3566	Chlorite
v(OH) stretching	3435	Chlorite
v(OH) stretching	2515	Calcite
v(Si-O) normal to the plane stretching	1090	Illite
v(Si-O) planar stretching	1031	Illite
(Al-Al-OH) deformation	916	Illite
(Al-Mgl-OH) deformation	833	Illite
M-OH stretching	817	Chlorite
v(OH) stretching	798	Quartz//Feldspar
v(OH) stretching	788	Quartz
v(OH) stretching	697	Quartz
(Si-O-Al) deformation	543	Chlorite
(O-Si-O) bending	525	Illite
(O-Si-O) bending	470	Ilite
(Si-O-Mg) deformation	443	Chlorite

Table 4. Results of the FTIR spectrum analyses of the clay sample (Kk1) from Kolsuz-Ulukisla (Turkey) after thermal treatment.

Only v(Si-O) strecthing at 988 cm⁻¹ among the vibrational frequencies of chlorite at 3660, 988, 819, 543, and 441 cm⁻¹ was not observed after the thermal treatment. This is because that the organic molecules are forced into silicate layers during the thermal treatment. XRD data acquired from preheated sample (Kk1) between 350-550 °C, indicated that chlorite and illite peaks were not shifted but smectite peak was shifted towards 10 A°. In the same sample, kaolinite should also be present, beacuse the 7 A° peak dissappears at 550 °C. Based on FTIR and XRD (even at slow scan) measurements, kaolinite was never observed. Thus we conclude that Kk1 does not include any kaolinite. The v(OH) stretching vibrations of quartz at 798, 788 and 697 cm⁻¹ and v(OH) stretching vibration of calcite at 2515 cm⁻¹ remain the same after the heat treatment. As a result from FTIR analysis, we can say that the framework of silicate (T-O-T) structures of Kolsuz-Ulukisla-Nigde (Turkey) clay minerals samples has not been destroyed.

Before the thermal treatment, in Kk1 and Kk3 samples, palygorskite's only bands including δ(water-OH) scissoring vibrational bands at 1731 and 1673 cm^{-1} (Frost et al., 2001), OH deformation bands at 905 cm^{-1} and (O-Si-O) bending vibration band at 468 cm^{-1} were observed. While illite's ν(OH) stretching vibrational band appears as shoulder at 3680 cm^{-1} in Kk3 clay samples, despite (O-Si-O) bending vibration band at 468 cm^{-1} observed in Kk1 and Kk3 samples was assigned as belonged to palygorskite, it could belong to Na-montmorillonite's (O-Si-O) bending vibration band. Similarly, vibration band observed at 3680 cm^{-1} in Kk3 sample was assigned as ν(OH) stretching vibration band, it belongs to inner-layer OH, (Al-O...H) stretching band (Farmer and Russell, 1964). In samples Kk1 and Kk3, when bands assigned at 3622, 920, 805, 620, and 468 cm^{-1} were evaluated altogether; they appear well-matched with fundamental vibration bands of Na-montmorillonite standard. But vibration band at 920 cm^{-1} of the Na-montmorillonite standard clay was observed at 917 cm^{-1} in Kk3 sample.

6. Conclusions

1. Clay minerals of selected profiles in Nigde-Ulukisla area were determined both by the XRD and the FTIR.
2. Structural evaluation of these clay varieties was carried using the FTIR spectra.
3. Structural changes could be determined effeciently by the FTIR.
4. Findings are evaluated with the known data for depth-temperature-clay mineral transformations during burial processies in sedimentary basins.
5. Study area has potential to produce hydrocarbons as indicated by the presence of certain clay species.

Author details

Burhan Davarcioglu
Department of Physics, Faculty of Arts and Sciences, Aksaray University, Aksaray, Turkey

Acknowledgement

We would like to thank Turkish Scientific and Technological Research Council (TUBITAK-Turkey) for the financial support (project code: CAYDAG 2005-101Y067). Professor Dr. Cigdem Ercelebi (Department of Physics, Middle East Technical University, Turkey) is also gratefully appreciated for the FTIR.

7. References

Akyuz S, Akyuz T (2003). FT-IR spectroscopic investigation of adsorption of pyrimidine on sepiolite and montmorillonite from Anatolia. Journal of Inclusion Phenomena and Macrocylic Chemistry. 46(1): 51-55.

Bayar M, Turkay E, Gumrah, F (1987). Formation of clay with water injection in increasing oil production to investigate the effects of experimental. Proceedings of 3rd National Clay Symposium. Turkmenoglu and Akiman (ed.), September 21-27, 287-299, Ankara, Turkey.

Bishop J, Madejova P, Komadel P, Froschl H (2002). The influence of structural Fe, Al and Mg on the infrared OH bands in spectra of dioctahedral smectites. Clay Minerals. 37(4): 607-616.

Bozkaya O, Yalcin H (1996). Diagenesis-Metamorphic transition the methods used to determine. Journal of Engineering Geology-Turkey. 49: 1-22.

Brown G (1961). The X-ray Identification and Crystal Structures of Clay Minerals. Mineralogical Society (Clay Minerals Group), London, pp. 543.

Cradwick PD, Wilson, MJ (1972). Calculated X-ray diffraction profiles for interstratified kaolinite-montmorillonite. Clay Minerals. 9(5):395-405.

Davarcioglu B, Gurel A, Kayali, R (2005). Investigation of Central Anatolia region Nigde-Dikilitas (Turkey) clays by FT-IR spectroscopy. Proceedings of 12th National Clay Symposium. Yakupoglu, Aclan and Kose (ed.), September 5-9, 63-72, Van, Turkey.

Davarcioglu B, Kayali R (2007). Investigation of Central Anatolia Aksaray-Guzelyurt region kaolinitic clays by FT-IR spectroscopy. Journal of Faculty Engineering-Architecture Selcuk University. 22(1-2): 49-58.

Davarcioglu B, Gurel A, Kayali R (2007). Investigation of Eastern Black Sea Rize-Findikli-Camlihemsin region clays by FTIR spectroscopy. 6th International Industrial Minerals Symposium. Kemal, Batar, Kaya and Seyrenkaya (ed.), February 1-3, 87-95, Izmir, Turkey.

Davarcioglu B, Kayali R, Gurel A (2008). FT-IR spectroscopic study of Arapli-Yesilhisar-Kayseri clays from Central Anatolia region. Journal of Clay Science and Technology. 1(3): 163-173.

Davarcioglu B, Ciftci E (2009). Investigation of Central Anatolian clays by FTIR spectroscopy (Arapli-Yesilhisar-Kayseri, Turkey). International Journal of Natural and Engineering Sciences. 3(3): 154-161.

Davarcioglu B (2009). Investigation of eastern black sea region clays by FTIR spectroscopy (Rize-Findikli-Camlihemsin, Turkey). Colloquium Spectroscopium Internationale XXXVI. August 30-September 3, Book of Abstract, 74, Budapest, Hungary.

Demirtasli E, Bilgin AZ, Erenler F, Isiklar S, Sanli DY, Selim M (1986). The general geology of the basin with the Eregli-Ulukisla Bolkardaglari. Report of MTA-Turkey. No: 8097, pp. 54.

Dunoyer de Segonzac G (1970). The transformation of clay minerals during diagenesis and low-grade metamorphism: a review. Sedimentology. 15(1-2): 281-346.

Dypvik H (1983). Clay mineral transformations in Tertiary and Mesozoic sediments from North Sea. American Association of Petroleum Geologists Bulletin. 67(1): 160-165.

Falaras P, Lezou F, Seiragakis G, Petrakis D (2000). Bleaching properties of alumina-pillared acid-activated montmorillonite. Clays and Clay Minerals. 48(5): 549-556.

Farmer VC, Russell JD (1964). The infrared spectra of layer silicates. Spectrochimica Acta. 20(7): 1149-1173.

Frost RL, Locos OB, Ruan H, Kloprogge JT (2001). Near-infrared and mid-infrared spectroscopic study of sepiolites and palygorskites. Vibrational Spectroscopy. 27(1-2): 1-13.

Gundogdu MN (1982). Investigation of Geological, Mineralogical and Geochemical of Bigadic Cretaceous Sedimentary Neogene. PhD Thesis, Hacettepe University (Turkey), pp. 386.

Gurel A (1999). Determination of sedimentalogical, mineralogical and chemical properties of Ulukisla-Kolsuz area (Nigde-Turkey). Proceedings of 9th National Clay Symposium. Yeniyol, Ongen and Ustaomer (ed.), September 15-18, Istanbul, Turkey.

Heroux Y, Chagnon A, Bertrand R (1979). Compilation and correlation of major thermal maturation indicators. American Association of Petroleum Geologists Bulletin. 63(12): 2128-2144.

Hunt JM (1995). Petroleum geochemistry and geology. W.H. Freeman and Company, New York, pp. 745.

Kayali R, Gurel A, Davarcioglu B, Ciftci E (2005). Determination of qualitative and quantitative properties of industrial raw materials clays and diatomites in Central Anatolia by spectroscopic methods. TUBITAK-Turkey (project code: YDABCAG-1001Y067), pp. 137.

Kok MV (2006). Effect of clay on crude oil combustion by thermal analysis techniques. Journal of Thermal Analysis and Calorimetry. 84(2): 361-366.

Kok MV, Smykatz-Kloss W (2008). Characterization, correlation and kinetics of dolomite samples as outlined by thermal methods. Journal of Thermal Analysis and Calorimetry. 91(2): 565-568.

Murray HH (1991) Overview-clay mineral applications. Applied Clay Science. 5(5-6): 379-395.

Murray HH (1999). Applied clay mineralogy today and tomorrow. Clay Minerals. 34(1): 39-49.

Oktay FY (1982). Stratigraphy and geological evolution of Ulukisla and its surroundings. Geological Bulletin of Turkey. 25(1): 15-24.

Perry G, Hower J (1970). Burial diagenesis in Gulf Coast pelitic sediments. Clays and Clay Minerals. 18(3): 165-177.

Smith CR (1934). Base exchange reactions of bentonite and salts of organic bases. Journal of the American Chemical Society. 56: 1561-1563.

Suchy V, Sykorova I, Melka K, Filip J, Machovic V (2007). Illite 'crystallinity', maturation of organic matter and microstructural development associated with lowest-grade metamorphism of Neoproterozoic sediments in the Tepla-Barrandian unit, Czech Republic. Clay Minerals. 42(4): 503-526.

Teichmuller M (1987). Organic material and very low-grade metamorphism. In Low Temperature Metamorphism. Frey M (ed.), Blackie, Glasgow and London, pp. 161.

Weaver CE (1960). Possible uses of clay minerals in search for oil. Bulletin of American Association of Petroleum Geologists. 44(9): 1505-1518.

Wilson MJ (1987). A Handbook of Determinative Methods in Clay Mineralogy. Blackie-Son Ltd., London, pp. 308.

Yener N, Onal M, Ustunısık G, Sarikaya Y (2007). Thermal behavior of a mineral mixture of sepiolite and dolomite. Journal of Thermal Analysis and Calorimetry. 88(3): 813-817.

Clay Minerals in Soils

Chapter 6

The Impact of Clay Minerals
on Soil Hydrological Processes

Milan Gomboš

Additional information is available at the end of the chapter

1. Introduction

In literature clay and clay-loamy soils are sometimes called heavy soils. The origin of this term lies in agriculture – they are difficult to cultivate and thus increasingly energy demanding. One of the basic characteristics of heavy soils is their capacity to change their volume, which is induced by swelling and shrinking processes. These processes occur in three dimensions. In horizontal plane they are represented by formation of cracks and in vertical plane by vertical movement of soil surface. With the formation of cracks, soil environment becomes a two-domain structure, cracks being one domain and soil matrix being the other one. Shrinkage cracks are distributed through the unsaturated zone of a soil profile. In general, unsaturated zone is a three-phase system pertaining to litosphere and limited by land surface upwards and water table downwards. Simultaneously, it is a sub-system of a wider system formed by atmosphere - plant cover - soil – groundwater. All these sub-systems are interconnected by interaction (hydrological) processes and together they constitute soil water regime. Hydrological processes as a part of soil water regime are called water regime elements. Water regime elements can cause temporal and spacial changes in soil water storage, which are denoted soil moisture regime.

Soil volumetric changes and related formation, duration and termination processes of two-domain soil structure have a great impact on the dynamics of soil hydrological processes. Volumetric changes depend on the content of clay minerals in soil and their response to contact with water. Clay minerals content in soils is relatively stable and it is characteristic of every soil type. Soil moisture level varies with the changes in hydrometeorological conditions, depth of active root zone and the intensity of anthropogenic activity in the area. Soil water regime and related issues have been well monitored and examined in rigid soils, i.e. soils which do not shrink. Water transport in unsaturated rigid soil have been described by Richards` equation [1]. It is based on the theory of laminar flow in small pores. Many mathematical models have been developed to be used for numerical solution of this problem.

In this regard, heavy soils are similar to rigid soils, but only until reaching the point of first crack formation. The presence of shrinkage cracks considerably alters hydrodynamic properties of rigid soils and consequently the methods used for the description of rigid soils water regime cannot be used with cracking soils. At the time of shrinkage crack formation, heavy soils profile contains two distinct elements: cracks (also called macropores) and soil matrix (containing, inter alia, micropores). Each of these elements is characterised by very different conditions for water transport and retention. In the study of a two-domain soil structure, unlike in the study of rigid soils, it is necessary to consider the phenomena caused by soil volumetric changes, such as definition of crack network geometry, soil surface vertical movement, shrinkage characteristics, water flow to cracks, water flow within cracks and water flow from cracks to soil matrix.

The aim of this chapter is to quantify the impact of clay minerals on soil hydrological processes. The presented results were gained from the field measurements on Eastern-Slovak Lowland and from numerical simulation of heavy soils water regime.

2. Clay minerals in soil

2.1. Properties of clay minerals

Crystals of clay minerals are formed by sheets (formed by silicon tetrahedrals and aluminium octahedrals) of varying number, from 1 (montmorillonite) theoretically up to infinite. The individual sheets are extremely thin (5 – 10 Å) and their specific surface is very large (15 $m^2.g^{-1}$ – kaolinite, 80 $m^2.g^{-1}$ – illite; 800 $m^2.g^{-1}$ – montmorillonite). Sheets specific surface is very closely linked to the volumetric changes [2]. The larger specific surface is, the better is the capacity of soil to expand. The surface of clay sheets carries negative charge and thus it is able to bond molecules of water. As a result, the crystals of some clay minerals can bond water within their sturcture and expand. They can re-gain their original volume by drying. The ability of clay minerals to bond water in their structure and expand is different:

- kaolinite clays – relativelly inactive (in this group only halloysite has a very restricted ability to bond water);
- illites – low to medium – expanding clays
- Montmorillonite clays – extremely expanding clays. In their original form (under laboratory conditions) they can expand by 1 400 up to 2 000 % (Na-montmorillonite), [3].

In nature, clay minerals do not occur in pure form but they combine into mixed structures composed of various clay and other minerals. In literature, the word "clay" is used for both clay soil and clay minerals. "Clay", however, denotes a material that contains clay minerals smaller than 0.002 mm, as well as material whose structure contains predominantly clay minerals [4]. That means the material can be formed not only by clay minerals but also by other minerals smaller than 0.002 mm, such as silica, carbonates or metal oxides. With regard to the classification of clay minerals by their grains, clay minerals pertain to the category I. of Kopecký division (< 0.01). This is further divided into two sub-categories: I. colloidal clays (< 0,001 mm) and II. physical clays (0.001 – 0.002 mm) and very fine silt (0.002 – 0.01 mm). Thus when measuring clay minerals content in soil, clay percentage can be considered the same as the percentage of particles smaller than 0.002 mm (colloidal clay and physical clay). The limit value of 0.002 mm, which divides clay particles from fine silt, is the value used most worldwide [5]

2.2. Characteristics of the observed area

Research into the quantification of clay minerals impact on soil hydrological processes was carried out in the Eastern-Slovak Lowland fig. 1. ESL forms part of a neogene basin, which was created by irregular tectonic subsidence of crust within Carpathian arc during Neogene and Post-tertiary. Because of the subsidence, accumulation processes dominated in the area and a flat lowland surface was formed, mainly composed of fluvial sediments, loess and windblown sands.

Figure 1. Localization of the observed area

From the climatic point of view, ESL lies on the boundary between oceanic and continental climate. The area is characterised by a high changeability of metheorological elements in time. Precipitations formation is heavily influenced by circulation factors. Long-term avg sum of precipitations per year is 600 – 900 mm. Annual sum of potential evapotranspiration is 619 – 687 mm with the maximum value of 115 – 125 mm in July and minimum values in January. Bearing in mind real moisture conditions of the upper soil layer, the actual evaporanspiration is lower than the potential. It reaches 450 – 480 mm.

Soil conditions on ESL correspond to the hydro-geological development in the area fig.2. There is 209 518 ha of farmland. Only 3.2% of the farmland is composed of light soils. Up to 20.6% of the total farmland in the ESL is formed by very heavy soils which contain more than 75% of clay particles < 0,01 mm. The average depth of unsaturated zone is up to 1m. It can be found on 23.36% of the area whereas the unsaturated zone reaches 2m depth on 45% of the area and 3m depth on 68% of the area.

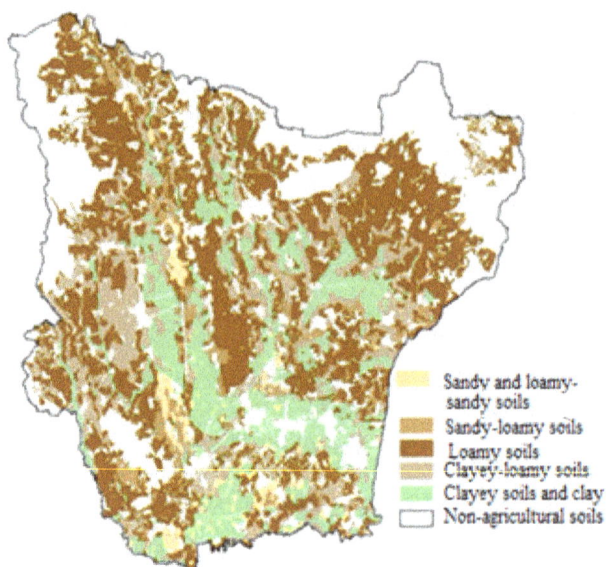

Figure 2. Planar distribution of soil types on East Slovakian lowland

2.3. Clay minerals in the soils of Eastern-Slovak lowland

The overall study of mineralogical soil structure in Slovakia was performed in 1970s. The results were published in Čurlík's work [6]. In spite of time distance between the year of publication and the present time, the work and the data published within are still topical and applicable. Due to demanding character of determining soil clay minerals in laboratory and unlike in other analysed areas, where pedological reasearch was performed, selected methotodology was based on determining the mixed layer clay structures in the individual soil types. This methodology presumes that pure forms of clay minerals are extremely rare in nature. More often than not they form combinations of diferent types. The study showed that predominant clay mineral in ESL is montmorillonite, others are kaolinite, chlorite, mica clay and mixed structures. The following table 1 illustrates the results of the research and presents the data from Svätuša probe (48⁰25′ 14,00″, 28⁰55′02,70″, 95m). This area is typical of ESL. The analysed soil profile was divided into six horizons from the surface to the depth of 1.50m. The data in the tab.1 concern clay and silt fraction which are typical of ESL.

2.4. Specific features of heavy soils water regime

2.4.1. Vertical and horizontal hydraulic conductivity of cracks

In our latitudes, heavy soil are severly dried in the summer half-year (April - September). As a result, cracks are formed and soil has a two-domain structure - domain of micropores and domain of macropores fig. 3. The first domain is formed by micropores in soil matrix.

Water movement here can be described by Richards` equation. Rate of water movement in this domain is relatively low. In case of macropores (cracks), Richards` equation cannot be used. Provided that large amount of water gets into cracks in a short time and the cracks are not properly closed, rainwater movement in cracks is analogous to the water movement in an open ditch. Rate of water movement is higher by orders in comparison with the water movement through micropores. Crack networkwork is a system of preferred water flow paths both in horizontal (horizontal crack conductivity) and vertical (vertical crack conductivity) direction. Cracks geometry changes in time and therefore hydraulic parameters of the preferred paths change too. In cracks, open water level can occur only if the cracks are open. The presence of water in cracks rapidly increases soil moisture and thus boosts the crack closure. Consequently, water movement in cracks is restricted by soil response time to moisture changes. Water shall move within cracks only if the cracks are filled with large amount of water in a very short time (e.g. torrential rains). Under normal conditions, cracks usually close before open water level could be formed.

Fraction / Sample	Microstructure				
	Oriented preparations			Non-oriented preparations	
	1 µm	1 – 2 µm	2 – 5 µm	5 – 10 µm	10 – 50 µm
P – 88 0,0 – 0,20 m	montmorillonite, clay mica, kaolinitee, quartz	quartz, chloritee, clay mica, spar, kaolinitee	quartz, mica, feldspars, mixed structure	quartz, feldspars, chloritee, mica	quartz, feldspars, mica, chloritee
mass.%	69,4	11,6	6,2	6,4	3,8
P – 89 0,21 – 0,40 m	montmorillonite, clay mica, kaolinitee, quartz	quartz, chloritee, clay mica, feldspars, kaolinitee	quartz, mica, feldspars, mixed structure	quartz, feldspars, chlorite, mica	quartz, feldspars, mica, chlorite
mass.%	75,9	9,6	4,7	2,2	4,8
P – 90 0,41 – 0,55 m	montmorillonite, clay mica, kaolinite, quartz	quartz, chlorite, clay mica, feldspars, kaolinite	quartz, chlorite, mica, feldspars	quartz, mica, feldspars, chlorite	quartz, feldspars, mica, chlorite
mass.%	73,8	7,8	6,8	2,0	8,2
P – 91 0,56 – 0,70 m	montmorillonite, clay mica, kaolinite, quartz, chlorite	quartz, chlorite, clay mica, feldspars, kaolinite, montmorillonite	quartz, chlorite, mica, feldspars	quartz, mica, feldspars, chlorite	quartz, feldspars, mica, chlorite
mass.%	78,4	8,8	2,0	2,4	7,2
P – 92 0,71 – 1,00 m	montmorillonite, clay mica, kaolinite, quartz, chlorite	quartz, chlorite, clay mica, feldspars, kaolinite, montmorillonite	quartz, chlorite, mica, feldspars	quartz, mica, feldspars, chlorite	quartz, feldspars, mica, chlorite, calcite
mass.%	68,1	4,6	8,0	11,3	6,0
P – 93 1,01 – 1,50 m	montmorillonite, clay mica, quartz, chlorite	quartz, chlorite, clay mica, feldspars, kaolinite, montmorillonite	quartz, chlorite, mica, feldspars, gypsum	quartz, mica, feldspars, chlorite, calcite	quartz, feldspars, mica, chlorite, calcite
mass.%	46,7	7,6	13,1	15,0	15,2

Table 1. Clay and silt characteristics on Svätuš locality. Mineralogical phase analysis of soils is difficult.

Figure 3. Double-domain soil structure and detail of a crack, photo Gomboš, locality Milhostov

With regard to the intensity of rainfall and rate of water interception, the following situations can occur fig.4. The symbols have the following meanings: I_c – intensity of crack infiltration [m.s^{-1}], I_{c1} –infiltration ratio from rainfall exceeding maximum infiltration intensity of soil matrix surface [m.s^{-1}], I_{c2} – crack infiltration of rainwater fallen directly to cracks [m.s^{-1}] , I_{in} – interception intensity [m.s^{-1}], I_m – soil matrix infiltration intensity of [m.s^{-1}], I_{po} – surface runoff intensity [m.s^{-1}], P_{in} – rainfall intensity [m.s^{-1}], S_c – inner crack surface on a unit of soil surface area [m^2.m^{-2}], S_m –soil matrix surface on a unit of area [m^2.m^{-2}], S_{pr} – cracks sectional area on a unit of soil surface area [m^2.m^{-2}].

1. $P_{in} - I_{in} \leq I_{m,max}$, $I_{c1} = 0$, $I_{c2} = S_{pr} . P_{in}$, $I_c = I_{c2}$, $I_m - S_m . (P_{in} - I_{in})$

   In the first and simplest case the intensity of rainfall is lower than or equal to the rate of water infiltration to soil matrix ($I_{m,max}$). The ratio of rainwater absorption by soil matrix surface and the area of crack sectional surface is the same as the ratio of their surfaces (fig.4a).

2. $P_{in} - I_{in} > I_{m,max} \wedge I_{po} = 0$, $I_{c1} = S_m . (P_{in} - I_{in} I_{m,max})$, $I_{c2} = S_{pr} . P_{in}$, $I_c = I_{c1} + I_{c2}$, $I_m = S_m . I_{m, max}$

   In the second case the intensity of rainfall is higher than water infiltration in soil matrix. Rainwater that is not absorbed by the surface of soil matrix (I_{c1}) drains to the nearest crack and thus increases water content absorbed by cracks (fig.4b).

3. $P_{in} - I_{in} > I_{m,max} \wedge I_{po} > 0$, $I_c = S_c . I_{m,max}$, $I_m = S_m . I_{m, max}$, $I_{ho} = (P_{in} - I_{in}) . (S_{pr} + S_m) - I_c - I_m$

   In the third case the intensity of rainfall exceeds the maximum rate of water infiltration in soil matrix while crack capacity to absorb another water flow has been exhausted and cracks are now wholly filled with water (fig.4c). There is surface run-off.

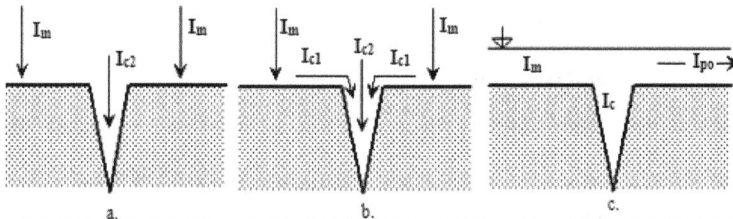

Figure 4. Different ways of water penetration through cracks and soil matrix surface depending on the rainfall intensity.

During water transfer from the surface to lower soil horizons or ground water level (GWL, rainwater can permeate directly under the root zone. As a result no efficient water contact with root zone occurs. This effect reduces the efficiency of irrigation and fertilizing. At the same time, cracks can accelerate soil water evaporation to atmosphere and crack networkwork anticipates the permeation of pollutants through soil to GWL.

2.4.2. Water balance in two-domain soil structure

The basic equation for determining the water balance in aeration zone is [7]:

$$W_t = W_0 + I_k + I_i + Q_{hp} - E - T - Q_{ho} - I_h \tag{1}$$

W_0 - initial overall soil content in aeration zone [m], W_t - overall soil water content in aeration zone at time t [m], I_k - capillary water inflow [m], I_i - water inflow by infiltration from rainfall [m], Q_{hp} – subsurface lateral water inflow [m], E - water losses by evaporation [m], T - water losses by transpiration [m], Q_{ho} – subsurface lateral water drainage [m], I_h - infiltration into GWL or lower horizons [m].

For the purposes of cracking soils, it is necessary to adapt some members of the balance equation, mainly I_i, representing water inflow by infiltration from precipitations. This shall be split to I_c – water inflow by infiltration through cracks, and I_m – water inflow by infiltration through soil matrix surface ($I_i = I_c + I_m$).

Similarly, water losses by evaporation E shall be split to the evaporation from soil matrix surface E_m and evaporation from sectional area of cracks E_c (E = E_m + E_c). However, the evaporation from the sectional area of cracks is disputable. On one hand, it could be objected that water does not evaporates merely from the narrow area of cracks sectional surface but from the whole surface of crack walls, which is larger. On the other hand, evaporation from the surface of crack walls is very limited due to high moisture levels and little air circulation inside a crack. Air inside the cracks that reach the depth of groundwater table is supposed to be saturated with water vapours. Evaporation from plant cover (transpiration) T occurs only on the area of soil matrix surface $T_m = T . S_m$. Modified balance equation shall be:

$$W_t = W_0 + I_k + I_c + I_m + Q_{hp} - E_m - E_c - T_m - Q_{ho} - I_h \tag{2}$$

In lowland areas, subsurface inflow and drainage (Q_{hp}, Q_{ho}) can be disregarded.

3. Methods and material

3.1. Texture of the analysed soil profiles

The research in heavy soils water regime on ESL was performed on 11 soil profiles. In this paragraph, results from three typical areas on ESL are presented: Senné (48°39'48,19"; 22°02'53,90"; 97 m), Milhostov (48°40'11,08"; 21°44'18,02"; 100 m) and Sírnik (48°3033,01"; 21°48'51,18"; 98 m). Grain-size analysis to various depths was performed on each of these three soil profiles- in Senné to 0.8m, in Milhostov to 2.0m and in Sírnik to 1.2m, in the layers 0.10m thick. The results were processed into USDA soil textural triangles [8], which are shown at fig.5.

It is obvious that the heaviest soil is in Senné, where clay is the predominant soil constituent. Soil profile is homogeneous and the fraction of sand is significantly lower compared to the other two areas. In Milhostov, where in addition to clay, silt can be found as the second most dominant soil type, ranging from clay silt to silty loam. Upper horizons of soil profile to 0.60m are heavier (36% of clay) than the lower horizons. Sírnik soil profile shows two markedly different layers along the vertical line. In the depth of 0.40m and downwards to 1.20 m clay fraction doubles from 25% to 50%. Apart from clay, the second more dominant constituent is silt.

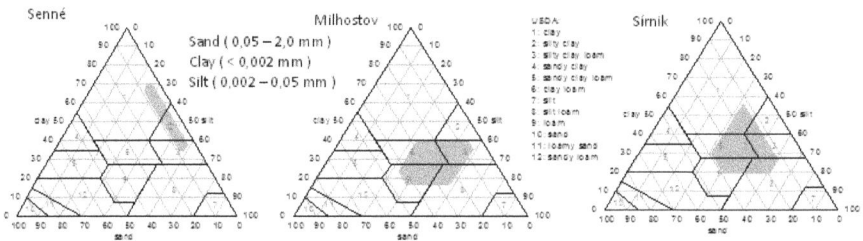

Figure 5. Classification of selected soil profiles according to the triangular classification diagram USDA

3.2. Field measurements and experiments

Field measurements included continual monitoring of the observed areas once every week. During the winter, field measurements were not performed because of possible imprecisions due to snow cover and frozen upper layer of soil.

The following parameters were monitored fig.6.:

- GWL position,
- volumetric soil moisture along the vertical line to the depth of 0.8m in the layers 0.1m thick.
- vertical movement of soil surface – these measurements were performed by the method of surface levelling in three points situated in the shape of equilateral triangle with 2m sides. For the details of one such point see fig.6. Casing pipe is used as a fixed point. It is supposed that due to deep embedding, the pipe is stable with regard to the vertical movements of soil surface through the active layer of a soil profile.

- lump-sum takings of undisturbed soil samples by the method of dug probe fig.7, in Kopecký cylinders, for the purpose of determining moisture retention curves, hydraulic conductivity, COLE values (Coefficient of Linear Extensibility) and PLE values (Potential Linear Extensibility). Undisturbed soil samples were taken as well to perform grain-size analysis. Furthermore, in Milhostov area, geometric parameters of crack networkwork such as length, width, specific length and soil matrix area were measured.

3.3. Laboratory measurements and experiments

Laboratory measurements were based on the evaluation of soil samples taken during the regular monitoring and lump-sum takings. The following analyses were performed:

- Grain-size analysis – on disturbed soil samples by hydrometer-method. On the ground of this, soils were classified according to USDA texture diagram.
- Analysis of soil shrinking properties – on undisturbed soil samples. Firstly, the samples were fully saturated with water and on the basis of geometric parameters measurements their original volume and measured weight were calculated. The samples were then dried at laboratory temperature and their volume and weight were regularly measured. When the weight loss had almost reached measurement error, samples were dried up in a laboratory dryer at 105°C and then measured again for their final volume and weight. Measured parameters were evaluated from the point of view of dependencies between soil volume, grain soil structure and volumetric moisture, shrinking properties in the form of shrinkage curve and COLE and PLE values and geometric factor r_s.

Figure 6. Sampling area (Senné area) and a detail of a point where vertical water movements are measured

Figure 7. Collection of untouched soil samples

- Analysis of volumetric soil moisture – gravimetric method was used.
- In the IH SAS laboratory in Bratislava, overpressure method was used for measuring moisture retention curves and courses of hydraulic conductivities on the solid soil samples.

3.4. Determination of the selected heavy soils characteristics

3.4.1. Determination of the soil shrinkage basic characteristics

Heavy soils shrinkage is often described with regard to the relationship between the volume of the soil and its moisture. This relation has been subject to numerous studies worldwide. The shrinkage characteristic used most is based on the relation between a soil water ratio - v and a void ratio – e, where v = soil water content / volume of soil solid phase [-] e = volume of pores/volume of soil solid phase [-]. Soil water ratio and void ratio are determined on the basis of volumetric soil moisture - θ and porosity – P. The following equations are used:

$$P = \frac{e}{1+e}; \quad \theta = \frac{v}{1+e} \, [-] \tag{3}$$

General relationship between soil water ratio and void ratio is figured in the graph in fig.8. Auxiliary diagonal line in the fig.8 represents the state of saturation when soil water ratio equals void ratio and the total volume of pores is filled with water. Curve divides shrinkage process to four domains.

- **Strucural domain** – occurs in saturated soils. Water is drained from macropores. Soil volume remains unchanged or changes very slightly;
- **Normal domain** – changes in soil volume are equal to the changes in water content and micropores remain saturated with water (part of the curve paralel to the diagonal line);
- **Residual domain** – changes in soil volume are lower than changes in water content , air starts to enter the micropores (curved part of the curve);
- **Zero domain** – there are no changes in volume (horizontal part of the curve).

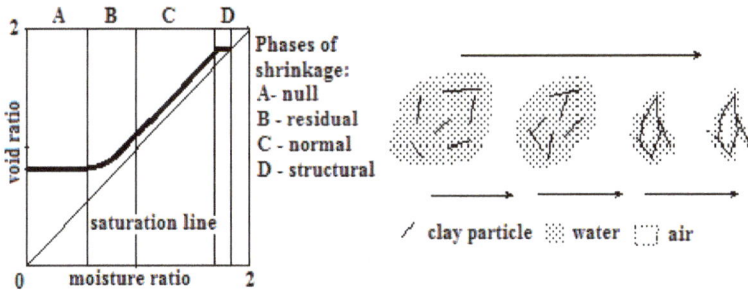

Figure 8. General form of expression of shrinkage characteristics of heavy soils and position of clayey particles during drying [9].

3.4.2. Shrinkage potential of heavy soils

When studying shrinkage characteristics of heavy soils, it is practical to quantify the potential of this phenomena. For this purpose two parameters were introduced in soil science: COLE (Coefficient of Linear Extensibility) and PLE (Potential Linear Extensibility). COLE was introduced by Grossman [10] and it is used for quantifying shrink-swell potential of soil:

$$COLE = \left(\frac{V_{wet}}{V_{dry}} \right)^{\frac{1}{3}} - 1 \, [-] \qquad (4)$$

where V_{wet} is the volume of wet soil and V_{dry} is the volume of dry soil.

The second parameter expressing shink-swell soil potential is PLE, which is the potential for soil swelling and shrinking in field conditions. It considers swelling and shrinking properties of the individual soil horizons in the studied soil profile.

$$PLE = COLE_{(1)} \cdot Z_{h(1)} + COLE_{(2)} \cdot Z_{h(2)} + \dots + COLE_{(n)} \cdot Z_{h(n)} \qquad (5)$$

where $COLE_{(n)}$ is a COLE value for n- horizon, $z_{h(n)}$ is the width of n-horizon [cm], and $z_{h(1)} + z_{h(2)} + \dots + z_{h(n)} = 100$.

PLE is the value of maximum (potential) change in length of a 100cm thick soil profile due to swelling and shrinkage process. After COLE and PLE has been set, a soil can be classified according to COLE value [11], tab.2 or PLE values [12], tab.2.

shrinkage - swelling potential	COLE	Shrinkage according to the PLE potential	PLE value [cm]
low	< 0.03	high shrinkage	> 14
medium	0.03 - 0.06	medium shrinkage	9 - 14
high	0.06 - 0.09	low shrinkage	< 9
very high	> 0.09		

Table 2. Classification of shrink-swell potential by COLE and PLE values

3.4.3. Formulation of relationships between volumetric changes and vertical subsidence of soils

Soils volumetric change is a three-dimensional process. In nature, drying of soils is partly reflected in cracks formation and partly in the soil surface subsidence. Therefore soil changes are horizontal, caused by opening and closure of cracks, and vertical as soil surface movement. In laboratory, soil volumetric changes are visible as changes in geometric dimensions of an undisturbed specimen of soil. Calculations are based on the following equations fig.9.:

$$\Delta V = \Delta Vv + \Delta Vh \tag{6}$$

where

$$\Delta Vv = z_s^2 . \Delta z \; [m^3] \tag{7}$$

Mathematical expression of the relationship between soil volumetric change and vertical subsidence is as follows [13]:

$$\Delta z = z_s - \left[\left(\frac{V}{V_s} \right)^{\frac{1}{r_s}} \right] z_s , [m^3] \tag{8}$$

where

ΔV total volumetric change of a soil specimen $[m^3]$;
ΔV_v vertical volumetric change of a soil specime $[m^3]$;
ΔV_h horizontal volumetric change of a soil specimen $[m^3]$;
V soil volume after shrinkage $[m^3]$;
V_s volume of a saturated soil specimen $[m^3]$;
Δz change in height of a soil specimen $[m]$;
z_s height of a saturated soil specimen $[m]$;
z height of a soil specimen after shrinking $[m]$;
r_s geometric factor $[-]$.

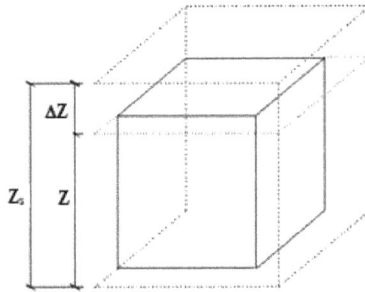

Figure 9. Volume change of isotropic soil sample in shaper of cube during drying process, (dashed line express the sample in saturated state (sample height Z_s), continuos line express the sample after shrinkage (sample height Z)

The equation above (8) is implemented in the mathematical simulation model FLOCR [14], which is designed to calculate the thickness of each layer. Horizotal volumetric change can be expressed by the following combination of equations (6,7,8):

$$\Delta V_h = V_s \left[\left(\frac{V}{V_s} \right)^{\frac{1}{r_s}} - \frac{V}{V_s} \right] \left[m^3 \right] \qquad (9)$$

Experimental measurements of the dependency $\Delta V = f(\theta)$ in the vertical line of a soil profile allow for the calculation of crack porosity volume and the overall subsidence or lift of soil surface induced by moisture changes. With the knowledge of geometric characteristics of crack networkwork, cross-sectional surface and avg thickness (rate of openness) of cracks can be calculated.

3.4.4. Geometric factor of volumetric changes

Non-dimensional geometric factor r_s is the ratio of participation in soil volumetric changes of both crack formation and vertical movements. It can be influenced by external load and possibly by terrain settling process, which occurs when the clay sheet particles are orientated in one predominant direction. r_s can reach the following values:

$r_s = 1$ no cracking process, all soil volumetric changes are vertical;
$1 < r_s < 3$ vertical movement predominates over crack formation
$r_s = 3$ isotropic shrinking;
$r_s > 3$ crack formation predominated over vertical movement;
$r_s \to \infty$ all soil volumetric changes are horizontal; i.e. only cracks are formed.

In nature, isotropic shrinking with $r_s = 3$ can occur in most soils. Provided that during drying the vertical change in height of a soil specimen and the volume of water saturated soil are measured, the equation (8) can be used to calculate the geometric factor r_s. This factor can be calculated from the equation mentioned previously in its analytical form:

$$r_s = \frac{\log(\frac{V}{V_s})}{\log\left[\frac{-(\Delta z) + z_s}{z_s}\right]} \quad [-] \tag{10}$$

3.4.5. Measurements of crack networkwork characteristics

Cracks represent horizontal volumetric changes of soil. On the soil surface they form a mosaic that reminds a network. Stability of soil matrix walls is ensured by cohesive forces in soil. It is enforced by humic and other organic substances. It is a well-known fact that cracks are formed in places where mechanical strength is lower, usually in the same area not only during one year but also in more consecutive seasons. It means that if a crack networkwork was created in an area, cracks shall open repeatedly, unless significant changes occurred in mechanical characteristics of the soil [15], [7]. Study of cracks geometry and determination of their characteristic has been increasingly important with regard to physical structure of soil which is characterised by physical soil properties. This relationship may be useful for formulating predictions about cracks formation in a soil type. With a view to define geometric properties of cracks the following characteristics can be used: L_c – length (total length of cracks on a measured surface) [m], d_c – width (avg crack openness) [m], z_c – depth [m], - specific length (length of cracks on a measured surface) [m.m^{-2}], R_c – specific density (number of cracks on a unit of length) [m^2.m^{-2}], $S_{pr} = L_c$. d_c, S_c – internal surface of cracks ("wall surface" of cracks on a unit of soil surface area) [m^2.m^2], $S_c \cong 2$. L_c . z_c, P_c – crack porosity (crack volume in a unit of soil volume) [m^3.m^{-3}], P_{ac} – crack porosity of aeration zone (crack volume on a unit of soil surface to GWL depth) [m^3.m^{-3}]. Geometric characteristics of crack networkwork were studied in Milhostov area. From 3m above the ground photodocumentation was made, on the ground of which geometric characteristics of crack networkwork were evaluated.

3.4.6. Numerical simulation of heavy soils water regime

Numerical simulation of water movement through the unsaturated soil is based on the interrelation of this sub-system with other sub-systems of the system: atmosphere – plant cover – unsaturated zone – groundwater. When coming in contact with water, impacted unsaturated zone responds by water regime changes in time and space. With a view to authentically simulate the results of such water movement, so that the model approaches reality as much as possible, very precise input data are vital. The input data are of five types: metheorological and climatic conditions, plant cover characteristics, hydrological conditions, topographic data of the observed area and initial and boundary conditions. In case of numerical simulation of water regime in heavy soils, shrinking and swelling characteristics must be included as well. Outputs from the model provide information on the development of soil water content in every horizon in unsaturated soil, on water flow through upper and lower boundary of unsaturated zone during the whole modelled period. The model can be verified by comparing the outputs from the model with the data monitored on-site.

In this particular case of heavy soils water regime, FLOCR model was used (FLOw in CRacking soils) [14]. The model was developed in The Netherlands and it simulates one-dimensional vertical water flow in aeration zone of a two-domain soil profile. It assumes that water flow through soil matrix in unsaturated zone can be calculated by Richards'equation and volumetric changes by the formula (6). A soil profile is divided in layers, there can be max. 30 layers in a soil profile. For each layer, the model calculates water inflow and drainage, volumetric changes, crack volumes, changes in layer width, moisture potential as a pressure height and volumetric moisture. For a soil profile as a whole, the model calculates total volume of cracks, vertical movements of soil surface, GWL, drainage, actual evapotranspiration and surface runoff. The model can be used also for the soils with more horizons of different hydrophysical properties. Max. Number of horizons is 5.

The input data entering the model are: hydrophysical properties of a soil profile (moisture retention curve, saturated and unsaturated hydraulic conductivity, shrinking characteristics), upper and lower boundary condition, precipitations, potential evapotranspiration, spacial determination of a soil profile, setting of a computational step.

4. Results

4.1. Quantification of volmetric changes in ESL soils

Altogether, 90 samples of different gran-size structure were analysed. Time-course of volumetric changes of soils with varying clay particles ratio is pictured in fig. 10. The graph shows considerable impact of clay particles ratio in soil on the rate of volumetric changes with regard to the conditions of saturation. When the content of particles of the I. fraction (I. fraction (I.fr.) is colloidal clay - particles < 0,001 mm, I. kat.- particles < 0,01 mm) exceeds 25%, this increase shoots up other volumetric changes (difference between upper and lower line). Therefore water desorption from soil during drying requires different time interval for every soil sample.

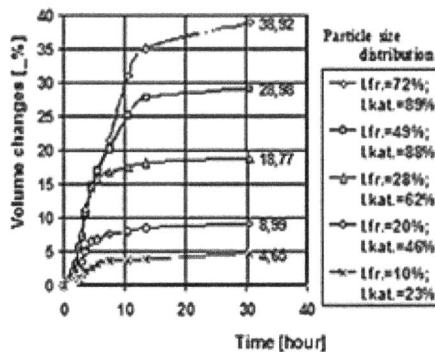

Figure 10. Processing the courses of the volume changes of the soil with the different particle size Distribution, I.fr. < 0.001 mm.

Fig. 11 shows the course of volumetric changes of soils of various grain-size structure with regard to volumetric moisture. The differences in volumetric changes in each part of I. fraction are obvious. It can be seen that volumetric changes can reach 40 % of the saturated soil volume. In fig.11 volumetric changes at different moisture levels for different soil samples are considered. Thus the lines in the graph represent volumetric changes measured simultaneously in various soil samples with different moisture values. Initially, all the lines of the first fraction are in one bundle.

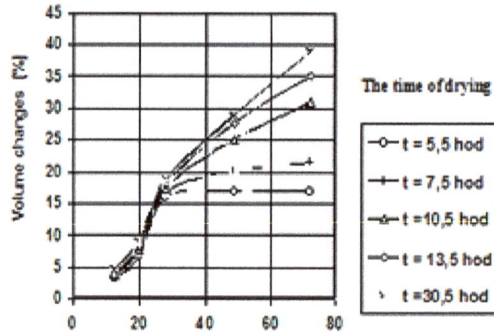

Figure 11. The course of the volume changes of the soils with the different particle size distribution during drying depending on the bulk water content. Differences at individual levels of 1. Fraction content are obvious.

When the particle content in the I. fraction is less then 20 %, important volumetric changes occured during the first 5.5h and in the following 25h there were just slight changes in soil volume.

Fig. 12 shows the course of volume changes in soils of different textural composition, with regard to soil volumetric moisture. The differences in volumetric changes in soils with various ratio of fraction I. are evident. On the grounds of the stated results, it is obvious that physical clay content and soil moisture have a decisive impact on soil volumetric changes.

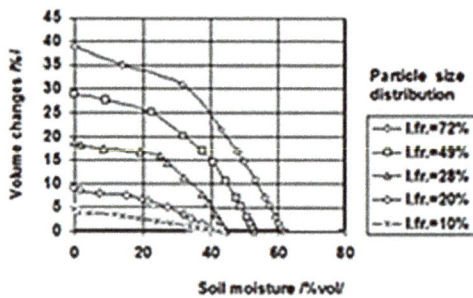

Figure 12. The soil volume changes at different water contents (bulk water content θ is expressed by the time of drying) with the different content of particle of 1. fraction

Fig.13 shows a three-dimensional visualisation of the measured results. It is a three - dimensional axionometric representation of 190 measurements of dependencies between volumetric soil changes with regard to the content of physical clay and volumetric moisture. Fig. 13 on the left is a graphical representation of the measured data. On the right there is a graphical representation of the same values by means of 3rd order polynomials.

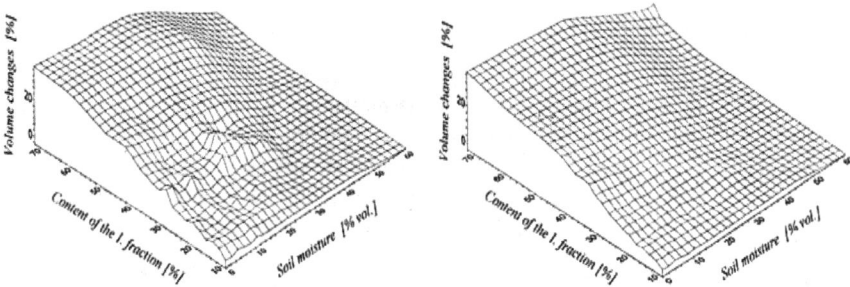

Figure 13. The three-domain representation of the measured volumetric changes of soil in dependence on the humidity and the grain structure and polynomial representation of this dependency.

4.2. Shrinkage characteristics

In the soil profiles mentioned in paragraph 3.1, shrinkage properties were determined in layers 0.1m thick by means of the method described in paragraph 3.4.1. Fig. 14 shows the comparison between the soil profiles with different clay minerals content, which has considerable impact on the course of shrinkage characteristics. Shrinkage process in clay soils comprise three distinct shrinkage phases. In light soils (on the right) shrinkage is very slight. Knowledge of shrinking characteristics is vital for precise numerical simulating of heavy soils water regime.

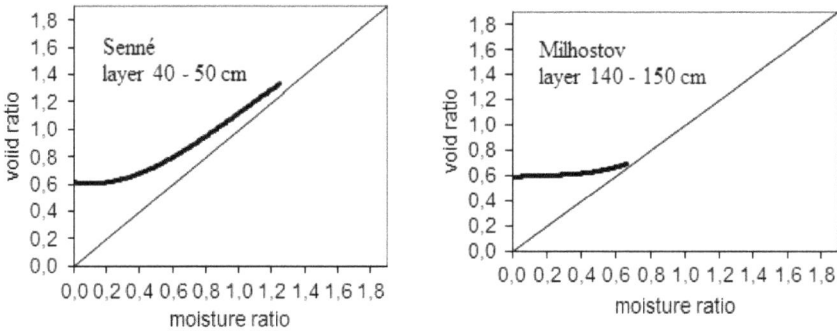

Figure 14. Comparison of shrinkage characteristics of heavy clay (Senné, clay 49%) and lighter loam (Milhostov, clay 23%).

4.3. Evaluation of shrink-swell potential of ESL soils

The quantification of volumetric change potential was performed in laboratory on the examined profiles by means of COLE and PLE coefficients.

Value of PLE=15,30cm in Senné classifies the soil profile as "High shrinkage" in Reeve classification. It means that soil profile can potentially change its thickness by 15,30cm into the depth of 1m. However, it is only theoretical assumption which is very unlikely to occur in nature. The highest vertical movement measured in Senné locality during research works was 5,51cm.

Dependency between COLE and clay minerals measured on ESL was compared to the results gained in the Netherland and USA [9], [11], [16] fig. 15. Comparison indicates that examined dependency measured on ESL has slightly steeper slope comparing to the data from literature. It is probably related to spatial variability of mineral structure of clay particles, mainly by presence of illite and montmorillonite group of clay minerals.

4.4. Field measurements of vertical soil surface movements

Vertical soil surface movement is one of the signs of volumetric changes. Fig. 16 shows the results of vertical water movement in 2003. Results indicate that the largest soil surface movements occured in Senné locality (5.51cm). It was 4.08cm in Sírnik area and 3.18cm in the Milhostov area. Results comply with the analysis based on PLE and COLE values. It is possible to calculate water storage in active layer of soil profile by vertical movements. Value of the geometric factor – r_s (that is dependent on the ratio of horizontal and vertical part of shrinkage) was laboratory determined. In every analysed profile the value of r_s was dentified for 0.10 m thick layers. In the Senné avg value $r_s = 2.85$ was measured. It means that the vertical soil surface movement dominates over creation of soil cracks. In Milhostov the avg measured value was $r_s = 3.1$ is. Horizontal changes slightly dominate over vertical changes with regard to the total volumetric change. In Sírnik profile to 0.80m depth r_s value was isotropical, $r_s=3$.

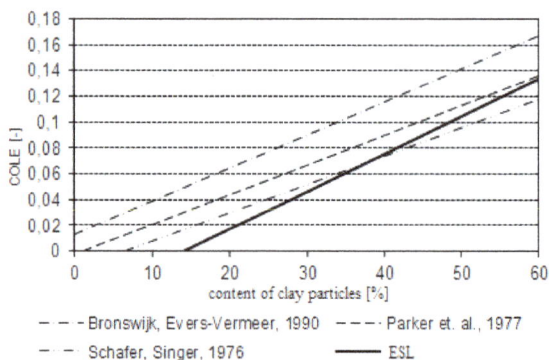

Figure 15. Comparison between cole dependency on clay particles content on ESL and dependencies measured in heavy soils in Netherlands and USA.

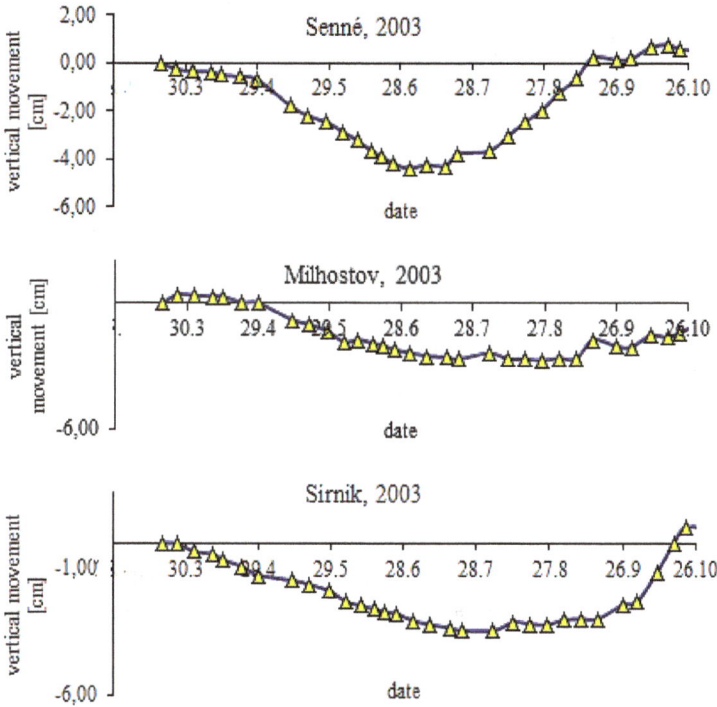

Figure 16. Results of vertical movements of soil surface measurements in 2003

4.5. Structure of crack network in the locality of Milhostov

In Fig.3 you can see the types of crack network structure identified in this work on the basis of common particular features [7]. The most common structure is *characteristic structure of crack network*, fig. 17a, which occurs in the soils cultivated by common agrotechnology. Another structure is a sporadic crack structure of *linear typ*, fig. 17b. It is an accidental structure occurring on seedbeds or on paths between them. It is a separate element on the surface of agriculturally cultivated fields. The third type is a sporadic structure of *cluster type*, fig. 17c. It is formed by small clusters of cracks on small areas which are caused by enormous soil drying in these areas. Crack network structure of *anthropogenic character*, fig. 17d, is a network which is caused by sowing mechanisms blades. It has a regular square structure. In comparison with preceding structure it is characterised by relatively smaller values of specific length and crack density.

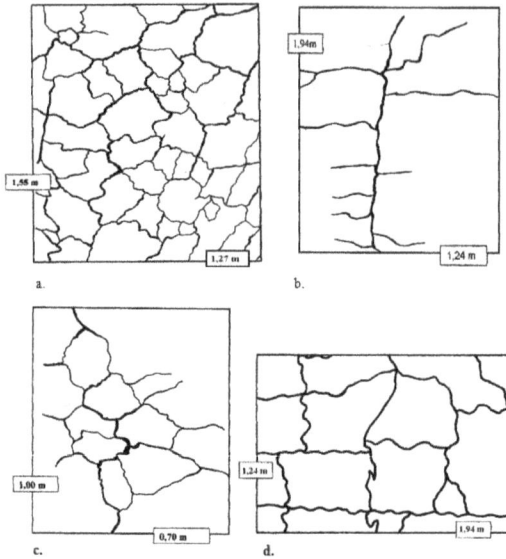

Figure 17. Structure of heavy soil crack net (locality Milhostov)

The analysis of photographs from Milhostov area showed that total length of cracks in the observed areas L_c was 123m, specific length was L_{mc} = 5.65 m.m^2, specific density was R_c = 3.1 cracks.m^{-1}. Depth of the cracks was estimated on 1.20m based on a dug probe. Therefore the average area of internal walls with neglected turtuzoity is 13.56 m^2.m^{-1}. At the occurrence of sudden and extreme rainfall, the infiltration is realised not only by soil matrix surface but also by water permeation through cracks as the cracks are filled with water. Fig. 18 shows the evaluation of 234 soil matrices. The distribution indicates that the most occurring soil matrix area is within the interval 0.05-0.10 m^2. The average area of the soil matrix is 0.121m^2.

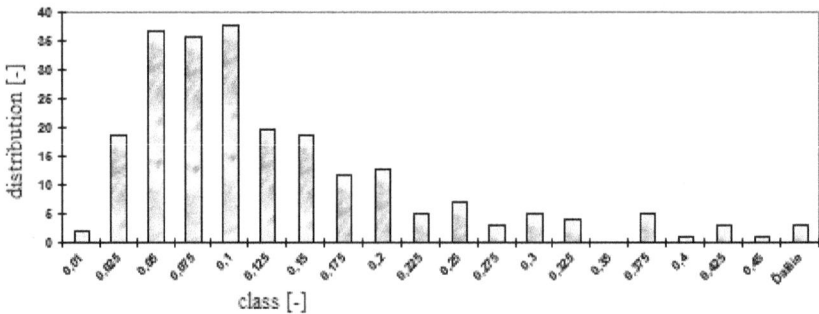

Figure 18. Distribution abundance of soil matrix area in the locality Milhostov

4.6. Selected results of numerical simulation of soil water regime on ESL

In the figure below, the results of 2003 numerical simulation of water regime are described. This year has been one of the driest years in Senné area in terms of soil water storage since 2007. Therefore the impact of the clay minerals on hydrological processes in soil environment can be proved. The schemes graphically represent the results of moisture volume measurements in the examined profiles. The depth to which the soil profiles are dried can be seen.

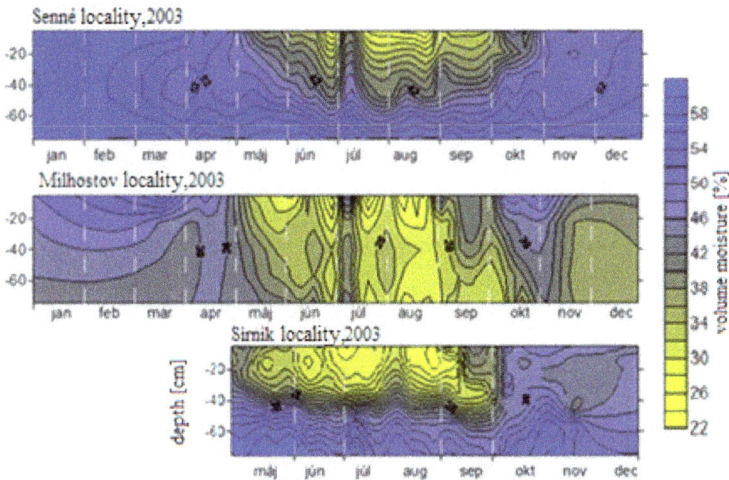

Figure 19. Monitored moisture regime of examined soil profiles in 2003 in Senné, Milhostov and Sírnik locality.

After the verification of the numerical simulation results, the simulation of water regime into the depth of 3.0 metres could be started. The targets of the verification were the measurements of the soil moisture regime, GWL course and soil surface vertical movement. The results of moisture regime numerical simulation of the observed profiles into the depth of 3m during 2003 vegetation period are shown in Fig. 20.

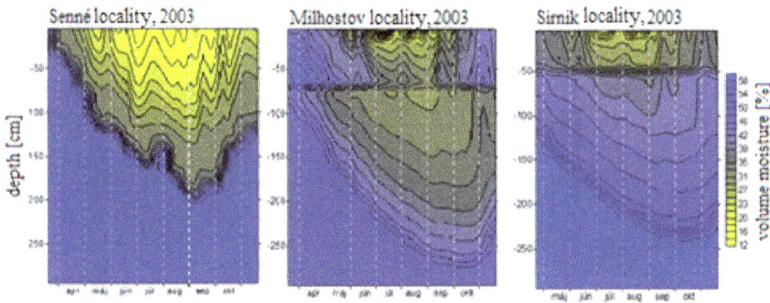

Figure 20. Numerical simulation of moisture regime of examined profiles

In Milhostov and Sírnik the influence of material interface along the vertical line of soil profiles is clearly manifested. In Milhostov the upper soil horizons are heavier with higher content of clay compared to the lower horizons. In Sírnik the situation is just the other way round and in Senné the soil profile is homogenous. Balance values illustrated in tab. 3 (based on fig. 21) confirm that the heavy soil profile in Senné can absorb rainwater through the cracks better then the Milhostov profile. Surface runoff did not occur in 2003. Its occurrence was observed only once, in Milhostov in 2001. It was induced by heavy rain in July 2001 when the daily rainfall accrual was 82mm. This rain event provoked 17mm surface runoff. The rest was absorbed by soil matrix and cracks.

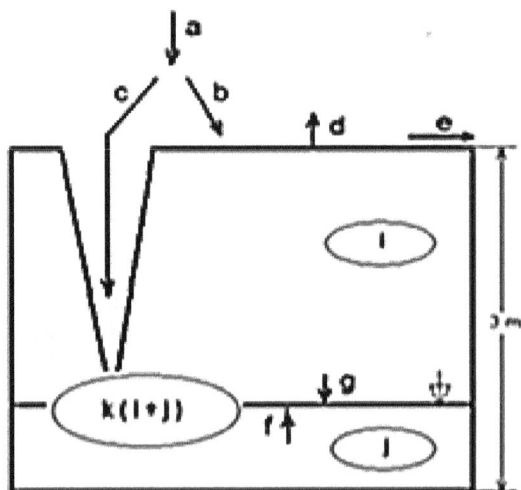

Figure 21. Balance scheme for simulated soil profile (explanations are listed in table 3).

4.7. Simulation of extreme rainfall influence on the water regime in unsaturated zone

The network of soil cracks represents a retention volume which is available for rainwater retention in case of extreme rainfall incidents. This effect is manifested most under extreme rainfall incidents. 2001 water regime of soil a profile (from February 14, 2001 to December 18, 2001) was simulated by numerical simulation using the mathematical model FLOCR (FLOw in Cracking soils). The profile is a specimen of extremely heavy soil with a two-domain structure. During the simulation the soil profile to a 3 m depth was defined. The course of groundwater level, volumetric moisture of the observed soil profile and the values of surface runoff were obtained by the simulation. Monitoring process, including the measurements of groundwater level and takings of soil samples in order to define volumetric moisture, took place in the observed locality. Measured data were used to verify the results of the simulation.

Profile water balance			Senné 2003	Milhostov 2003	Sírnik* 2003
precipitation	(b+c)	(a)	300	315	295
rainfall absorbed bzysoil matrix		(b)	287	305	293
rainfall absorbed by cracks		(c)	13	10	2
actual ET		(d)	462	504	473
Surface runoff		(e)	0	0	0
flow from GW (to the zone of areation)		(f)	177	56	155
flow to GW (from the zone of areation)	(h-c)	(g)	92	30	90
flow to GW (from the zone of areation+from the cracks)		(h)	105	40	92
water content change in the zone of aeration	(b-d-g+f)	(i)	-90	-173	-115
water content change in the zone of saturation	(h-f)	(j)	-72	-16	-63
water content change in the soil profile	(i+j)	(k)	-162	-189	-178

*values are listed for the Sírnik locality for period from 1.5. to 30.9.

Table 3. Balance table (values according to the model (mm) for vegetal period: 1.4 to 30.9.

The other step was to analyse a 30-year series of daily rainfall accrual, from which the rainfall event with the periodicity of 0.01 was calculated, i.e. the probability of its repetition is once in every 100 years. The value of extreme rainfall daily accrual was 76mm, fig. 22. This datum was incorporated to the inputs entering the model on the day when the soil profile was the driest, which was calculated to September 5th, 2001. Then he consequent output from the model, based on the added fictitious rainfall event, was evaluated and its influence on the soil profile water regime was analysed. The model showed a very sensitive response of the soil profile to the added fictitious rainfall. It proves the differentiation of volumetric moisture states along the vertical line of the soil profile and the course of groundwater level fig. 22, 23, 24. Surface runoff had a zero (0) value when the simulation considered real rainfall (no extreme rainfall). The addition of extreme rainfall resulted in surface runoff while the soil profile absorbed 56mm of water. At the very limited infiltration ability of the soil matrix surface it means that the significant part of this water content was absorbed immediately by soil cracks. Crack network represents an important retention volume available in heavy soils mainly during dry periods.

Maximum rainfall accrual that the soil profile would be able to absorb is app. 56 mm, provided the periodicity of rainfall is 0.02, fig. 25. This value disregards surface runoff. From the retention point of view, crack network of a two-domain soil structure has a significant influence on the water regime in immediate high rainfall accruance. This fact is of great importance for the future investigation of two-domain soil structures influence on the water regime as the periodicity of heavy rainstorms occurence continually rises at present.

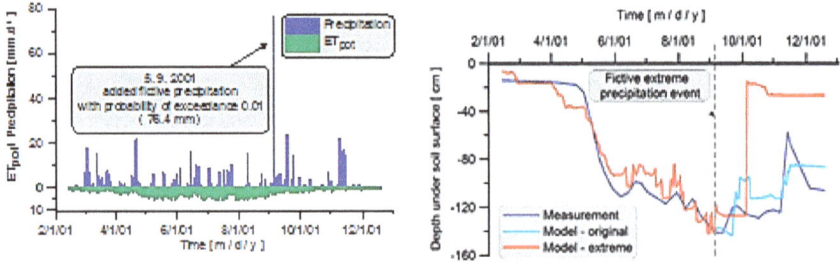

Figure 22. Meteorological inputs to the model with the added fictious extreme rainfall and observed and calculated courses of groundwater level

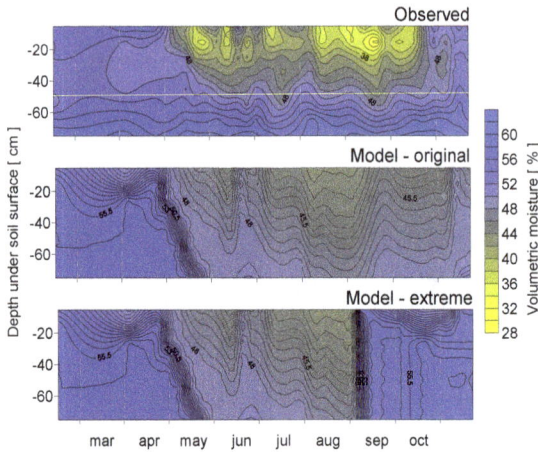

Figure 23. Redistribution of volume moisture in soil profile

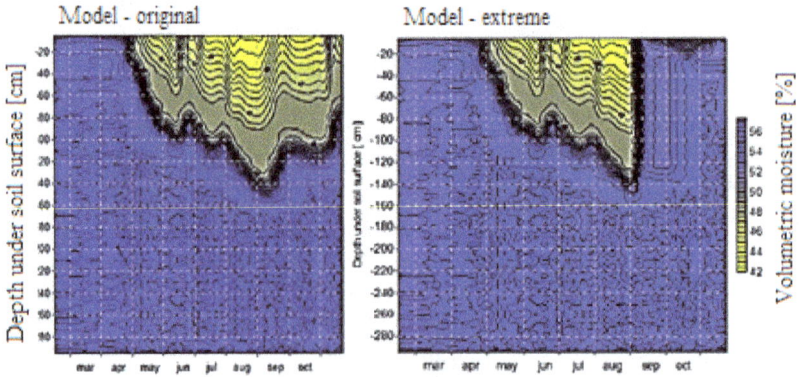

Figure 24. Redistribution of volume moisture in soil profile (model – original), (model – extreme).

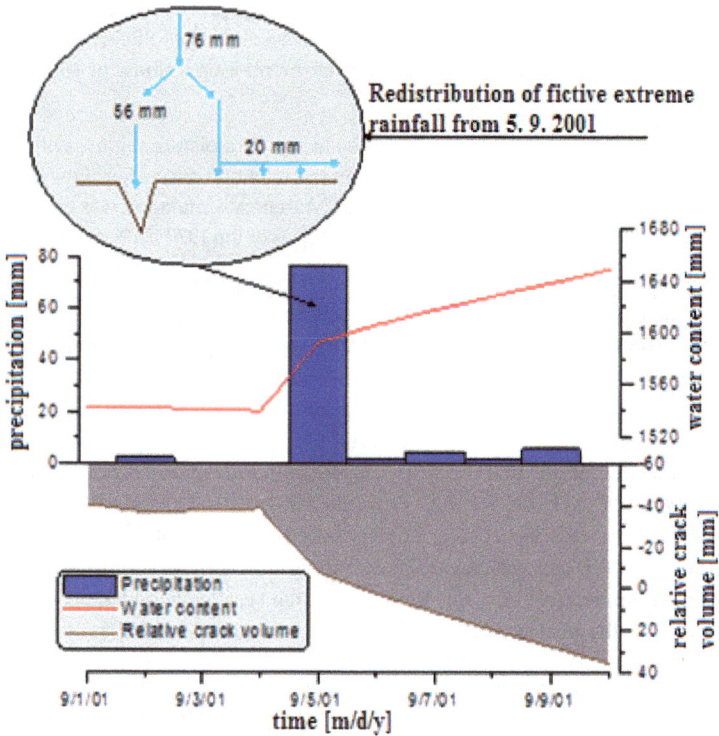

Figure 25. The reaction of model at the time of fictious rainfall event

4.8. Creation of soil cracks as a soil drought indicator

One of the evidence of an extreme hydrological processes is the creation of soil drought. In general, soil drought is defined as a shortage of the soil water used by plant cover for its growth. During water shortage plants get into the state of stress and all their physiological processes are focused on plant survival. To assess soil water storage available for plant cover, we conventionally use the following characteristics: moisture retention curve (soil-water content); wilting point (WP) - represents the value of pF = 4.18; threshold point (TP) - represents the value of pF = 3.3; field water capacity (FWC) - represents the value of pF = 2.0 to 2.7. Water drought starts when the water supplies in the root zone of a soil profile are on the TP level. TP is a soil moisture state when the physiological processes in plant cover are restricted by water shortage. WP is a soil moisture state when plant cover is in a constant lack of water what finally leads to wilting [17], [18],

Heavy soils contain higher percentage of clay minerals. With the changes in soil moisture, these minerals cause the changes in soil volume. During soil profile drying, volumetric changes are manifested by cracks creation. Volume of cracks depends on soil moisture. Soil moisture is thus an indicator of a soil profile moisture conditions. The aim of the presented part is to quantify one of the stress factors on the selected soil profile. [19], [20]. In other words, we focus on quantifying water drought by monitoring volume of soil cracks [21], [22], [23], [24].

The study of this problem is based on the analysis of soil moisture regime with regard to cracks volume and the position of TP and WP on the retention curve. For thew purposes of the analysis, Milhostov area soil profile was used. Numerical simulation was calculated with a computational time of 1 day step in 38 year time series of the 1970-2007 vegetation periods.

Fig. 26 shows shrinkage characteristics. These are expressed in the form of a dependency between a void ratio and a moisture ratio and between the overall soil volumetric changes and moisture. Both expressions include TP and WP values. The development of the dependency between a void ratio and a moisture ratio it is assumed that plants get into stress when shrinkage characteristics shifts from normal domain to residual domain. During the shrinkage process non-linearity is observed. Decrease in soil volume is lower than water reduction. That means that the air penetrates into micropores. Moisture ratio has the value of 0.56 for TP and 0.26 for WP. From the development of overall soil volumetric change (shown in the second part at fig. 26) it is assumed that overall soil volumetric change at TP is 6.7% and at WP 14% with regard to the saturated state. This shrinkage induces crack formation in the upper layer of a 1m- thick soil profile i.e. double domain soil structure is formed. Volume of the created cracks at TP moisture is 43mm and at WP moisture 85mm.

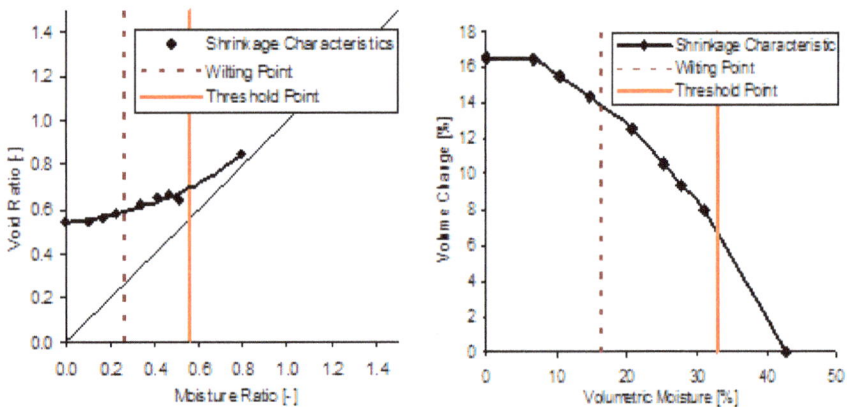

Figure 26. Basic shrinkage characteristics of heavy soil (Milhostov, Slovakia)

5. Conclusion

Sources of clay minerals impact on soil hydrological processes consist in their capacity to bind water and subsequently change their volume. Changes in soil hydrological processes result in the change in soil water regime.

This phenomena was studied in the soils of the East Slovakian Lowland. Research of mineralogical structure in ESL soils showed a predominant presence of illite and montmorillonite type of clay minerals. Therefore it was possible to identify percentual content of clay with percentual content of particles smaller than 0,002mm (colloidal+physical clay) without a need to distinguish between individual types of clay minerals during determination of percentual content of clay in soil.

Significant dependency of volume changes on clay minerals was demonstrated. Maximum volumetric changes reached up to 40% of saturated soil volume. Varying time intervals of water desorption from the soil samples were identified depending on content of clay particles. The intervals varied between 5 and 80 hours. In shrinkage characteristics of ESL heavy soils all shrinkage phases are contained very distinctly. The soils in Senné area show the highest shrink-swell potential on ESL. The potential compared to Reeve classification is of a high value: PLE = 15.3cm. Maximum values of 5.51cm were identified by field measurements of vertical movements. Assessment of soil isotropy in relation to their volumetric changes and based of the geometric factor r_s showed that its values are about 3.0. This value represents isotropy of volumetric changes. Research of geometric characteristics identified 4 types of crack network. It shall be noted that specific length of cracks on a unit of area was $L_{mc} = 5.65$ m.m^2 and avg soil matrix surface is 0.121 m^2.

Selected components of soil water regime were monitored in the field. Numerical simulation results were verified by comparison with the monitored results. This paper presents assimilation measurements of heavy soils moisture regime in homogeneous profile and two materially heterogenous profiles. Moisture distribution at the interface of material layers are illustrated as well.

The impact of water retention in a two-domain soil structure on rainfall-runoff processes during sudden and intensive rainfall in dry periods was quantified. There is a theory in which cracks are used as an indicator of soil drought creation in the end.

Clay minerals presence in soils and consecutive review of their influence on hydrological processes course in the soils is a very extensive issue. Due to limited extent of the present paper only basic approaches to the quantification of clay minerals impact on hydrological processes could be studied and illustrated in this chapter.

6. Further research

Clay minerals content in soil is stable. Their impact on hydrological processes is manifested in changes in soil water storage. This is most significant during extreme hydrological incindents such as soil drought or extreme rainfall.

In the last years, the increased occurence of crack network formation has been observed. It can be probably accounted for the ongoing climatic changes whose primary manifestation is in the change in rainfall distribution throughout the year- more frequent dry periods and subsequent extreme rainfall. With regard to this, the study of clay minerals impact on the dynamics of soil hydrological processes has an increasing importance.

The future research should be focused on the prognostics of the occurence of a two-domain soil structure in an area and on the study of its impact on soil water regime. The prognostics should be based on the numerical simulation and the outputs of the available climatic scenarios. It is advisale that the occurence of a two-domain soil structure be quantified. Mathematical models can be used to numerically simulate the effects of cracks retention volume on water drainage from an area during extreme rainfall incidents. In addition to this, further research into the knowledge of cracks temporal and spacial characteristics, water flow within a crack, into a crack, and flow between cracks and soil matrix should be developed. On the grounds of numerical simulations and field observations, the quantification of crack network influence on the transfer of water, other substances and solutions in soil should be performed. The results of this can be used for proposing the measures for the mitigation of cracks negative impacts and, on the contrary, for using their positive sides in water management in the countryside.

Author details

Milan Gomboš
Slovak Academy of Sciences/Institute of Hydrology, Slovak Republic

Acknowledgement

The authors would like to thank for the kind support of the project APVV-0139-10, and project VEGA 2/0125/12.

7. References

[1] Richards, L. A.: Capillary conduction of liquids through porous mediums. In: Physics 1, 1931, pp.318-333.

[2] Ross, G. J.: 1978. Relationships os specific surface area and clay content to shrink – swell potential of soils having different clay mineralogical composition. In: Canadian Journal Soil Science. 1978, č. 58, s. 159-166.

[3] Olive, W. W. – Chleborad, A. F. – Frahme, C. W. et al.: 1989. Swelling Clays Map Of The Conterminous United States. U.S. Geological Survey. 1989.

[4] Velde, B.: 1995. Composition and mineralogy of clay minerals. In: Origin and mineralogy of clays. New York : Springer-Verlag. 1995, s. 8-42.

[5] Tall, A.: 2002. Porovnanie klasifikačných systémov pre určovanie textúry pôd so zameraním na ťažké pôdy. In: Acta Hydrologica Slovaca, roč. 3, 2002, č. 1, s. 87-93, (in Slovak).

[6] Čurlík, J. 1977. Pedograficko-mikroskopické a mineralogické štúdium pôd Slovenska : Kandidátska dizertačná práca. Bratislava : Katedra mineralógie a kryštalografie PriFUK. 1977, 301 s.,(in Slovak)

[7] Šútor, J. – Gomboš, M. – Mati, R. - Ivančo, J.: 2002. Carakteristiky zóny aerácie ťažkých pôd Východoslovenskej nížiny. Monografia, ISSN 80-968480-8, ASCO pre ÚH SAV a OVÚN, , 215 s., (in Slovak).

[8] Soil Survey Division Staff.: 1993. Soil survey manual : USDA Handb. 18. Washington, DC. : U.S. Gov. Print Office.

[9] Bronswijk, J. J. B. – Evers-Vermeer, J. J.: 1990. Shrinkage of Dutch clay soil aggregates. In: Netherlands Journal of Agricultural Science. 1990, č. 38, s. 175-194.

[10] Grossman, R. B. – Brasher, B. R. – Franzmeier, D. P. et al. 1968. Linear extensibility as calculated from natural-clod bulk density measurements. In: Soil Science Society of America Proceedings. 1968, č. 32, s. 570-573.

[11] Parker, J. C. – Amos, D. F. – Kaster, D. L.: 1977. An evaluation of several methods of estimating soil volume change. In: Soil Science Society of America Journal. 1977, č. 41, s. 1059-1064.

[12] Reeve, M. J. – Hall, D. G. M. – Bullock, P.: 1980. The effect of soil composition and environmental factors on the shrinkage of some clayey British soils. In: Journal of Soil Science. 1980, č. 31, s. 429-442.

[13] Bronswijk, J. J. B.: 1989. Prediction of actual cracking and subsidence in clay soils. In: Soil Science. 1989, č. 148, s. 87-93.

[14] Oostindie, K. – Bronswijk, J. J. B.: 1992. Flocr – A simulation model for the calculation of water balance, cracking and surface subsidence of clay soils : Report 47. Wageningen : The Winand Staring Centre for Integrated Land, Soil and Water Research. 1992, 65 s.

[15] Novák, V. – Šimunek, J. – Genuchten Van, M. Th.: 2000. Infiltration of Water into Soil with Cracks. In: Journal of Irrigation and Drainage Engineering, roč. 126, 2000, č. 1, s. 41-47.

[16] Schafer, W. M. – Singer, M. J.: 1976. Influence of physical and mineralogical properties on swelling of soils in Yolo County, California. In: Soil Science Society of America Journal. 1976, č. 40, s. 557-562.

[17] Kotorová, D. - Mati, R.: 2008. Properties and moisture regime of heavy soils in relation to their cultivation. Cereal Research Communications 36. 5: 1751-1754.

[18] Stehlová, K.: 2007. Assessment of the soil water storage with regard to prognosis of the climate change at lowlands. Cereal Research Communications, 35. 2: 1093-1096

[19] Velísková, Y. (2010) Changes of water resources and soils as components of agro-ecosystem in Slovakia. Növénytermelés, Vol. 59, suppl., pp. 203-206, ISSN 0546-8191, http://www.gabcikovo.gov.sk

[20] Nagy, V. – Sterauerova, V. – Neményi, M. – Milics, G. – Koltai, G.: 2007. The role of soil moisture regime in sustainable agriculture in both side of river Danube in 2002 and 2003. Cereal Research Communications, 35. 2: 821-824.

[21] Šoltész, A. - Baroková, D.: 2004. Analysis, prognosis and control of groundwater level regime based on means of numerical modelling. In: Global Warming and other Central European Issues in Environmental Protection: Pollution and Water Resources, Columbia University Press, Vol.XXXV, Columbia, pp.334-347, ISBN 80-89139-06-X

[22] Tall, Andrej - Pavelková, D.: 2010, Dana. Predpokladaný vplyv klimatických zmien na hladinu podzemnej vody na Východoslovenskej nížine. In Acta Hydrologica Slovaca, 2010, roč. 11, č. 1, p. 162-166. ISSN 1335-6291.

[23] Šútor, J. - Gomboš, M. - Mati, R. - Tall, A. – Ivančo, J. : 2007. Voda v zóne aerácie pôd Východoslovenskej nížiny. ISBN 80-89139-10-8, Bratislava:Michalovce, Ústav hydrológie SAV, 279 s.

[24] Kandra, B.: 2010. The creation of physiological stress of plants in the meteorological conditions of soil drough. Növénytermelés, Vol. 59, ISSN 0546-8191, Supplement, p. 307-310.

Chapter 7

Soil Moisture Retention Changes in Terms of Mineralogical Composition of Clays Phase

Markoski Mile and Tatjana Mitkova

Additional information is available at the end of the chapter

1. Introduction

The mineral part of the soil is inherited from the parent material but it modifies under the influence of different factors and processes.

The Earth's crust contains almost 100 elements and only 8 of them (O, Si, Al, Fe, Ca, Na, K, and Mg) form 98.5% of the crust and compose the base of the soil body. Most of the soil types contain 60% of all existing minerals where the silicates and aluminosilicates are the most dominating ones. They are present in a form of primary and secondary minerals. The primary minerals originate as a result of igneous, sedimentary, and metamorphic rocks' weathering while the secondary minerals originate as a result of chemical weathering of the primary minerals.

Theoretically, the soil contains all the different types of minerals but the real number and type of the minerals that constitute one type of a soil is quite limited.

The clay minerals are large group with common silicate characteristics. According to (Ilić, et al., 1975), these minerals evolve from a surface weathering of the aluminosilicates in the parent rock or a deposit of surface water.

According to (Kostić, 2000), phyllosilicates (clay minerals) dominate in the clay fraction (<0.002mm or 2 microns) of many soils which affect the physical, physical-chemical, water-physical and physical-mechanical properties of the soil (plasticity, stickiness, swelling, shrinkage, cohesion) and the soil structure and moisture retention as well.

The degree the mineralogical composition affects the soil moisture retention depends on the percentage amount and fraction of the clay minerals present in a soil type. The clay particles represent the most active part of the fine earth because of their large external and internal active surface, cation-exchange capacity (CEC) and mineralogical composition (Skorić, 1991).

Many authors such as Barteli and Peters, Salter et al, Petersen et al. cit. by (Pelivanoska, 1995), state that the increase content in the slightly smaller particles increases the surface of tangency between the solid phase and soil moisture.

In addition to the mineralogical composition of the clay explanation presented before, the paper researches the influence of the moisture retention given the different levels of tension in soil (starting with 0.1 (pF-2) and going to 15 bar or (pF-4.2), which corresponds to the wilting point). The remaining soil moisture above 15 bar is not available to the plant (Bogdanović, 1973).

According to (Filipovski et al., 1980) the presented model research is based on three moments: (1) the unexplored research in influence of the mineralogical composition of soil retention curves in the recent years; (2) their practical and theoretical importance; and (3) the great importance in the periods with aridity conditions and irrigation conditions.

The clay mineral content and the soil retention curves represent a significant characteristic for each soil horizon, subhorizon or layer and the soil profile as a whole. As a result of the laboratory investigation in the mineralogical composition of many soil types the moisture retention curves are created for each horizon, subhorizon or layer.

We are presenting our research and research made by other authors on the mineralogical composition of clay and how it influences the moisture retention in 6 soil types: *Fluvisol, Chernozem, Vertisol, Chromic Luvisol on saprolite, Albic Luvisol* and *Molic Vertic Gleysol (WRB - 2006,* World Reference Base for Soil Resources).

The water retention curves are very different because the soils have heterogeneous mineral content.

2. Material and methods

The Republic of Macedonia is situated in the central part of the Balkan Peninsula. Its total territory is only 25,700 km², but it shows a great diversity of natural conditions. Here, as a natural museum, can be seen almost all relief forms, geological formations, climatic influences, plant associations and soil which appear in European. The relief is very heterogeneous, with numerous relief forms, with different expositions and inclinations, and with great differences altitude (from 40 to 2764 m above sea level). In Macedonia, there are numerous geological formations of a very heterogeneous petrographic-mineralogical composition. The mountains are composed of non-calcareous hard rocks, including quartzite and various silicate rocks: acidic, neutral, basic and ultrabasic rocks; as well as of calcareous rocks such as pure limestones, marbles and dolomites. Basins are composed of loose and lightly cemented sediments, and a small quantity of young volcanic rocks. Undulating-hilly terrains in the basins are composed of sea and lake sediments (Mesozoic, Paleogenic, Neogenic, diluvial (superficial) deposits. The sloping terranians consist of colluvial and some fluvioglacial deposits. Macedonia is under the influence of two zonal climates (Mediterranean and temperate – eastern continental) and one local (mountain) climate (Mitkova et. al., 2005).

A great heterogeneity appears in the soil cover, too (40 soil types and even more subtypes, varieties and forms).

The places where the soil profiles have been dug are marked with GPS coordinates (Figure 1). Fluvisol 41°51, 08' 55" N and 22°26, 14' 15" E; Chernozem 41°49, 38' 58" N and 22°00, 45' 29" E; Vertisol 41°49, 06' 42" N and 22°11, 59' 71" E; , Chromic Luvisol on saprolite 41°54, 00' 31" N and 22°15, 57' 59" E; Albic Luvisol 41°42, 41' 65" N and 22°49, 04' 56" E; Molic Vertic Gleysol 41°20, 45' 63" N and 21°25, 38' 49" E.

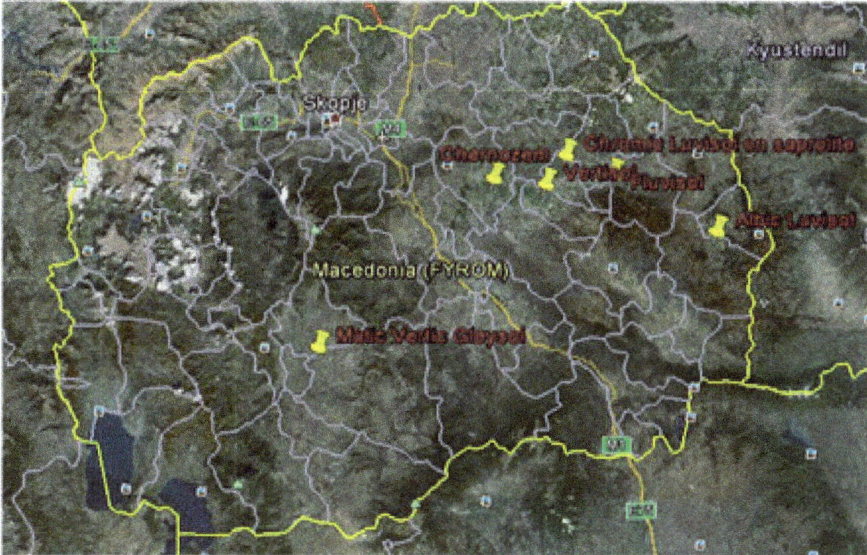

Figure 1. Location of Profile pits

The following methods were used for analysis of soil materials: mechanical composition was determined with the dispersion of the soil with 1 M solution of $Na_4P_2O_7$ x 10 H_2O. First, 10 g soil sample was put in Erlenmeyer flask, the 25 ml of 1 M $Na_4P_2O_7$ x 10 H_2O solution was poured into the flask, left for 12 hours, and the mixed for six hours. Finally the suspension was sieved through a 0.2 mm sieve into the 1000 ml cylinder and then pipetted - (Škorić, 1986). Fractionation of mechanical elements was done by the international classification of soil texture by Scheffer & Schachtschabel (Mitrikeski & Mitkova, 2006). The pH of the soil solution was determined electrometrically in a water suspension and in a suspension of H_2O. The humus content was determined on the basis of the total carbon according to the Tjurin method, modified by Simakov (Orlov et al., 1981).

Mineralogical composition of clay was determined by diffractometer, brand PHILIPS, type PW 1051 in range $2\theta = 50 \div 60^0$, at wavelenght radiation (Cu, Kα). Mineralogical composition analysis was done at Mining and Geology Faculty at University of Belgrad, Republic of Serbia. To determine the clay minerals in each soil sample, two products were made, one of which was untreated, and the filled with glycerin, while the other was annealed at temperatures of 480^0 C for the determineted type of clay minerals in the range $2\theta = 30\div14^0$ (methods are described by (Đurič, 1999 and 2002).

The soil moisture retention at pressures of 0.1 bar (pF-2), 0.33 bar (pF-2.54) and 1 bar (pF-3) was carried out with application of pressure by Bar Extractor (Fig. 2), while the soil moisture retention at higher pressures 2.0 bar (pF-3.3); 6.25 bar (pF-3.90); 11 bar (pF-4.04) and 15 bars (pF-4.2) was carried out with Porous plate extractor (Richards, 1982), described by (Resulović et al., 1971).

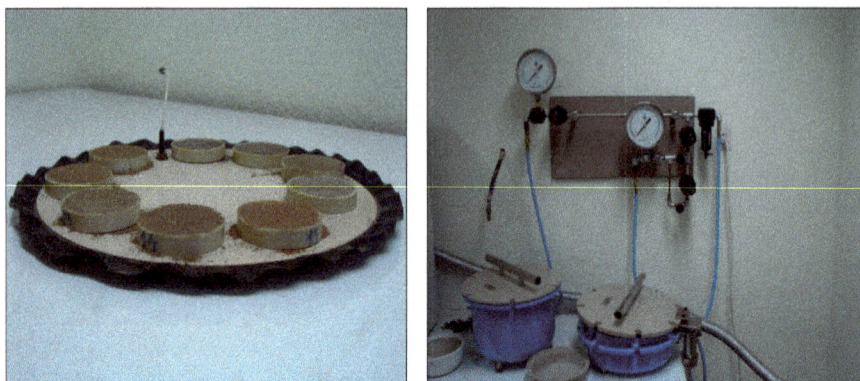

Figure 2. Preparing soil and placing samples on Bar extractor and Porous plate extractor

3. Review of references and discussion

The soil water retention is a result of two forces: adhesion (soil particles attract the water molecules) and cohesion (mutual attraction of the water molecules). The adhesion is much stronger than the cohesion. The force which retains the water in the soil is called capillary potential and is closely related to the water content. The free flowing water in the soil has a capillary potential equal to zero, a condition where all the soil pores, both capillary and non-capillary, are filled with water. Soil water potential can be determined indirectly by recourse to measurements of soil water content and soil water release or soil moisture characteristic curves that relate volumetric or gravimetric content to soil water potential. The measurement of water potential is widely accepted as fundamental to quantifying both the water status in various media and the energetics of water movement in the soi l- plant-atmospheric continuum (Livingston, 1993). Mukaetov (2004) underlines that by decreasing the water content in the soil the value of the capillary potential is increased. In order to assess the humidity of the soil using the capillary potential Schofiled 1935 quot. (Vučić,1987) has proposed the pF values where the force of the water in the soil is expressed on the height of a water column in cm (1 bar = 1063 cm water/cm^2). The pF values are affected by the mechanical content and according to the same author, the bigger the participation of the fine fractions the greater the pF values, especially under pressure of 0.33 bars.

It is known that clay minerals that are characterized with swelling (smectites, vermiculite) have stronger moisture retention ability than other clay minerals. Warkentin & Meada (1974), Meada & Warkentin (1975), Rausseaux & Warkentin (1976) concluded that amorphous materials such as allophanes (hydratated aluminosilicate minerals) are characterized with high moisture retention ability.

All the factors mentioned show a strong influence in the specific surface area of the soils. The water retention capacity is defined by the total osmotic surface activity (Shainberg et al., 1971). The loamy soils and the clay soils, especially if their clay fractions contain swelling minerals are characterized with large active surface and fine pores (Voronin, 1974).

According to Hillel, (1980) not only the mineralogical composition but also the mechanical and organic matter content of the soil affects the water- physical relations.

In Table 1 we have a presentation of the mechanical composition of soils.

Profile	Soils (WRB – 2006)[a]	Depth in cm	Total sand 0.002-2mm	Silt 0.002-0.02mm	Clay < 0.002mm	Silt + Clay <0.02mm	Textural class[b]
P1	Fluvisol	0-20	46.30	40.20	13.50	53.70	Loam
		20-35	49.60	37.30	13.10	50.40	Loam
		35-50	48.10	38.60	13.30	51.90	Loam
		50-70	53.10	35.00	11.90	46.90	Sandy loam
P2	Chernozem	0-40	55.80	13.60	30.60	44.20	Sandy clay loam
		40-60	53.20	13.70	33.10	46.80	Sandy clay loam
		60-90	42.50	15.00	42.50	57.50	Clay
		90-110	44.30	18.60	37.10	55.70	Clay loam
P3	Vertisol	0-26	45.90	15.40	38.70	54.10	Loam clay
		26-54	46.40	18.00	35.60	53.60	Loam clay
		54-105	40.90	15.60	43.50	59.10	Loam clay
		105-150	34.10	23.50	42.40	65.90	Loam clay
		150-165	46.90	15.60	37.70	53.10	Loam clay
P4	Chromic Luvisol on saprolite	0-15	38.10	38.10	23.80	61.90	Loam
		15-30	25.10	40.10	34.80	74.90	Clay loam
		30-55	21.50	41.30	37.20	78.50	Clay loam
		55-70	25.00	45.80	29.20	75.00	Clay loam
P5	Albic Luvisol	0-8	45.20	41.90	12.90	54.80	Loam
		8-23	46.90	36.60	16.50	53.10	Loam
		23-33	33.80	35.00	31.20	66.20	Clay loam
		33-44	52.30	29.60	45.10	47.70	Clay loam
		44-65	35.20	26.60	38.20	64.80	Clay loam
P6	Molic Vertic Gleysol	0-30	24.40	23.60	52.00	75.60	Clay
		30-65	26.20	18.70	55.10	73.80	Clay
		65-85	21.50	16.50	62.00	78.50	Clay
		85-124	29.70	14.80	55.50	70.30	Clay

[a](WRB – 2006) World Reference Base for Soil Resources
[b]Scheffer & Schachtschabel textural class

Table 1. Mechanical composition of soils

(Maclean & Yager, 1972), (Jamison & Kroth, 1958), (Shaykewich & Zwarich, 1968) and (Heinonen, 1971) studied the effect of the organic matter and the mechanical content over the water retention in several soils throughout U.S., Europe and Asia. The research conducted by (Hollist et al., 1977) confirms that the soil water retention in the West Midlands (United Kingdom) depends mostly on the organic matter and the mineralogy content of the soil.

According to Filipovski, (1996), soil water retention in different tension is in tight correlation with humus, clay, and silt content and mineralogical composition of the clay.

The data for the mineralogy content of the clay within the fluvisol (profile 1, table 2) are showing that it consists of a more minerals. This is a result of both the different mineralogy content of the rocks which when weathering result in clay minerals and the conditions where the weathering takes place (Filipovski et al., 1980).

Results in the Tables (2, 3, 4, 5, 6, 7 and 8) are given semi quantitatively, thus the presence of certain minerals is indicated by plus (+), which indicate the presence of the individual minerals (+++++) > 70%; (++++) = 50-70%; (+++) = 30-50%; (++) = 10-30% and (-) = traces (tr.).

Depth cm	Chemical properties			Clay mineralogical composition						
	Organic matter %	CaCO₃ %	pH in H₂O	Montmorillonite	Vermiculite	Illite	Chlorite	Kaolinite	Quartz	Feldspar
0-20	1.40	0.00	6.40	+ +	-	+ + + +	+	+ + +	-	-
20-35	0.85	0.32	7.30	+	-	+ + +	tr.	+ + + +	-	+
35-50	0.57	0.47	7.60	+	-	+ + +	tr.	+ + + +	tr.	-
50-70	0.26	0.55	7.80	tr.	-	+ +	tr.	+ + + +	-	-

Table 2. Chemical properties and clay mineralogical composition of Fluvisol

In this profile, according to the same authors is deposited detritus that comes from the silicate rocks consisting of albite, quartz, sericite, chlorite schists and from the soils over them with an acid reaction. Therefore the fluvisol is richer in kaolinite and illite arising from the sericites. Parts of the slates contain ferromagnesian minerals and their decomposition result in montmorillonite. The clay minerals in profile 1 are represented with the association: kaolinite-illite-montomorillonite.

If we are continuously measuring the tension of the soil moisture and for each tension we are measuring the amount of moisture expressed in volume percentages and then we put the data on coordinates for each horizon, we come to the retention curves. The express the ratio between the attracting forces (tension) and the amount of soil moisture. The soil water retention curves are sometimes called desorption curves, indicating that the measurement determines how much water is retained by the soil at each successively lower matric potential (Topp et al., 1993). They indicate the availability of water capacity in which the upper limit is field water capacity while the lower limit is wilting point. Using the given limits, it is easier to understand when the irrigation would take place and how much water the plant requires. Furthermore, the retention curves are also used to explain the water movement in the soil. Soil water retention curves are useful directly and indirectly as indicators of other soil behavior traits, such as drainage, aeration infiltration, and rooting patterns. Retention curves are the relationship between soil water content and soil water potential (matric potential) for soil during a drying phase.

As a result of the values received from the determination of the soil moisture with ten different tensions (0.1; 0.33; 0.75; 2, 6.25; 11 and 15 bars) we get the retention curves (Fig. 3) of the soil moisture for profile 1.

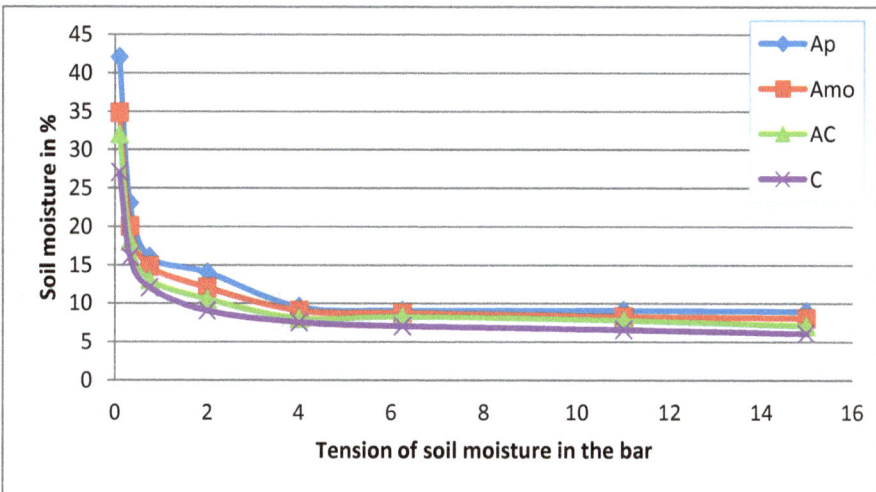

Ap - tillage mineral horizon; Amo- mollic horizon; AC – transitional horizon; C – substrate (parent material)

Figure 3. Soil moisture retention curves for Fluvisol

The retention curves of the profile show that the retention moisture is low (small content of clay and humus). The curves up to tension of 2 bars show quite big slope and afterwards, they are almost horizontal at a tension of 15 bars. Due to the similar mechanic and mineralogy content of the soil, the retention curves do not differ essentially (Filipovski et al., 1980). As a result of the smallest clay content, silt, smectites and humus in the lowest layer of the soil, it shows the lowest retention.

In Table 3 we have a presentation of the clay mineralogical composition in the chernozem (profile 2). The clay mineralogical composition depends on the stratification of the substrate. Until the depth of 90 cm in the chernozem there is almost the same amount of illite and smectites (montmorillonite) and small contents of vermiculite and kaolinite. The illite prevails until the depth of 40 cm, and the smectites (montmorillonite) prevail deeper, at 90 cm (Markoski, 2008).

Depth cm	Chemical properties			Clay mineralogical composition						
	Organic matter %	CaCO₃ %	pH in H₂O	Smectite	Vermiculite	Illite	Kaolinite	Chlorite	Quartz	
0-40	1.64	0.0	7.4	+	+	+ +	+	+	-	
40-60	1.61	0.0	7.4	+ +	+	+ +	+	+	-	
60-90	1.33	0.0	7.3	+ +	+	+ +	+	+	-	
90-110	1.17	17.36	8.2	+ + +	+	+ +	+	tr	-	

Table 3. Chemical properties and clay mineralogical composition of Chernozem

In conditions of layers and lithological discontinuity it is hard to explain the increase of smectities beneath 90 cm. It is most common in other processes (transformation of other minerals or transfer of the minerals). None of the mineral is absolutely dominant, but the bigger representation of the minerals with the type of layers 2:1 (illite, smectite, vermiculite) compared to the minerals with type of layers 1:1 (kaolinite). The 2:1 layers of the mineral in the group of smectities have a bond weak and thus, when moisturized, water molecules and different ions enter between them and they become distant. The montmorillonite has active an internal areas between the layers and the ions and water molecules are adsorbed in the inside and outside (extramicellar and intramicellar adsorption). Due to this the montmorillonite has a high cation-exchange capacity (CEC), high degree of hydration, dispersion, plasticity, viscosity and swelling in humid conditions. When dried, the layers come nearer and the particles are contracting. Montmorillonite under commonly prevailing pH conditions converted to non-expanding hydrous mica – illite. The structure of illite in comparison with montmorillonite has a higher layer charge arising from isomorphous substitution of Al^{3+} for Si^{4+} in the tetrahedral sheet and interlayer is occupied by non-hydrated K^+ ions. The illite has a lower cation-exchange capacity (CEC) and does not manifest signs of swelling, plasticity or viscosity. Vermiculite is a high-charge 2:1 phyllosilicate clay mineral. It is generally regarded as a weathering product of micas in which the potassium ions between the molecular sheets are replaced by magnesium and iron ions. Vermiculite is also hydrated and somewhat expansible though less so than smectite because of its relatively high charge. As a result, the vermiculite has a bigger cation-exchange capacity (CEC) and similar to the smectites, it manifests swelling, plasticity and viscosity.

The data for the water retention capacity in the chernozem are showing that the biggest value is found in the mollic Amo horizon (bigger content of humus, clay and colloids).

Furthermore, high values for wilting point are received (pressure of 15 bars). The highest value is found in the Amo horizon (high content of the fraction of physical clay (clay and silt), as well as the presence of bigger number of micro-pores). The retention curves of the profile show that in all the horizons starting from tension of 2 bars, are almost horizontal (Fig. 4). A gradual change in the retention forces can be noticed coming with the change of moisture without jumps. This shows that the division of the soil moisture in different forms cannot be justified with the retention curve because the decrease of the amount of water does not have big jumps under different tensions.

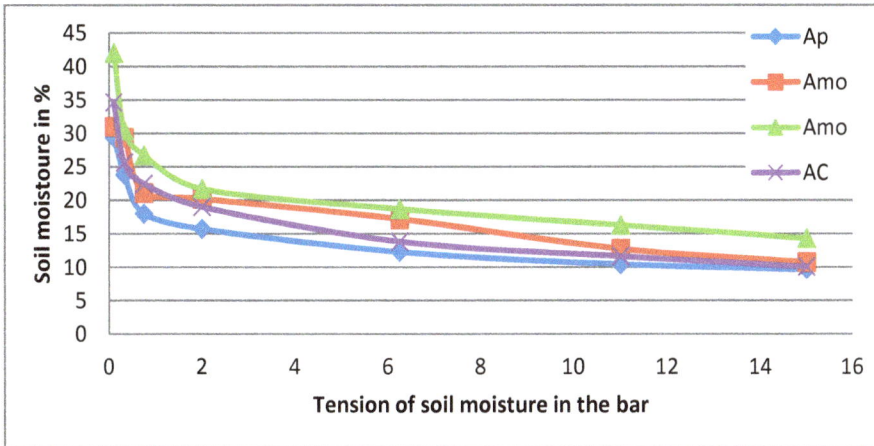

Ap - tillage mineral horizon; Amo- mollic horizon; AC – transitional horizon

Figure 4. Soil moisture retention curves for Chernozem

When speaking about the mineralogy content of the vertisol we can say that it is equal in the whole depth of the profile (Table 4). It is formed on lacustrine clay sediments and it has inherited the clay minerals from the substrate where all the secondary minerals do not represent products of weathering of the basic rocks only. Between the substrate, the AC horizon and the humus horizon there is no difference; thus the processes of pedogenesis did not change the composition of clay (Jovanov, 2010). It is equaled by the pedoturbation. The equality of the composition of clay in the substrate and the soil is yet another proof for the "lithogenous" character of the vertisols i.e. for its genesis related only to certain substrates rich in smectites (montmorillonite) or such that during weathering were converted mainly to montomorillonite. According to (Dudal, 1956) in the vertisols in Mississippi (U.S.), the smallest fraction (the clay has 2-3 times bigger cation - exchange capacity) contained more montmorillonite than the larger fractions. The presented profile of vertisol is characterized by dominant presence of the montmorillonite and mixed layer minerals (illite - montmorillonite) where it prevails. Afterwards, illite, chlorite and kaolinite which are a result of the weathering of different rocks made up the fields above the lacustrine sediments.

Depth	Chemical properties			Clay mineralogical composition						
cm	Organic matter %	CaCO₃ %	pH in H₂O	Montmorillonite	Vermiculite	Ilite	Chlorite	Kaolinite	Quartz	Amorphous minerals
0-15	3.32	0.00	5.8	+ + + +	-	+ +	-	+	Tr	-
15-77	2.19	0.00	6.3	+ + + +	-	+ +	-	+	Tr	-
77-94	1.72	0.79	7.2	+ + + +	-	+ +	-	+	Tr	-
94-117	0.94	5.06	7.6	+ + + +	-	+ +	-	+	Tr	-
117-145	0.33	3.95	7.5	/	/	/	/	/	/	/

Table 4. Chemical properties and clay mineralogical composition of Vertisol

The vertisol is characterized with several more important characteristics of retention curves (Jovanov, 2010), (Fig. 5).

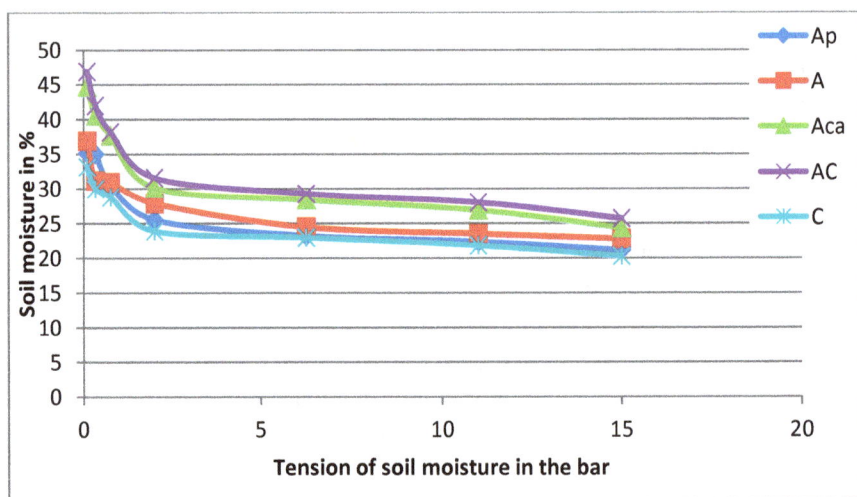

Ap - tillage mineral horizon; A – humus accumulative horizon; Aca – carbonate/calcareous humus accumulative horizon; AC – transitional horizon; C – substrate (parent material)

Figure 5. Soil moisture retention curves for Vertisol

Firstly, due to the high content of clay and its montomorillonite (smectite) character, the moisture retention is very high in the whole depth of the profile. The second characteristic is that the release of moisture during low tension (not bigger than 2 bars) is slow and therefore the curves do not have a big slope, and with a further increase of the tension (15 bars), the curves are almost horizontal, with an insignificant slope. It can be explained with a high percentage of micro-pores through which the water is slowly released. Therefore, with the increase of the tension i.e. decrease of the moisture, the plants in these soils are suffering the lack of moisture respectively spending more energy for its reception because it is held with much bigger retention forces. The third characteristic of the curves is that they are relatively close to each other meaning that there is no big difference in the moisture retention between the separate horizons (equal mineralogical content of the soil, homogeneity of the profile).

The profile of the chromic cambisol is composed of andesite breccia rich in volcanic ash and contains a few clay minerals in crystal form. The secondary minerals are a result of the weathering of the primary minerals in the andesites. The volcanic ash mixed with the breccia resulted in amorphous minerals (Filipovski et al., 1980).

In the biggest fractions there are more primary minerals, but only two are dominant – the quartz and the feldspar (together more than 90%), (Table 5). Mostly, quartz is found in the fraction of 0.05 – 0.10 mm. There is a tendency for its decrease in the depth of the profile. Parts of the quarts and feldspar are connected with the fragments of rocks. According to (Filipovski et al., 1980) this points out to the fact that the chromic cambisols usually contain the primary minerals such (quartz) which is resistant to weathering or (feldspars) which are hardly weathers.

Depth	Chemical properties			Clay mineralogical composition						
cm	Organic matter %	CaCO₃ %	pH in H₂O	Montmorillonite	Vermiculite	Ilite	Chlorite	Kaoliite	Quartz	Feldspar
0-15	2.65	0.00	6.3	+	-	+	-	+	-	+ +
15-30	1.72	0.00	6.1	+	-	+	-	+	+	+ +
30-55	1.00	0.00	6.4	+	-	+	-	+	+	+ +
55-70	1.12	1.42	7.5	+	-	+	-	+	-	+ +

Table 5. Chemical properties and clay mineralogical composition of Chromic Luvisol on saprolite

Practically, the mica and ferromagnesian minerals are weathered into secondary minerals. In the lower part of the horizon (B) and the horizon (B)C there are corrections. If we review the mineralogical content of the fraction directly larger than the clay (0.002 – 0.05 mm, Table 6), we can see that the secondary minerals are amorphous and they are dominant throughout the whole profile and are equally distributed in the whole depth.

There are no crystal secondary minerals, and the primary ones have the same amount of quartz and feldspar because the breccia is rich in volcanic ash. In the clay fraction the primary minerals are lost meaning that it only contains secondary minerals. It is a result of the weathering of the primary minerals of the substrate, from the basic and neutral rocks. The amorphous minerals are dominant (of the volcanic ash in the substrate) equally distributed throughout the depth. In all the horizons, including the substrate, there are more crystal secondary minerals: illite, kaolinite and montmorillonite (smectite). Mostly there are amorphous minerals and illite equally distributed throughout the depth. In the profile the following associations of the minerals occur: in horizon A, amorphous minerals-illite-kaolinite-smectite, and in horizon (B) amorphous minerals-illite-smectite, with insignificant presence of quartz.

Depth cm	Mineralogical composition of the soil separate from 0.002 to 0.05 mm							
	Kaolinite	Hidrobiotit	Montmorillonite	Vermiculite	Chlorite	Quartz	Feldspar	Amorphous minerals
0-15	Tr	0	0	0	0	+ +	+ +	+ +
15-30	Tr	+ +	0	0	0	+ +	+ +	+ +
30-55	Tr	+ +	0	0	0	+ +	+ +	+ +
55-70	Tr	+ +	0	0	0	+ +	+ +	+ +

Table 6. Mineralogical composition of the soil separate from 0.002 to 0.05 mm

For the retention curves of this soil the most significant is the fact that it is rich in clay and mostly fine clay (under 0.001 mm). The textural differentiation is very clear and this has reflected in the retention curves where the on the graph (Fig. 6) the retention curves are high meaning that the moisture retention is big (a lot of clay and amorphous minerals).

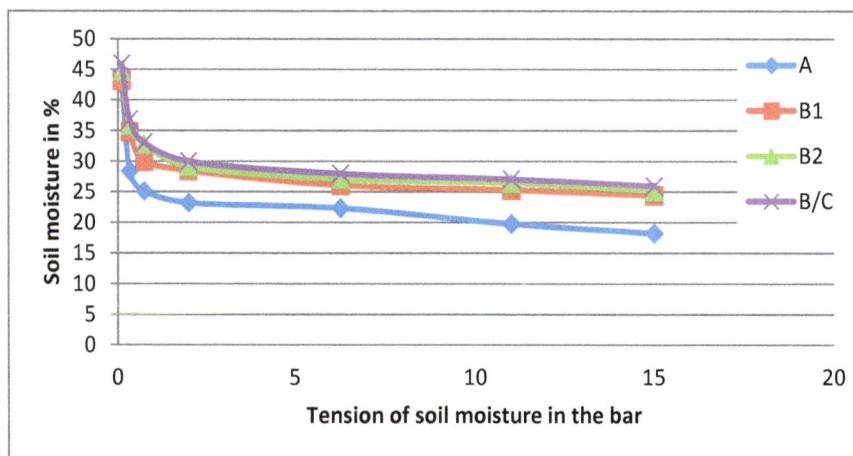

A- humus accumulative horizon; B1- cambic horizon; B2- cambic horizon; B/C – transitional horizon

Figure 6. Soil moisture retention curves for Chromic Luvisol on saprolite

Having in mind that the mineralogy content of the clay is equal throughout the depth, the retention depends on the clay content: the retention differences between separate horizons are followed by the content of the clay. The fall of the retention curves starting from the lower to the upper tensions is big and this, according to (Filipovski et al., 1980) means that a great part of the water is held with a big force by the soil particles and they are hard to be released. The textural differentiation is clearly reflected on the moisture retention: the lowest retention can be seen in horizon A because it contains the least silt, fine clay and clay. The other probes of the horizon (B) have a bigger retention because they contain more than those fractions. The curves on these probes are approximated because they contain almost equal amount of clay and have the same claymineralogical composition.

The albic luvisol is formed on clastic sediments. One part of the clay minerals is inherited by the substrate and in that regard the inherited minerals in the upper layer (horizons A and E) are more equal than those in the lower layer (horizon B). Besides, by leaching the clay not all the clay minerals are leached equally. The montmorillonite (smectite), that has finer particls and is easily peptized, can be leached in bigger amounts compared to other clay minerals (Table 7).

Depth cm	Chemical properties			Clay mineralogical composition						
	Organic matter %	CaCO₃ %	pH in H₂O	Montmorillonite	Vermiculite	Ilite	Chlorite	Kaolinite	Ilite vermiculite mixture mix.mineral mineral	Amorphous minerals
0-8	5.44	0.00	6.0	Tr	+ +	+	-	+	-	+ +
8-23	1.79	0.00	5.3	Tr	+ +	+	-	+ +	+	+ +
23-33	0.80	0.00	4.9	+	+	Tr	+	+	+	+ +
33-44	0.45	0.00	4.6	+ +	Tr	Tr	-	Tr	+	+ +
44-65	0.33	0.00	4.6	+ +	Tr	Tr	-	Tr	-	+ +

Table 7. Chemical properties and clay mineralogical composition of Albic Luvisol

The profile can be divided into three parts depending on the mineralogical content of the clay: upper part (horizons A and E), middle part (upper part of the horizon Bt) and lower part (the lower part of the horizon Bt). It is obvious that the mineralogical content, especially in (horizons A and E) and the upper part of the horizon B is more complex – it contains 6 different minerals, but in lower amount. In the upper part of the profile is dominated we have the prevail of the vermiculite followed by the kaolinite and the illite, and we can find only traces of montmorillonite. It is characteristic that there are many amorphous minerals (sesquioxides of Fe and Al), nearly as the vermiculite. The middle part has the most complex content: it is dominated by the amorphous minerals and followed by almost equal amounts of kaolinite, vermiculite, illite-vermiculite, chlorite and it also appears in the montmorillonite. In the lower part of the profile the composition is less complex: it contains mostly amorphous minerals and montmorillonite (due to the bistratified and the upper translocation), followed by the illite-vermiculite, and there are only traces of the other minerals. The content of the illite decreases and of the smectite is increased in the depth of the profile.

According to the data for the content of the organic matter the content of the total and fine clay and its mineralogical composition (Filipovski et al., 1980) we are discussing a bistratified profile where the horizons A and E are formed by one, and the horizon Bt by another layer.

The retention curves (Fig. 7) show the strong differentiation of the profile. The content of the amorphous minerals does not influence this differentiation because their content is equal throughout the depth of the soil. (Filipovski et al., 1980) explain this with the bistratified profile and with the pedogenesis processes.

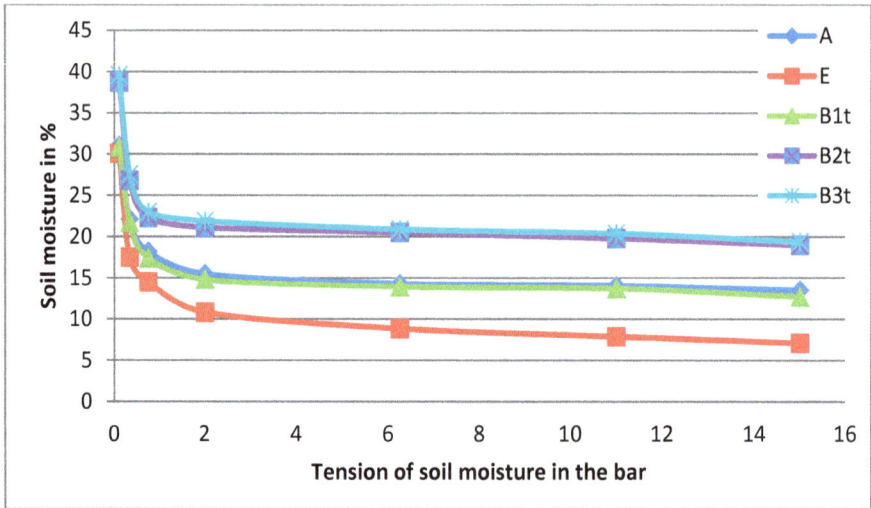

A – humus accumulative horizon; E- albic (eluvial) horizon; B1t-illuvial horizon, modif. (argillic); B2t-illuvial horizon, modif. (argillic); B3t-illuvial horizon, modif. (argillic);

Figure 7. Soil moisture retention curves for Albic Luvisol

The retention curves are grouped in mostly three groups. The lowest retention can be found in the E horizon (it contains the lowest amount of clay and much less humus compared to horizon A). In the second group we have the upper part of the horizon Bt (it contains more clay compared to horizon E), and here we can mention horizon A (the same characteristics as the horizon E, but with bigger content of humus). It is familiar that the humus has high retention. In the third group, the biggest retention is reserved for the lower part of the horizon Bt (it contains mostly clay dominated by the montmorillonite that is characterized with a high retention of moisture). The ration of the water retention in horizon E to that one in horizon Bt, under tension of 15 bars is 1 : 2.84.

When describing the mineralogical content of the clay in the molic vertic gleysol we used data by (Mukaetov, 2004) and we would like to underline that the sediments in the valley where the soil were formed have been created during the Quarter. This means that they are very young and the geo-chemical processes did not reach the high degree of weathering of minerals. Besides, the same author claims that the proluvial - deluvial deposits that derive from the surrounding tops are heterogeneous in their composition and consist of granite, gneiss, schists, as well as marbles and limestone rocks (limestone and lesser degree of dolomite).

The molic vertic gleysol is a homogenous soil that in the bigger part of the year are saturated with the maximum water capacity. In its biochemical activity in neutral and alkali environment, the plants take up the bigger part of the potassium of mica and illite. The pushing of the potassium from the interlayer of illite leads to creation of the so-called "open illite". This is a phase of minerals arising from vermiculite or smectites. According to (Kostić, 2000)replacing K^+ with Mg^{2+} ions which have a smaller dimension, but are significantly more hydratated led to creation of vermiculite. During the formation of smectites the balance is established by introducing water molecules and protons (Hydronium H_3O^+ ion) and the so-called transformative smectite is received. The partial leaching of the potassium from the interlayer of mica within the chlorites leads to creation of the inter-stratified minerals in the soil. This can be proved with the presence of mixed layer of layered silicates of the type chlorite - smectite, as well as of illite –smectite. The presence of chlorite is confirmed. The chlorite is slowly weathering in the low reductive conditions of the environment. Furthermore, there is a significant presence of plagioclase of the type of albite (Na-plagioclase). The biggest part of orthoclase (K - feldspar) was alternated into kaolinite as a result of the pushing out of the potassium with the biochemical processes of the swamp plants in the low drained conditions of the environment and with water of the saturated gley soil. The presence of the amphiboles and chlorite (ferromagnesian minerals) which are easily weathered in well - drained soils at oxygen conditions of the environment confirm the previous statement for the low aeration of the soil.

We can see a dominant presence of the montmorillonite (smectites) in the upper part of the soil. In the lower part the content of the montmorillonite and amorphous minerals is almost equal (Table 8). It is obvious that this soil contains a lot of minerals that are characterized with high degree of water retention.

Depth	Chemical properties			Clay mineralogical composition					
cm	Organic matter %	CaCO₃ %	pH in H₂O	Montmorillonite	Vermiculite	Illite	Chlorite	Kaolinite	Amorphous minerals
0-10	3.22	0.0	6.6	+ + +	-	Tr	-	Tr	+
10-17	1.22	0.0	6.9	+ + + +	-	Tr	-	Tr	+
17-34	1.59	0.0	6.6	+ +	-	+	-	Tr	+ +
34-80	1.38	0.0	6.4	+ +	-	+	-	Tr	+ +

Table 8. Chemical properties and clay mineralogical composition of Molic Vertic Gleysol

We received high values for the content of moisture that is retained in the molic vertic gleysol during the wilting range as a result of the high content of silt, clay, organic matter, presence of clay minerals with high retention (smectites and amorphous phase) as well as the presence of a large number of fine capillary pores. For the influence of the mechanical content over the retention in the soil we can take the high positive correlation between the moisture retention at 0.33 - 15 bars and the content of physical clay and clay (Table 1) and the high negative correlation between the retention of moisture at 0.33 - 15 bars and the content of sand.

The data presented in the graph (Fig. 8) show a decrease of the retention curves and it is most obvious at lower pressures. The reason for this is the high content of clay and clay minerals with high retention (smectites, amorphous minerals), afterwards the retention curves are stabilized and gradually decreasing.

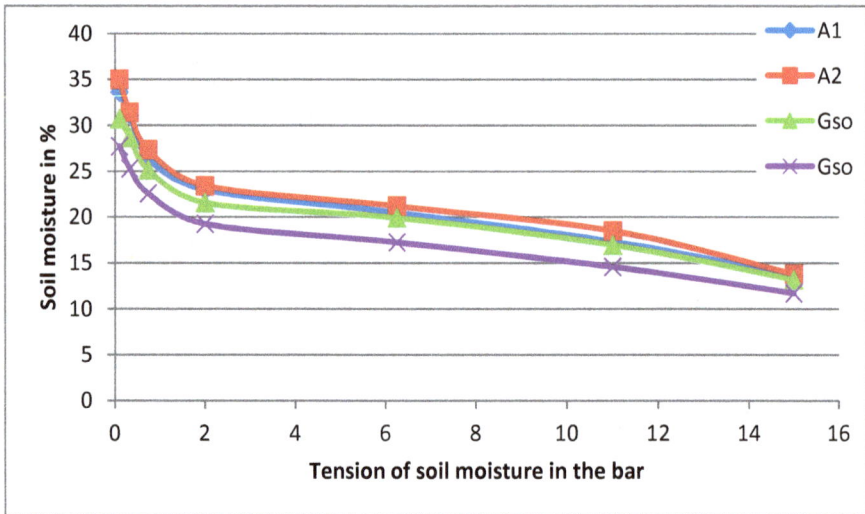

A1- humus accumulative horizon; A2- humus accumulative horizon; Gso – gleyic horizon/sub horizon Gso

Figure 8. Soil moisture retention curves for Molic Vertic Gleysol

The small difference between the retention curves at the upper three horizons according to (Filipovski et al., 1980) is a result of the combined influence of the clay, humus and mineralogical content of the clay. The highest retention can be seen in the upper two depths (richest in silt, clay, humus and presence of smectites, then the AC horizon and the lowest values can be found in horizon Gso that has the lightest mechanical content and the least amount of humus and smectites.

4. Conclusion

The analyzed mineralogical content and the retention curves of these types of soil show that the water retention depends on the content and the type of the clay minerals.

In general, due to the high content of clay and its smectite character, the water retention within some soil types is big throughout the whole depth of the profile. The others soil types showed, a lower retention as a result of the lower content of clay and other minerals (kaolinite and illite).

The soil water retention curves were successfully used in the prediction of the flow of water and other fluids in soil. Also many attempts have been made to define adequate and yet useful mathematical functions for the retention curve.

The measurement of soil water retention curves and the reliability of the resulting data are very important for adequate characterization of a number of other soil phisical attributes. The information can be used to indicate the soil pore size distribution interpreted in relation to aeration and water availability for plants, for water flow and infiltration in soil, and frost susceptibility.

Author details

Markoski Mile and Tatjana Mitkova
Department of Soil Science, Faculty of Agricultural Sciences and Food, Sc. Cyril and Methodius University in Skopje, Republic of Macedonia,

5. References

Bogdanović, J., Bogdanović, M. (1973): Uporedna ispitivanja metoda za odrđivanje vlažnosti venenja kod različitih tipova zemljišta. Zemljište i biljka, Beograd.Vol. 22, No.3, p.p. 325-334.

Dudal, R. (1965): Dark clay soils of tropical and subtropical regions. FAO. Rome.

Filipovski, G., Pradhan, S. (1980): Moisture retention characteristics in the soils of the Socialist Republic of Macedonia. Macedonian Academy of Sciences and Arts. Skopje, p.p. 1-87.

Filipovski, Đ . (1996): Soil of the Republic of Macedonia.Vol II. Macedonian Academy of Sciences and Arts. p. p.175-209.

Heinonen, R. (1971): Soil management and crop water supply. Lantbruk - shögskolans kompendienämnd.

Hollis, J. M., Jones, R. J. A., Palmer, R. C. (1977): The effects of organic matter and particle size on the water retention properties of some soil in the west Midland of England. Geoderma. 17. p. p. 225-238.

Hillel, D. (1980): Application of Soil Physics. Department of Plant and Soil Sciences. Massachusetts, Academic press.

Ilić, M., Karamata, S. (1975): Specijalna mineralogija. Prvi deo. Izdavačko-Informativni centar studenata (ICS), Beograd.

Jamison, V. C., Kroth E. M. (1958): Available moisture storage capacity in relation to textural composition and organic content of several Missouri soils. Soil Sciences. Am. Poc. 22. p.p. 189-192.

Jovanov, D. (2010): Water, physical and physical-mechanical properties of vertisols of the Stip, Probistip and Sv. Nikole region. Master Degree Thesis. p. p. 1-50.

Kostić, N. (2000): Agrogeologija. Beograd. str. 132-149.

Livingston, N. J. (1993): Soil water potential. Soil sampling and methods of analysis. Canadian Society of Soil Sciences. p. p. 559-567.

Maclean, A. H., Yager, T. V. (1972): Available water capacities of Zambian soils in relation to pressureplate measurments and particle size analysis. Soil Scinces.113. p.p. 23-29.

Meada, T. Warkentin, B. P., (1975): Void changes in allophane soils determining water retention and transmission. Soil Sciences. Am. Proc. 398-403.

Mitrikeski J., Mitkova, T. (2006). Practicum on pedology, second edition. University "Ss. Cyril and Methodius", Faculty of Agricultural Sciences and Food, Skopje, p. 63-67.

Mukaetov, D. (2004): Conditions for formation, genesis and characteristics of the hidromorpfhic black soils spread out in Pelagonija. Doctoral Degree Thesis, Faculty of Agriculture, Skopje. p. p. 1-162.

Markoski, M. (2008): Physical and physical-mechanical properties of chernozems of the Ovće Pole region. Master Degree Thesis. p. p. 1- 64.

Mitkova, T., Mitrikeski J. (2005): Soils of the Republic of Macedonia: Soil Resourches of Europe, secound edition, European Soil Bureau Research Report, N⁰9, 225-234, EUR 20559 EN, 2005.

Orlov, S.D. Grishina, A.L. (1981): Practicum on Chemistry on humus. Moscow University Press, p. p.1- 271.

Pelivanoska, V. (1995): Physical and physical-mechanical characteristics of cinnamon forest soils in the Prilep region. Master Degree Thesis, Faculty of Agriculture, Skopje. p. p. 1-60.

Resulović, H. red. (1971): Methods of investigation the physical properties of the soil, JDZPZ, Belgrade.

Rousseaux, J. M., Warkentin, B. P. (1976): Surface properties and forces holding water in allophane soils. Soil Sciences Am. Proc. 40. p. p. 446-451.

Richards, L.A. (1982): Soil water and planth grow. Soil phyisical conditions , New York.

Shaykewich, C. F. Zwarich, M. A. (1968): Relationships between soil physical constants and soil physical components of some Manitoba soils. Can. J. Soil Sciences. 48. p. p. 199-204.

Shanberg, J., Bresler, E., Klausner, Y. (1971): Studies on Na/Ca montmorillonite system. I Swelling pressure. Soil Sic. III, p. p. 214-220.

Topp, G.C., Galaganov, Y.T., Ball., Carter, M. R. (1993): Soil water desorption curves. Soil sampling and methods of analysis. Canadian Society of Soil Sciences. p.p. 569 - 579.

Voronin, A. D. (1974): The energy state of soil moisture as related to soil fabric. Geoderma. 12. p. p. 183-189.

Vučić, N. (1987): Water, air and thermal regime of soil, VANU. Department of natural sciences. Volume 1. Novi Sad. p. p. 1-320.

Warkentin, B. P., Meada, T. (1974): Physical properties of allophane soils from the West Indies and Japan. Soil Sciences. Am. Proc. 39. p.p. 372-377.

WRB, (2006): Diagnostic Horizons, Properties and Materials. Chapter 3.World Reference Base for Soil Resources. FAO, ISSS-AISS-IBG, IRSIC, Rome, Italy.

Đurić, S. (1999): Determination of the clay mineral X-ray diffraction. Published 99th 53 (1997) 7,8,9, Belgrade, p.p. 224-227.

Đurić, S. (2002): Methods of research in crystallography. Technical Faculty, Čačak. ITN. SASA. Belgrade, p.p. 207-213.

Škorić, A. (1986): Priručnik za pedološka istraživanja. Sveučilište u Zagrebu, Fakultet poljoprivrednih znanosti, Zagreb.

Škorić, A. (1991): Composition and soil properties. Textbook, Faculty of Agricultural Sciences, University of Zagreb.

Chapter 8

Fougerite a Natural Layered Double Hydroxide in Gley Soil: Habitus, Structure, and Some Properties

Fabienne Trolard and Guilhem Bourrié

Additional information is available at the end of the chapter

1. Introduction

In soils, the clay fraction defined by particles with a size less than 2 μm, contains numerous minerals, which have fascinated many people to determine their structure and study their properties in the environment. For example, very soon after the time of the discovery of the Bragg's rule and the development of X-ray diffraction since 1918, Tamm (1922) developed a chemical pre-treatment of the samples to separate the well crystallized clays from the "undefined mixture of Si, Fe, Al" contained in soils. Then in 1960 – 1980, a large community in the World has studied the well crystallized clay - minerals, such as kaolinite, nontronite, montmorillonite, vermiculite, bentonite... from mining deposits. It has established the structures, explained the negative charges of the layer in relationship with substitution of structural cations III by cations II in the layer, neutralized by interlayered cations. This kind of structures induced particular properties in soils such as cationic exchange capacity or swelling... In the 1990-years, a new family of phyllites appears, the Layered Double Hydroxides (LDHs), named "anionic clays", that has layered structure with an anionic exchange capacity and particular electrochemical and magnetic properties. Easily synthetized, LDHs are essentially studied in the laboratory for industrial applications.

In soils, the study of LDHs emerges, because these constituents well crystallized, have particle size often less than 500 nm, develop a large reactivity with the soil solution and interfer with many biogeochemical cycles, e.g. Fe, Si, Al, Mg, Ca, K, N.... In soils, fougerite, is a LDH, responsible for the blue-green colour of gleysols. It can be identified simply in the field by its colour that changes to ochre when in contact with oxygen from the atmosphere.

The purpose of this chapter is to expose some thermodynamic properties deduced from the study of the natural LDH, fougerite, and propose their extension to other LDHs.

2. The natural LDH: fougerite

Fougerite (IMA 2003-057) is the natural green rust mineral responsible for the bluish to greenish colours expressing *reductomorphic properties* (Driessen *et al.* 2001, Annex 2, p.314). As early as in the original definition of gley by Vyssotskii (1905, [1999]), the colour of gley was considered as indicating the presence of Fe "protoxide", *i.e.* of ferrous oxide (*s.l.*). This colour has been ascribed to "green rust" by Taylor (1981). Green rusts are intermediate compounds in the corrosion of steel first evidenced by Girard and Chaudron (1935). The first evidence for green rust, as a natural mineral, was provided by Trolard *et al.* (1996, 1997). The mineral was identified in a gleysol developed on granite in Fougères (Brittany, France), from which the name fougerite was proposed. To determine the occurrence, the reactivity and the structure of fougerite, selective dissolution techniques, Mössbauer, Raman and EXAFS spectroscopies and thermodynamic modelling were used both on natural samples and synthesized compounds. The mineral was homologated by the International Mineralogical Association in 2004 (Trolard *et al.* 2007).

The originality of fougerite is that instead of the other LDHs, such as Ca – Al(III) or Ni(II) – Al(III) LDH, Fe(II) and Fe(III) can exchange electrons in the layer between each other. Though it is generally a nano-mineral, it is however not poorly ordered, but well crystallized (trigonal system).

2.1. Habitus

Iron is the main biogeochemical marker of gleyey soils. To observe fougerite in soils, different conditions with regard to hydric variations, biological and biochemical processes must be simultaneously fulfilled. They are:

- an excess of water;
- moderate reductive conditions;
- temperature conditions favourable to microflora activity;
- the presence of chemical elements that are able to record more or less irreversibly the variations of aerobic or anaerobic conditions, named geochemical markers (Trolard et al., 1998).

Soil colour is closely related to the nature of iron oxides, more specifically to their degree of hydration and their amount (Vyssotskii, 1905[1999]; Taylor, 1981; Cornell and Schwertmann, 2003). Moderately reduced waterlogged soils are characterized by the blue-green colour, which turns into ochre when the soils are open to the outer atmosphere (Vyssotskii, 1905[1999]; Ponnamperuma et al., 1967). It has been often ascribed to the occurrence in the milieu of mixed Fe(II) – Fe(III) compounds with a likely structure of green rust (GR). This assumption has been formulated since the 1960s and largely discussed in the literature (e.g. reviews by Taylor, 1981 or Lewis, 1997).

2.2. Structure

Green rusts belong to the large group of LDH compounds (table 1). The LDH structure is built of stacked layers of edge-sharing metal octahedra containing divalent and trivalent metal ions separated by anions between the interlayer spaces. The range of composition of LDHs is very narrow and they differ mainly by the nature of the interlayer anion (Allada et al., 2002). The general structural formula can be expressed as $[Me(II)_{1-x} Me(III)_x (OH)_2][x/n \, A^{-n}, mH_2O]$, where, for example, $Me(II)$ could be Mg(II), Ni(II), Co(II), Zn(II), Mn(II) and Fe(II) and $Me(III)$ is Al(III), Fe(III) and Cr(III) (Sparks, 2003). As all sites of the layer $[Me(II)_{1-x} Me(III)_x (OH)_2]$ are occupied, this exhibits a net positive charge x per formula unit, which is balanced by an equal negative charge from interlayer anions A^{-n}, such as Cl^-, Br^-, I^-, NO_3^-, OH^-, ClO_4^- and CO_3^{2-}. Water molecules are present too in the interlayer (figure 1).

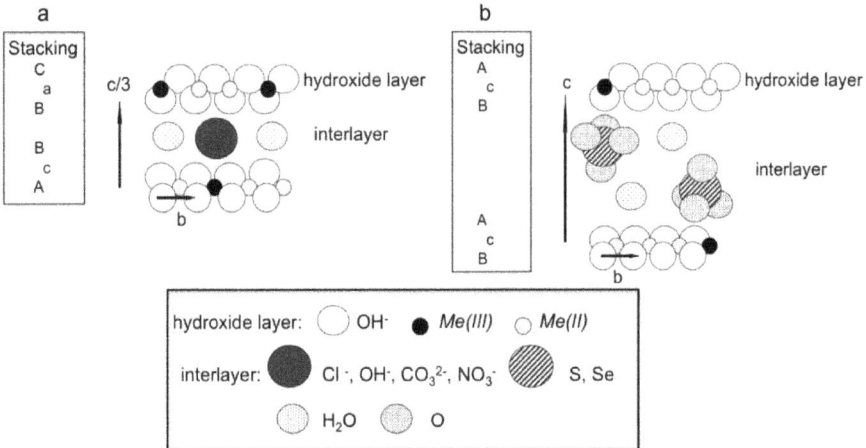

Figure 1. Stacking and structures of Green Rusts 1 (a) and 2 (b). $Me(II)$ = Fe or Mg and $Me(III)$ = Fe (from Trolard and Bourrié, 2008).

mineral	Structural formula	Interlayer anion
fougerite	$[Mg_y Fe^{II}_{1-x} Fe^{III}_x (OH)_2][x/n \, A^{-n}, mH_2O]$	Possible : OH^-, Cl^-, CO_3^{2-}...
meixnerite	$[Mg_6 Al_2 (OH)_{16}] [(OH^-)_2, 4 \, H_2O]$	OH^-
woodallite	$[Mg_6 Cr_2 (OH)_{16}] [(Cl^-)_2, 4 \, H_2O]$	Cl^-
iowaite	$[Mg_4 Fe^{III} (OH)_{10}] [(Cl^-), 2 \, H_2O]$	Cl^-
takovite	$[Ni_6 Al_2 (OH)_{16}] [(CO_3^{2-}, OH^-), 4 \, H_2O]$	OH^-, CO_3^{2-}
hydrotalcite	$[Mg_6 Al_2 (OH)_{16}] [(CO_3^{2-}), 4 \, H_2O]$	CO_3^{2-}
pyroaurite	$[Mg_6 Fe^{III} (OH)_{16}] [(CO_3^{2-}), 4 \, H_2O]$	CO_3^{2-}

Table 1. Structural formula of some natural layered double hydroxides

Fougerite was first characterized by selective dissolution techniques, Mössbauer and Raman spectroscopies (Trolard et al., 1997). By EXAFS, the structure was confirmed, but it appeared that in addition to Fe, Mg was present in the natural mineral (Refait *et al.* 2001).

The generic formula of synthetic green rusts is $[Fe(II)_{1-x} Fe(III)_x (OH)_2][x/n \, A^{-n}, mH_2O]$, where x is in the range [1/4 – 1/3]. In the synthetic compounds, the interlayer anion is largely variable: Cl⁻, Br⁻, I⁻, $CO_3{}^{2-}$, $C_2O_4{}^{2-}$, $SeO_4{}^{2-}$, $SO_3{}^{2-}$, $SO_4{}^{2-}$... With Cl⁻, Br⁻, I⁻, $CO_3{}^{2-}$ and $SO_3{}^{2-}$, that are small sized, spherical or planar anions, there is only one layer of water molecules in the interlayer, and the symmetry group is trigonal (GR1 structure) (Refait *et al.* 1998), while with $SeO_4{}^{2-}$, $SO_4{}^{2-}$, that are tetrahedral, there are two layers of water molecule, and the layer stacking is different (GR2 structure) (Simon *et al.* 2003). Green rusts can be easily synthesized in the laboratory (Murad, 1990), form by oxidation of Fe(II) in solution (Lewis, 1997), by partial oxidation of $Fe(OH)_2$ (Génin *et al.* 1994) or by bacterial oxidation of Fe(III) oxides (Fredrickson *et al.* 1998).

As smectites can accommodate different cations in the interlayer, fougerite, green rusts and other LDHs can accommodate different anions. Indeed, natural minerals have been described with the same structure, in which OH⁻, Cl⁻, $CO_3{}^{2-}$ are the interlayered anions (figure 1). In Fougères, the most likely anion is OH⁻, as evidenced by considering soil/solution equilibria (Feder *et al.*, 2005) and by XRD decomposition (Trolard and Bourrié, 2008). In other environments, Cl⁻ or $CO_3{}^{2-}$ could be present too with a GR1 structure, and $SO_4{}^{2-}$ with a GR2 structure, but this was not yet found in the field.

2.3. Some properties

2.3.1. Ionic exchange capacity

Ion exchange involves electrostatic interactions between a counter-ion in the boundary layer between the solution and a charged particle surface and counter-ions in a diffuse cloud around the charged particle. It is usually a rapid, diffusion – controlled, reversible, stoichiometric reaction, and, in most cases, there is some selectivity of one ion over another by the exchanging surface (Sparks, 2003).

Ionic exchange capacity can be studied on bulk soil sample by measuring CEC (cationic Exchange Capacity) or AEC (Anionic Exchange Capacity). In 1:1 clay minerals, such as kaolinite, in metal oxides, amorphous materials and organic matter, ion exchange is due to surface sites that dissociate (Lewis acid, carboxylic or phenolic moieties), so that the net electric charge is pH dependent (Stumm, 1987; Sposito, 1989). In 2:1 clay minerals, in addition to Lewis acid sites, there exist isomorphic substitutions of different charge ions in the lattice, so that the net charge is the sum of a constant charge defect, always negative, and a pH dependent charge. The net resulting charge is negative, and is compensated by interlayer cations, more or less hydrated. The magnitude of CEC in soils is usually larger than AEC, but highly weathered soils in acidic environment can exhibit a substantial AEC.

As 2:1 clay minerals, LDHs contain constant positive charge due to the presence of both divalent and trivalent metals in the octahedral layer, and both external and internal exchanges sites, but the net resulting charge is positive, and the interlayer ions are anions. This is the basis of their designation as "anionic clays", though strictly speaking, LDHs are not clay minerals, as they are not phyllosilicates. Anions present in the interlayer sheet can be exchanged.

The property of exchange is often used to obtain some synthetic compounds that cannot be precipitated directly. This is the case for green rust with interlayer $CO_3{}^{2-}$ or $C_2O_4{}^{2-}$, because presence of carbonate or oxalate in the initial solution inhibits the formation of $Fe(OH)_2$

sheets, which is the first step to build up green rusts by precipitation. These green rusts are then obtained by exchanging the SO_4^{2-} from a sulphate - green rust with CO_3^{2-} or $C_2O_4^{2-}$ (Drissi et al., 1994; Refait et al., 1998).

Much more basically, the properties of cationic or anionic exchange of clays and LDHs in soils contribute to the capacity of these soils to retain or release ions in a form available for plant uptake and contribute then to the supply of nutrients for plant growth. For example, this is the case for Ca^{2+}, Mg^{2+} or K^+ from clays and NO_3^- or SO_4^{2-} from LDHs.

2.3.2. Redox reactivity

Redox reactivity is a particular property of Fe-LDH (green rust) and fougerite due to possible transition from ferric and ferrous state inside the hydroxide sheets. Thus two kinds of reactions can be observed: one affects only the hydroxide sheets and the second interacts with the interlayer anion.

The redox reactivity of the hydroxide sheets determines the stability and the transformation of Fe-LDH in other minerals. Concerning the stability, Refait *et al.* (2001) showed that in a same layer each Fe^{3+} ion must be surrounded only by bivalent, either Fe^{2+} ions or Mg^{2+} ions for the natural mineral fougerite. This is due to the fact that if the Fe^{III}/Fe_{total} is up to 1/3, the electrostatic repulsion of two neighbouring Fe^{3+} induces an oxolation reaction and the formation of Fe-O-Fe bonds. In the laboratory, this reaction permits to obtain a range of intermediate compounds between green-rusts and well crystallized Fe^{III}- oxides such as lepidocrocite, goethite or hematite (*e.g.* Schwertmann and Fechter, 1994; Lewis, 1997). The final products depend on the kinetics of oxidation and the nature of the interlayer anion. In the field, *in situ* Mössbauer spectra can be obtained every two days and at different depths (Feder *et al.*, 2005). The results show that Fe^{III}/Fe_{total} ratio changes with the dynamics of pH and Eh in the soil solution of the waterlogged soils, indicating that the proportion of Fe^{II} and Fe^{III} in the mineral changes with time and depth (figure 2).

Other works underlined that oxidation of Fe(II) from GRs was the most relevant abiotic reaction of reduction pathway in natural environments, soils and sediments. This is the case for reduction of selenate Se(VI) into Se(0) or Se(-II) (Myneni et al., 1997), Ag(I), Au(III), Cu(II) and Hg(II), respectively into Ag(0), Au(0), Cu(0) and Hg(0) (O'Loughlin et al., 2003) or Cr(VI) into Cr(III) (Loyaux-Lawniczak et al., 2000).

The redox reactivity due to the interactions between the hydroxide sheets and the interlayer anion can induce reactions which are not able without the contribution of the particular position of electron donors and acceptors. For example, while in solution NO_3^- cannot be reduced by Fe^{2+}, Hansen et al. (2001) show that Fe-LDHs reduced stoichiometrically NO_3^- into NH_4^+ and the kinetics of the reaction depend on the type of the interlayer anion, the hydroxide layer charge and the relative content of Fe(II) in the hydroxide layers. But the reduction of NO_3^- into NH_4^+ in one step as observed, requires a simultaneous transfer of 8 electrons from Fe^{2+} to N. When NO_3^- is in interlayer position in a green rust, the hydroxide layer reacts by oxolation of Fe(II) – OH – Fe(II) bonds into Fe(III) – O – Fe(III) releasing $8e^-$ and $8H^+$ (figure 3). The GR oxidizes into hematite or goethite. For GR(SO_4), the half-reactions and net reaction can be written as follows:

$$2Fe^{II}_4Fe^{III}_2(OH)_{12}SO_4 \rightarrow 6\ Fe_2O_{3,\ hematite} + 6\ H_2O + 2\ SO_4^{2-} + 8e^- + 12\ H^+$$

$$NO_3^- + 8e^- + 10\ H^+ \rightarrow NH_4^+ + 3\ H_2O$$

$$2Fe^{II}_4Fe^{III}_2(OH)_{12}SO_4 + NO_3^- \rightarrow 6\ Fe_2O_{3,\ hematite} + 9\ H_2O + 2\ SO_4^{2-} + NH_4^+ + 2\ H^+$$

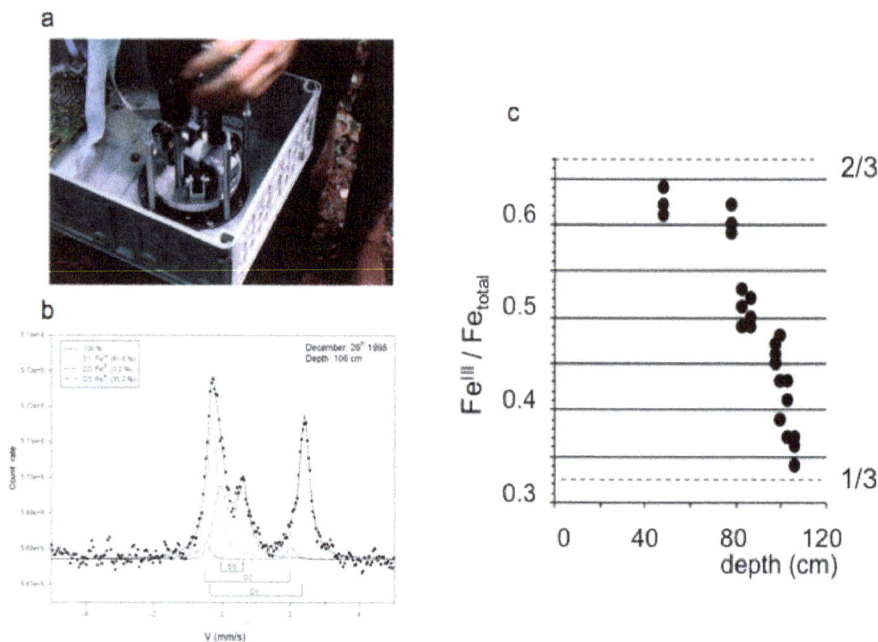

Figure 2. *In situ* Mössbauer monitoring in Fougères's soils (France): a. MIMOSII spectrometer used; b. Example of spectra obtained after two days of measurement in one depth; c. Variations of the ratio Fe^{III}/Fe_{total} in soil with depth and time (from Feder *et al.*, 2005).

3. Thermodynamic modelling of stability of LDHs

Thermodynamic properties of minerals mainly depend on the chemical composition of the mineral, and at a lesser degree on its structure. Evidence for this is the fact that nesosilicates include very soluble compounds such as olivine and quasi-insoluble minerals, *e.g.* zircon. Considering isostructural compounds, thermodynamic properties can thus be related with a suitable parameter depending on the chemical properties of ions. As the interaction between layer and interlayer is mainly electrostatic, we proposed to consider the electronegativity of the ion (Bourrié et al., 2004). Jolivet (1994) developed a partial charge model based upon Allred and Rochow's electronegativity scale.

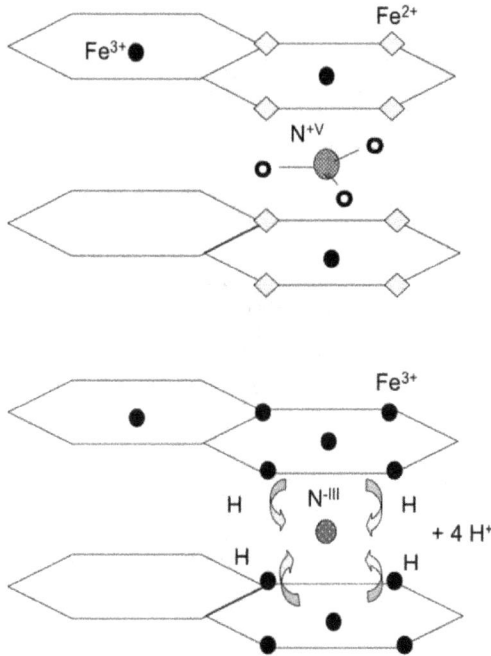

Figure 3. Reaction mechanism during the reduction of the interlayer NO_3^- with Fe^{II} from hydroxide sheets in fougerite

3.1. Thermodynamic data

To develop thermodynamic modelling, the first difficulty is to obtain a homogeneous data base. Tables 2 and 3 contain the reference data that are used for all thermodynamic calculations presented here.

3.2. Allred - Rochow electronegativity: A predictive parameter of the stability of hydroxides, oxyhydroxides and green rusts

The Allred - Rochow electronegativity scale is based upon the interaction energies between one molecule or one ion and its electrons (Allred and Rochow, 1958). It is a function of bulk charge Z of this molecule or ion.

$$\chi = (\partial E / \partial N)$$

where E is the energy of the molecule or ion and N the number of electrons.

The electronegativities of the elements, χ_i^* are derived from the electronic affinity and first ionization energy of the element and are independent of P and T. Following the partial charge model, developed by Jolivet (1994), the electronegativity of any molecule or ion can

R	8.31451
F	96485.309
T°	298.15
ln10	2.302585093
J	4.184

Reaction of formation of basic species	Nom	S°(T°) element (J/mol/K)	n	E° (V)	∂E°/∂T (mV/K)	∂²E°/∂T² (µV/K²)	$\Delta_r G°(T°)$ (kJ/mol)	$\Delta_r S°(T°)$ (J/mol/K)	$\Delta_r Cp(T°)$ (J/mol/K)	log K (T°)	$\Delta_r H°(T°)$ (kJ/mol)	$\Delta_f G°(T°)$ (kJ/mol)	$\Delta_f H°(T°)$ (kJ/mol)	S°(T°) species (J/mol/K)	Ref
$Na^+ + e = Na$	Na^+	51.21	1	-2.714	-0.757	-1.13	261.89	-73.04	-32.51	-45.88	240.11	-261.89	-240.11	58.96	1
$Mg^{2+} + 2e = Mg$	Mg^{2+}	32.68	2	-2.360	0.199		455.41	38.40		-79.78	466.86	-455.41	-466.86	-136.30	1
$Al^{3+} + 3e = Al$	Al^{3+}	28.33	3	-1.677	0.533	-0.23	485.42	154.28	-6.62	-85.04	531.42	-485.42	-531.42	-321.81	1
$K^+ + e = K$	K^+	64.18	1	-2.936	-1.074		283.28	-103.63		-49.63	252.39	-283.28	-252.39	102.52	1
$Ca^{2+} + 2e = Ca$	Ca^{2+}	41.42	2	-2.868	-0.186		553.44	-35.89		-96.96	542.74	-553.44	-542.74	-53.26	1
$Sc^{3+} + 3e = Sc$	Sc^{3+}	34.64	3	-2.09	0.41		604.96	118.68		-105.98	640.35	-604.96	-640.35	-279.90	1
$Cr^{3+} + 3e = Cr$	Cr^{3+}	23.77	3	-0.740	0.44		214.20	127.36		-37.53	252.17	-214.20	-252.17	-299.45	1
$Mn^{2+} + 2e = Mn$	Mn^{2+}	32.01	2	-1.182	-0.129	-0.9	228.09	-24.89	-51.78	-39.96	220.67	-228.09	-220.67	-73.67	1
$Fe^{2+} + 2e = Fe$	Fe^{2+}	27.28	2	-0.474	-0.07		91.50	-13.51		-16.03	87.47	-91.50	-87.47	-89.79	2
$Co^{2+} + 2e = Co$	Co^{2+}	30.04	2	-0.282	0.065		54.42	12.54		-9.53	58.16	-54.42	-58.16	-113.08	1
$Ni^{2+} + 2e = Ni$	Ni^{2+}	29.87	2	-0.236	0.146		45.54	28.17		-7.98	53.94	-45.54	-53.94	-128.88	1
$Cu^{2+} + 2e = Cu$	Cu^{2+}	33.15	2	0.339	0.011		-65.42	2.12		11.46	-64.78	65.42	64.78	-99.55	1
$Zn^{2+} + 2e = Zn$	Zn^{2+}	41.63	2	-0.762	0.119	-0.9	147.04	22.96	-51.78	-25.76	153.89	-147.04	-153.89	-111.91	1
$Cd^{2+} + 2e = Cd$	Cd^{2+}	51.76	2	-0.402	-0.029		77.57	-5.60		-13.59	75.91	-77.57	-75.91	-73.22	1
$Pb^{2+} + 2e = Pb$	Pb^{2+}	64.81	2	-0.126	-0.395		24.31	-76.22		-4.26	1.59	-24.31	-1.59	10.46	1
$Sr^{2+} + 2e = Sr$	Sr^{2+}	52.3	2	-2.899	-0.237		559.42	-45.73		-98.01	545.79	-559.42	-545.79	-32.54	1
$1/2 F_2 + e = F^-$	F^-	202.67	1	2.890	-1.87	-4.75	-278.84	-180.43	-136.64	48.85	-332.64	278.84	332.64	-13.81	1
$Cu^+ + e = Cu$	Cu^+	33.15	1	0.518	-0.754		-49.98	-72.75		8.76	-71.67	49.98	71.67	40.61	1
$PO_4^{3-} + 4 H_2O + 5e = P + 8 OH^-$	$(PO_4)^{3-}$	41.09	5	-1.470	-0.89		709.17	-429.36		-124.24	581.15	-1018.77	-1724.48	-221.63	1
$H_2SiO_4 + 4e + 4H^+ = Si + 4H_2O$	H_2SiO_4	18.83	4	-0.931	-0.395		359.31	-152.45		-62.95	313.86	-1308.03	-1457.18	189.78	1
$NO_3^- + 6H^+ + 5e = 1/2 N_2 + 3H_2O$	$(NO_3)^-$	191.5	5	1.244	-0.347	1.77	-600.14	-167.40	254.59	105.14	-650.05	-111.40	-207.44	146.45	1
$1/8 S_8 + 2e = S^{2-}$	S^{2-}	254.4	2	-0.570	-1.34		109.99	-258.58		-19.27	32.90	-109.99	-32.90	-96.21	1
$SO_4^{2-} + 8H^+ + 6e = 1/8S_8 + 4H_2O$	$(SO_4)^{2-}$	254.4	6	0.353	-0.173	3.1	-204.36	-100.15	535.07	35.80	-234.22	-744.36	-909.11	19.88	1
$CO_3^{2-} + 3H_2O + 4e = C + 6OH^-$	$(CO_3)^{2-}$	5.74	4	-0.766	-1.225		295.63	-472.78		-51.79	154.67	-527.83	-677.12	-56.86	1
$1/2Cl_2 + e = Cl^-$	Cl^-	222.957	1	1.360	-1.248	-5.83	-131.26	-120.41	-167.71	23.00	-167.16	131.26	167.16	56.35	1
$H_2O^+ + e = 1/2H_2 + OH^-$		130.574	1	-0.828	-0.836	-7.78	79.89	-80.66	-223.81	-14.00	55.84				1
$1/4O_2 + 1/2H_2O + e = OH^-$		205.029	1	0.401	-1.6816	7.23	-38.70	-162.25	207.99	6.78	-87.08				1
e		65.287													
NO_2^-			2	0.017	-1.183		-3.28			0.57					
NH_4^+			8	0.880	-0.448		-679.26			119.00					

Ref: (1) : data from Brasch (1989); (2) Refait et al. (1999)

Table 2. Thermodynamical data used for calculation

Fe: dissolution reactions of minerals

Reaction	Mineral or compound	n	E^0 (V)	$\partial E^0/\partial T$ (mV/K)	$\Delta_r G^0$ (T⁰) (kJ/mol)	$\Delta_r S^0$ (T⁰) (J/mol/K)	log K (T⁰)	$\Delta_r H^0$ (T⁰) (kJ/mol)	$\Delta_r G^0$ (T⁰) (kJ/mol)	$\Delta_r H^0$ (T⁰) (kJ/mol)	S^0 (T⁰) (J/mol/K)	Rf.
$FeOOH + 3H^+ + e = Fe^{2+} + 2H_2O$	goethite	1	0.886		-85.46	-101.31	14.97	-115.66	-480.30	-543.47	86.06	2
$FeOOH + 3H^+ + e = Fe^{2+} + 2H_2O$	lepidocrocite	1	0.985		-95.06		16.65		-470.70			2
$Fe_2O_3 + 6H^+ + 2e = 2Fe^{2+} + 3H_2O$	hematite	2	0.720	-1.250	-138.94	-241.21	24.34	-210.86	-755.45			1
$1/2Fe_2O_3 + 3H^+ = Fe^{3+} + 3/2H_2O$	½ hematite				4.92	-233.98	-0.86	-64.84				1
$Fe(OH)_{3,(s0)} + 3e = Fe + 3OH$		3	-0.776	-1.062	224.62	-307.40	-39.35	132.97				1
$Fe(OH)_{3,(s0)} = Fe^{3+} + 3OH$					207.51	-407.26	-36.35	86.08				1
$Fe(OH)_{3,(s0)} + e + 3H^+ = Fe^{2+} + 3H_2O$	ferric hydroxide	3			-922.04		161.53		119.00			4
$Fe(OH)_3 + 2e = Fe + 2OH$	virtual Green Rust	2	-0.890	-1.090	171.74	-210.34	-30.09	109.03	486.23	-569.02	85.54	4
$Fe(OH)_2 + 2H^+ + 2e = Fe + 2H_2O$		2	-0.062	-0.254	11.96	-49.01	-2.10	-2.65				4
$Fe(OH)_2 = Fe^{2+} + 2 OH$	ferrous hydroxide				83.81	-208.77	-14.68	21.57	-489.80	-569.02	83.98	2
$Fe(OH)_2 + 2H^+ = Fe^{2+} + 2H_2O$	ferrous hydroxide				-76.06		13.33					2
$GR1\text{-}Cl + 8H^+ + e = 4Fe^{2+} + Cl^- + 8H_2O$	chloride – Green Rust				-249.29		43.67		-2145.00			2
$GR2\text{-}CO3 + 12H^+ + 2e = 6Fe^{2+} + CO_3^{2-} + 12H_2O$	carbonate – Green Rust				-334.36		58.58		-3588.00			2
$GR2\text{-}SO4 + 12H^+ + 2e = 6Fe^{2+} + SO_4^{2-} + 12H_2O$	sulfate – Green Rust				-354.08		62.03		-3785.00			2
$Fe_4(OH)_8 + 8 H^+ + 2e = 3Fe^{2+} + 8H_2O$	ferrosoferric hydroxide	2	1.373		-264.95		46.42		-1897.03			3

Fe: Reactions of equilibria between minerals

Reaction	n	E^* (V)	$\partial E^*/\partial T$ (mV/K)	$\Delta_r G^0$ (T⁰) (kJ/mol)	$\Delta_r S^0$ (T⁰) (J/mol/K)	log K(T⁰)
2 Goethite = Hematite + H_2O				-31.98		5.60
4 Lepidocrocite + Cl^- + $4H^+$ + 3e = GR1-Cl	3	0.452		-130.94		22.94
6 Lepidocrocite + SO_4^{2-} + 6 H^+ + 4e = GR2-SO4	4	0.561		-216.44		37.92
6 Lepidocrocite + CO_3^{2-} + 6 H^+ + 4e = GR2-CO3	4	0.611		-235.97		41.34

Ref.: (1) Bratsch (1989); (2) Refait et al. (1999); (3) Bourrié et al. (1999); (4) Bourrié et al. (2004)

Table 3. Thermodynamical data used for calculation (suite)

be computed from the stoichiometric formula, the electronegativities of the chemical elements, χ_i^* and the net electric charge of the ion. The relationship is given by the following equation:

$$\chi = \frac{\sum_i \nu_i (\chi_i^*)^{1/2} + 1.36\, Z}{\sum_i \nu_i \left[1/(\chi_i^*)^{1/2} \right]}$$

where Z is the global electrical charge of the molecule or ion, and the summation is carried upon all elements of the molecule or ion.

The electronegativity of any molecule or ion can thus be calculated. It is a microscopic term with the dimension of an energy. For example, to calculate the electronegativity of the hydroxyl group OH-, we used the electronegativity of oxygen and hydrogen $\chi^*(O) = 3.5$ and $\chi^*(H) = 2.1$, respectively and the charge of the anion is $Z = -1$; thus the electronegativity of the hydroxyl group, OH- , is obtained as $\chi(OH^-) = 1.6$. In this way, with $\chi_i^* = 2.5$ for C, 2.83 for Cl and 2.48 for S (Jolivet, 1994), the values obtained for the anions are χ (Cl-) = 0.5421, $\chi(CO_3^{2-}) = 2.007$ and $\chi(SO_4^{2-}) = 2.2856$ (Trolard and Bourrié, 2008). Table 4 summarizes the values of Allred-Rochow electronegativities of some chemical elements, anions and cations used in calculations.

element	χ_i^*	ion	χ_i
H	2.10	Cl-	0.54
C	2.50	OH-	1.60
N	3.07	SO$_4^{2-}$	1.86
O	3.50	CO$_3^{2-}$	2.00
Mg	1.29	C$_2$O$_4^{2-}$	2.33
Al	1.47	NO$_3^-$	2.76
Si	1.74	H$_3$SiO$_4^-$	2.37
S	2.48	Mn^{2+}	5.10
Cl	2.83	Fe^{2+}	5.29
Cr	1.59	Co^{2+}	5.35
Mn	1.63	Ni^{2+}	5.45
Fe	1.72	Cu^{2+}	5.35
Co	1.75	Zn^{2+}	5.16
Ni	1.80	Al^{3+}	6.42
Cu	1.75	Cr^{3+}	6.73
Zn	1.66	Fe^{3+}	7.07
		Co^{3+}	7.15
		Ni^{3+}	7.27

Table 4. Allred-Rochow electronegativity values of some elements (from Jolivet, 1994) and ions (this work).

The Gibbs free energy, i.e. $\Delta_f G^0$, and the enthalpy of formation, i.e. $\Delta_f H^0$, of minerals or solid compounds are the main terms of macroscopic energy of interest. They can be obtained by the following semi-empirical equations:

$$\Delta_f H^0(i) - \Delta_f H^0(j) = F\left[\chi(i) - \chi(j)\right]$$

or

$$\Delta_f G^0(i) - \Delta_f G^0(j) = F\left[\chi(i) - \chi(j)\right]$$

where F is the Faraday constant, i.e. $F = N\,e$ where N is the Avogadro Number et e the electron charge, i and j indexing the ions exchanged or substituted in isostructural compounds. The multiplication by F transforms the energy from the microscopic (atomic) scale to the macroscopic (molar) scale.

3.2.1. Hydroxides and oxyhydroxides

Thermodynamic data used here were taken from Bratsch (1989). The enthalpies of formation of hydroxides, with the general formula: $Me^{II}(OH)_2$, and oxyhydroxides, $Me^{III}OOH$, are plotted versus Allred-Rochow electronegativities of bivalent and trivalent metals respectively (figure 4). The considered divalent metals are: Mn, Zn, Fe, Co, Ni, Cu and the trivalent metals are: Al, Cr, Fe, Ni and Co. Linear correlations are observed both for these hydroxides and oxyhydroxides.

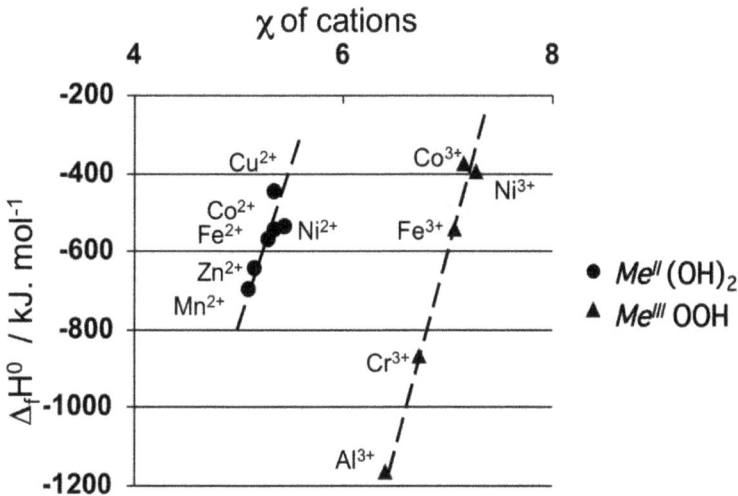

Figure 4. Relationship between enthalpies of formation of Me(II)-hydroxides and Me(III)-oxyhydroxides and electronegativity of divalent and trivalent cations respectively.

3.2.2. Green rusts

The Gibbs free energies of formation of synthetic green rusts are plotted versus Allred – Rochow electronegativities of the interlayer anions (figure 5). The value of the Gibbs free energy of Fe(OH)$_2$ is plotted at $\chi = 0$, as the interlayer is empty. Both Gibbs free energies and electronegativities are normalized to 2 structural OH per mole formula, *i.e.* for one third of the unit cell of a brucitic layer Mg$_3$(OH)$_6$ or Fe$^{II}_3$(OH)$_6$. A linear correlation is obtained for Green Rusts, from which the value of the Gibbs free energy for fougerite is derived, using χ = 1.6 for OH$^-$. This relationship has as consequence to show a differential selectivity between the different anions, following the sequence Cl$^-$ < OH$^-$ < CO$_3^{2-}$ < SO$_4^{2-}$ increasing the stability of the structure and permitting the interlayer anion exchanges.

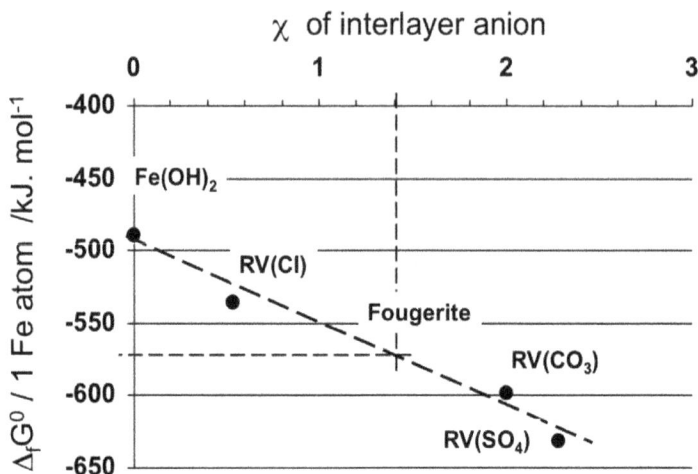

Figure 5. Relationship between Gibbs free energies of formation of green rusts (Fe LDHs) and the electronegativity of the interlayer anion.

3.3. Modeling of the LDH thermodynamic stability in presence of Si: a possible pathway to clay mineral formation.

3.3.1. Experimental data used

The study of the thermodynamic stability of LDH in presence of Si is based on experimental data obtained on hydrotalcite – like compounds with the general formula [Me(II)$_{1-x}$Al$_x$(OH)$_2$]. [x/n A^{-n}, mH$_2$O] with (0.2 < x < 0.4) and Me(II) = Ni (Allada et al., 2002; 2006).

The range of composition of LDHs is very narrow and they differ, in the first order, mainly by the nature of the interlayer anion (Allada et al., 2002, Bourrié et al., 2004).

Enthalpies of formation of Ni-Al LDHs were taken from Peltier *et al.* (2006) (table 5).

3.3.2. Results of the calculation

Same as Green Rust above, the value of the enthalpy of formation of Ni(OH)$_2$ is plotted at χ = 0, as the interlayer is empty. A large decrease is observed when LDHs form and figure 6 shows that the enthalpies of formation of synthetic Ni-Al LDHs are very close to each other when the compensating anion is sulphate, carbonate or nitrate. The enthalpy of LDHs when silicate anion is present in addition to carbonate shows a more negative value.

| compounds | Value of the stoichiometric coefficients in the Ni-Al LDHs | | | | | | | | $\Delta_f H^0$ | $\Delta_f G^0$ |
	Ni	Al	NO_3^-	CO_3^{2-}	SO_4^{2-}	$H_3SiO_4^-$	OH	H_2O	/kJ.mol^{-1}	/kJ.mol^{-1}
NO3-2	0.65	0.35	0.21	0.07	0	0	2	0.42	-942.41	-807.60
NO3-H	0.66	0.34	0.24	0.05	0	0	2	0.3	-904.03	-773.90
NO3-10	0.77	0.23	0.13	0.05	0	0	2	0.6	-908.42	-776.00
SO4-2	0.66	0.34	0	0.02	0.15	0	2	0.21	-940.30	-823.20
SO4-H	0.65	0.35	0	0.025	0.15	0	2	0.22	-957.52	-839.40
SO4-10	0.72	0.28	0	0.02	0.12	0	2	0.43	-952.78	-829.30
CO3-2	0.64	0.36	0	0.18	0	0	2	0.46	-987.30	-862.00
CO3-H	0.66	0.34	0	0.17	0	0	2	0.42	-950.57	-851.40
CO3-5	0.67	0.33	0	0.17	0	0	2	0.41	-930.47	-828.70
CO3-Si	0.65	0.35	0	0.1	0	0.15	2	0.08	-1132.27	-925.40
Ni(OH)$_2$	1	0	0	0	0	0	2	0	-540.34	-459.10
Al(OH)$_3$	0	1	0	0	0	0	3	0	-1293.00	-1155.00

Table 5. Enthalpies of formation and Gibbs free energies obtained by Peltier et al. (2006) on Ni-Al LDHs used in calculations

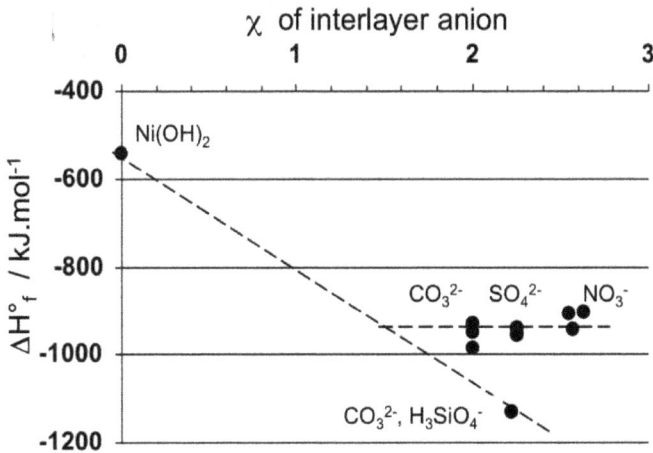

Figure 6. Relationship between enthalpies of formation of Ni-Al LDHs and average electronegativities of A^{-n} anion.

As the composition of the layer is quasi-constant, i.e. $x \in [1/4; 1/3]$, the variation of the enthalpy formation can be written:

$$\Delta_f H^0_{(LDH)} = \Delta_f H^0_{(hydr.)} - x\, F\, \chi(anion) + m\, \Delta_f H^0_{w,c}$$

where m is the number of moles of water in the interlayer under a crystalline state, x the electric charge of the hydroxide layer and F the Faraday constant (*cf. supra*). The minus sign comes from the negative charge of the anion.

The quantity $\Delta_f H^0_{(LDH)} + x\ F\ \chi(anion)$ is plotted versus the number of moles of water molecules per mole formula (figure 7). Two groups appear distinctly: all LDHs without silicate anion in the interlayer, and mixed silicate-carbonate Ni/Al-LDHs.

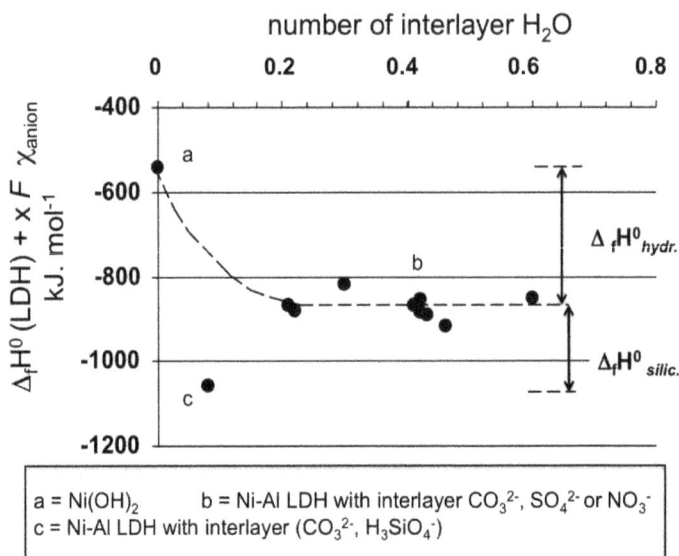

Figure 7. Relationship between enthalpies of formation of Ni-Al LDHs corrected from the contribution of electronegativities of A^{-n} anion and the number of moles of water in the interlayer.

The large variation of enthalpy from the pure *Me(II)* – hydroxide to the first stable LDHs can be ascribed to the opening and hydration of the interlayer as soon as the net charge of the layer is not zero, *i.e.* the enthalpy of hydration of the layers: $\Delta_f H^0_{hydr.}$. This explains the large deviation from ideality observed when modelling these compounds as solid solution (*e.g.* Bourrié et al., 2004 for Fe-LDHs or green rusts). As this term is large, the influence of the nature of the interlayer anion is masked. However, it would be interesting to check the influence of anions with larger electronegativities, such as Cl$^-$.

The silica anion exerts a large influence by expelling water molecules. This does not lead to a reversal to more positive values of $\Delta_f H^0_{(LDH)}$, but instead to more negative. It can be ascribed to condensation of silicate anion with the LDH layer with elimination of water molecules, hence the nucleation of Ni(II)-Al clay mineral by polymerization of silica within the interlayer following the anion exchange carbonate/silicate (Peltier et al., 2006).This implies that silica combines to the hydroxide sheets, as a first step towards new formation of phyllosilicate minerals. It is possible that the hydroxyl groups of the hydroxide sheets begin then condensate with hydroxyl groups of the silicate tetrahedra. This is not the case with sulphate, which give rise in green rusts to two monomolecular layers of water, instead of only one with planar anions.

4. Conclusions

Fougerite, as other layered double hydroxides (Sparks, 2003), plays an essential role in the formation of clay minerals, both phyllosilicates and iron hydroxides and oxides. They control both major elements, e.g. Mg, Fe, and trace metals (Co, Cr, Mn, Ni and Zn) in the environment.

Electronegativity of ions, computed following the model of partial charges by Jolivet (1994) is a suitable parameter to correlate macroscopic thermodynamic properties of isostructural compounds when interactions are mainly electrostatic. It is of special value for small sized natural minerals that cannot be isolated and purified, and to assess consistency between experimental data sets.

The introduction of silicate anion by anionic exchange in a LDHs structure leads to a reduction of the number of water molecules in the interlayer and increases the stability of lamellar double hydroxide. This transient structure can be considered as the first step towards the formation of a new clay mineral.

Author details

Fabienne Trolard and Guilhem Bourrié
INRA, UMR1114, Environnement Méditerranéen et Modélisation des Agro-hydrosystèmes, Avignon, France
UAPV, UMR1114, Environnement Méditerranéen et Modélisation des Agro-hydrosystèmes, Avignon, France

5. References

Allada R.K., Navrotsky A., Berbeco H.T. and Casey W.H. (2002) Thermochemistry and aqueous solubilities of hydrotalcite-like solids. Science, 296, 721-723.

Allada R.K., Peltier E., Navrotsky A., Casey W.H., Johnson C.A., Berbeco H.T. and Sparks D. (2006) Calorimetric determination of the enthalpies of formation of hydrotalcite –like solids and their use in the geochemical modeling of metals in natural waters. Clays & Clay Minerals, 54, 4, 409-417.

Allred A.L. and Rochow E.G. (1958) A Scale of electronegativity based on electrostatic force. J. Inorganic & Nuclear Chemistry, 5, 264-268.

Bratsch S.G. (1989) Standard electrode potentials and temperature coefficients in water at 298.15 K. J. of Physical Chemical Reference Data, 18, 1-21.

Bourrié G., Trolard F., Génin J.M.R., Jaffrezic A., Maître V. and Abdelmoula M. 1999. Iron control by equilibria between hydroxi-green rusts and solutions in hydromorphic soils. Geochimica et Cosmochimica Acta, 63, 3417-3427.

Bourrié G., Trolard F., Refait P. and Feder F. (2004) A solid solution model for Fe(II) – Fe(III) – Mg(II) green rusts and fougerite and estimation of their Gibbs free energies of formation. Clays & Clay Minerals, 52, 382-394.

Cornell R.M. and Schwertmann U. (2003) The iron oxides. 2nd edition. VCH, Weinheim, Germany.

Driessen P., Deckers J., Spaargaren O. and Nachtergale F. (2001) Lecture notes on the major soils of the world. World Soil Resources Report n°94, FAO, Rome.

Drissi S.H., Refait P., and Génin J.M.R. (1994) The oxidation of Fe(OH)$_2$ in the presence of carbonate ions: structure of carbonate green rust one. Hyperfine Interactions, 90, 395-400.

Feder F., Trolard F., Klingelhöfer G. and Bourrié G. (2005) In situ Mössbauer spectroscopy – Evidence for green rust (fougerite) in a gleysol and its mineralogical transformation with time and depth. Geochimica et Cosmochimica Acta, 69, 4463-4483.

Fredrickson J.K., Zachara J.M., Kennedy D.W., Dong H., Onstott T.C., Hinman N.W. and Li S.M. (1998) biogenic ion mineralization accompagnying the dissimilatory reduction of hydrous ferric oxide by a groundwater bacterium. Geochimica et Cosmochimica Acta, 62, 3239-3257.

Génin J.M.R., Olowe A.A., Resiak B., Confente M., Rollet-Benbouzid N., L'Haridon S. and Prieur D. (1994) Products obtained by mycrobially-induced corrosion of steel in a marine environment. Role of green rust two. Hyperfine Interactions, 93, 1807-1812.

Girard A. and Chaudron G. (1935) Sur la constitution de la rouille. Comptes Rendus de l'Académie des Sciences, Paris, 200, 127-129.

Hansen H.C.B., Gulberg S., Erbs M. and Koch C.B. (2001) Kinetics of nitrate reduction by green rusts: effects of interlayer anion and Fe(II): Fe(III) ratio. Applied Clay Science, 18, 81-91.

Jolivet J.P. (1994) De la solution à l'oxyde. Savoirs Actuels, Interéditions, CNRS, 387 pp.

Lewis L.W. (1997) Factors influencing the stability and properties of green rusts. In : soils and Environment (Auerswald K., Stanjek H., Bigham J.M. eds.), Adv. In GeoEcology, Catena Verlag, 30, 345-372.

Loyaux-Lawniczak S., Refait P., Ehrhardt J.J., Lecomte P., and Génin J.MR. (2000) Trapping of Cr by formation of ferrihydrite during the reduction of chromate ions by Fe(II)-Fe(III) hydroxysalt green rusts. Environmental Science & Technology, 34, 438-443.

Mineny S.C.B., Tokunaga T.K. and Brown G.E.J. (1997) Abiotic selenium redox transformation in the presence of Fe(II, III) oxides. Science, 278, 1106-1109.

Murad E. (1990) Application of 57Fe Mössbauer spectroscopy to problems in clay minerals and soil science: possibilities and limitations. Advances in Soil Science, 12, 125-157.

O'Loughlin E.J., Kelly S.D., Kemner K.M., Csencsits R. and Cook R.E. (2003) Reduction of Ag^I, Au^{III}, Cu^{II} and Hg^{II} by Fe^{II}/Fe^{III} hydroxysulfate green rust. Chemosphere, 53, 437-446.

Peltier E., Allada R.K., Navrotsky A. and Sparks D.L. (2006) Nickel solubility and precipitation in soils: a thermodynamic study. Clays and Clay Minerals, 54, 153-164.

Ponnamperuma F.N., Tianco E.M. and Loy T. (1967) Redox equilibria in flooded soils: I. The iron hydroxide systems. Soil Science, 103, 374-382.

Refait P., Charton A. and Génin J.M.R. (1998) Identification composition, thermodynamic and structural properties of pyroaurite-like iron(II)-iron(III) hydroxy-oxalate green rust. European J. Solid State Inorganic Chemistry, 35, 655-666.

Refait P., Bon C., Simon L., Bourrié G., Trolard F., Bessière J. and Génin J.M.R. 1999. Chemical composition and Gibbs standard free energy of formation of Fe(II)-Fe(III) hydroxysulphate green rust and Fe(II)hydroxide. Clay Minerals, 34, 499-510.

Refait P., Abdelmoula M., Trolard F., Génin J.M.R., Ehrhardt J.J. and Bourrié G. (2001) Mössbauer and XAS study of green rust mineral; the partial substitution of Fe^{2+} by Mg^{2+}. American Mineralogist, 86, 731-739.

Schwertmann U. and Cornell R.M. (2003) Iron oxides in the laboratory. Preparation and Characterization. VCH edition, Weinheim.

Schwertmann U. and Fechter H. (1994) The formation of green rust and its transformation to lepidocrocite. Clay Minerals, 29, 87-92.

Sparks D. (2003) Environmental soil chemistry. 2e edition, Academic press, Elsevier Science (USA).

Sposito G. (1989) The surface chemistry of soils. Oxford Univ. Press, New York.

Stumm W. (1987) Aquatic surface chemistry, Wiley, New York.

Tamm O. (1922) Eine Method zur Bestimmung der organisches Komponenten des Gelcomplexes in Boden. Medd. Statens Skogsfösöksanstalt, 19, 385-404.

Taylor R.M. (1981) Color in soils and sediments. A review. In: International Clay conference 1981 (ed. H. van Olphen and F. Veniale), Developments in Sedimentology, 35, 749-761 (Elsevier, Amsterdam).

Trolard F., Abdelmoula M., Bourrié G., Humbert B. and Génin J.M.R. (1996) Mise en évidence d'un constituant de type "rouilles vertes" dans les sols hydromorphes – Proposition de l'existence d'un nouveau minéral : "la fougérite". C.R. Académie des Sciences, Paris, 323(série IIa), 1015-1022.

Trolard F., Génin J.M.R., Abdelmoula M., Bourrié G., Humbert B. and Herbillon A.J. (1997) Identification of a green rust mineral in a reductomorphic soil by Mössbauer and Raman spectroscopies. Geochimica et Cosmochimica Acta, 61, 1107-1111.

Trolard F., Bourrié G., Soulier A., Maître V., Génin J.M.R. and Abdelmoula M. (1998) Dynamique de l'oxydo-réduction dans les zones humides. In: Agriculture intensive et qualité des eaux (C. Cheverry coord.), pp. 185-208. Collection Update, INRA publications, Paris.

Trolard F., Bourrié G., Abdelmoula M., Refait P. and Feder F. (2007) Fougerite, a new mineral of the pyroaurite-iowaite group: description and crystal structure, Clays and Clay Minerals, 28, 179-187.

Trolard F. and Bourrié G. (2008) Geochemistry of green rusts and fougerite: a reevaluation of Fe cycle in soils. Advances in Agronomy, 99, chap. 5, 227-287.

Vyssotskii G.N. (1905, [1999]) Gley. Pochvovedeniye, 4, 291-327 [Gley. An abridged publication of Vyssotskii 1905 on the 257th anniversary of the Russian Academy of Sciences. Eurasian Soil Science, 32, 1063-1068].

Chapter 9

Relations of Clay Fraction Mineralogy, Structure and Water Retention in Oxidic Latosols (Oxisols) from the Brazilian Cerrado Biome

Carla Eloize Carducci, Geraldo César de Oliveira, Nilton Curi, Eduardo da Costa Severiano and Walmes Marques Zeviani

Additional information is available at the end of the chapter

1. Introduction

In Brazil, Latosols are by far the main class of soils, mainly when one considers the soils potentially used for agricultural purposes. They cover approximately 50% of the Cerrado Biome, totaling about 200 million hectares, in [46, 52]. The clay mineralogy of these soils are very simple, basically composed by 1:1 clay minerals, mainly kaolinite, and varying proportions of iron- and aluminum-oxides (in this chapter this general term includes oxides, hydroxides and oxi-hydroxides). As the oxides content increases, they tend to be associated with the formation of granular structure, composed by very small and resistant micro-aggregates, occurring in both superficial and sub-superficial soil horizons.

To explain the formation of micro-aggregates in these highly weathered Latosols, in [30, 31] it is highlighted that iron- and mainly aluminum-oxides as aggregating agents of mineral particles by changing the arrangement of their components in relation to the plasma, resulting in granular aggregates with a diameter < 300 μm, in an agglutinated pattern, having a high pore volume, which is in turn organized into interconnected cavities, in [100]. Consequently, in these soils, the pore distribution by size is predominantly characterized by two distinct classes of pores: the first one is related to very large or structural pores (among micro-aggregates), which promote rapid internal drainage of the soil, however, they are very susceptible to alteration; and the second one is related to very small pores or textural pores (inside micro-aggregates), in which water is retained with very high energy, in [10, 21, 63]. This segregation of contrasting pores is typical of oxidic Latosols from this region, in which the increase of clay content is associated with higher total porosity and lower bulk density, in [31, 37, 46, 72, 88, 89].

Thus, the content and nature of the clay fraction are very important in the hydro-physical behavior of these highly weathered soils. Under natural conditions, these soils have a high total pore volume, one part being composed by drainable pores (approximately 2/3 of the total pore volume; those with diameters > 145 μm), which are of fundamental importance for the soil high permeability. However, these soils also have high volume of pores with very small diameter (< 2.9 μm; approximately 1/3 of the total pore volume). Therefore, in order to remove the residual water content a considerable amount of energy is required, in [63]. Small amounts of water have been observed to be adsorbed on the soil matrix under 300.000 kPa pressure, in [10].

In the current stage of evolution of the agricultural systems in Brazil, in which yield increments are searched without increasing the productive area, it is necessary to understand in details the hydro-physical behavior of these Latosols, taking into consideration the environmental sustainability, in which underground water recharge is fundamental for the maintenance of the most varied types of edaphical life. It should be mentioned that irrigated agriculture in this region is undergoing accelerated growth and it is not clear until now if the existing water resources are sufficient to support this expansion, in [78]. These soils are usually very deep, providing large reservoirs of water for crops, since there are no chemical constraints for the expansion of the crop root system. Even in the sub-superficial soil horizons, the residual water plays an important role in the maintenance of adequate thermal and physical conditions, which minimizes root death during the pronounced dry season, which is typical of this region.

2. Cerrado biome, mineralogy and structure of Oxidic Latosols

The Brazilian Cerrado, which is the second largest biome in the country, is located in the central part of the South America, including large portions between parallels 3 ° and 24 ° south and between parallels 41º and 63º west. In Brazil, the biome occupies approximately 23.92% of the territory, covering several states (Mato Grosso, Mato Grosso do Sul, Rondônia, Tocantins, Minas Gerais, Bahia, Maranhão, Piauí, Sao Paulo, and particularly Goiás and the Federal District, where this vegetation covers the landscape on a relatively more continuous way (Figure 1), but there is still remaining "islands" of this biome in Pará, Roraima and Amapá states, in [5]. The soils that support this biome are hydrologically important since the major basins (Amazônica, Platina and Sanfranciscana) have many of their springs in this region, in [29, 47].

The Cerrado can be defined as a formation composed of tropical vegetation, represented mainly by grasses, with sparse trees and shrubs, in other words, including floristic and physiognomic aspects of vegetation, constituting a unique biome, in [29], also called neotropical savanna. The soils are represented by Latosols (Oxisols) (50%), Argisols (Ultisols) (15%), Quartzarenic Neosols (Entisols) (15%), Cambisols (Inceptisols) (10%) and Plinthosols (Oxisols having drainage restrictions) (6%) and other soils (4%), in [78]. The Latosols and Quartzarenic Neosols are located in predominantly gentle relief associated with a very sparse hydrography.

Figure 1. Cerrado area in Brazil, highlighting Goiás state and Federal District (darker area) (Source: [85])

The Latosols are considered the oldest soils on earth. They go from deep to very deep, non hydromorphic, and show great textural variation with clay content ranging from 150 g kg^{-1} to more than 800 g kg^{-1} [88, 89]; exhibit low natural fertility due to the strong weathering-leaching, which contrasts with their excellent physical conditions favored by the strong and very small granular structure. Latosols also tend to present high acidity (pH 4.0 to 5.5), low cation exchange capacity, high anion adsorption capacity (especially phosphate and heavy metals) and low levels of available P (phosphorus) to plants [27, 46, 77].

The beginning of weathering of Latosols in this region dates from the Cretaceous and Tertiary, in [54]. They were formed under conditions of significant weathering-leaching, which contributed to their advanced degree of pedogenic development, resulting in a very simple mineralogy, in [76]. Their clay mineralogy consists basically of 1:1 clay minerals, mainly kaolinite ($Si_2Al_2O_5$ $(OH)_4$), iron oxides (hematite (Fe_2O_3) and goethite (FeOOH)) and aluminum oxides (gibbsite ($Al(OH)_3$)) in different proportions, as well as quartz and other resistant minerals, in [16, 38, 39, 44, 46, 54, 65, 73, 74, 75, 76, 77, 83]. There are also registers in the clay fraction of some Latosols formed from rocks richer in iron, of maghemite (Fe_2O_3) as well as magnetite (Fe_3O_4) and ilmenite ($FeTiO_3$) in the coarse fraction [93, 97]. The identification of hydroxi-interlayered vermiculite in the clay fraction of A and B horizons of some Latosols has been also registered, in [71].

Knowledge of Latosol genesis facilitates the identification of their corresponding classes in international soil classification systems: the Oxisols in Soil Taxonomy, in [92] and the Ferralsols in World Reference Base [43]. As peculiar characteristics of Latosols can be cited: the presence of latossolic B horizon (Bw = intense weathering), minimal differentiation between A and B horizons, color varying from reddish to yellowish, depending on the parent material and the factors and processes of soil formation, in [15, 54]. They exhibit weak macrostructure and strong microstructure [28, 30, 31], resulting in 50-300 µm size micro-aggregates, in [100]. These soils constitute the largest class in terms of territorial expression having high potential for agriculture, forestry and livestock purposes, in [46].

The pelitic rocks of the Bambuí Group which occur in Minas Gerais, Bahia and Distrito Federal states are important parent materials of many Latosols of the Cerrado Biome. These rocks are fine grained, resulting in clayey or very clayey soils. In these soils, the kaolinite is the mineral with higher expression in the clay fraction, in [66] and its presence in combination with low levels of iron and aluminum oxides favors the hard consistency when the soil dries, and higher bulk density, which is related to the blocky macrostructure, function of the face-to-face arrangement of the kaolinite plates in [30, 31].

Ferreira et al., in [30] relating the mineralogy and structure of Latosols in southeastern Brazil, stratified them into kaolinitic or gibbsitic soils: in kaolinitic Latosols the micromorphological evaluation showed that the distribution of quartz grains in relation to the plasma, is porphyric. In other words, the grains are enveloped in a dense and continuous plasma, with little tendency to develop the microstructure. This phenomenon is associated with the blocky structure, so that the soils are more compact, less permeable, with lower aggregate stability in water and have a greater susceptibility to sheet erosion. On the other hand, the gibbsitic Latosols show a more uniform distribution of the minerals in relation to the plasma, resulting in smaller granular and resistant aggregates (< 300 µm diameter), in an agglutinated pattern, influencing higher void ratio, which are in turn arranged into interconnected cavities, in [100], showing a greater susceptibility to gully erosion.

Consequently, in these soils the pore distribution by size is characterized by presenting predominantly two distinct classes of pores: the first one is related to very large or structural pores (among micro-aggregates), which promote rapid internal drainage of the soil being, however, very susceptible to alteration; and the second one is related to the very small pores formed among the mineral particles (inside micro-aggregates), in which water is retained with very high energy, characterizing it as hygroscopic water, in [19, 21, 63]. This segregation of contrasting pores is typical of the oxidic Latosols from this region, in [73]. Usually, increasing the clay content of these oxidic Latosols results in increased total porosity and lower bulk density, in [30, 31, 89].

Based on this knowledge it can be understood that in very weathered tropical soils the micro-aggregates are very resistant and play a prominent role in the formulation of the soil aggregation hierarchy hypothesis. An indication of this resistance is the difficulty of evaluating the clay content in the field, requiring more time for reliable estimates, in [13]. This micro-aggregates resistance also manifests itself in the laboratory analysis of particle size distribution, mainly during the chemical and mechanical soil dispersion, in [38, 39, 63, 99].

3. Bimodal pore distribution and water retention of oxidic Latosols

The development of a specific type of soil structure is usually a consequence of the parent material and soil formation processes and factors, and these will condition many of the physical properties of soil. Marshall, in [55] stated the soil structure is defined as the arrangement of soil particles and the associated voids, including shape, size and arrangement of the aggregates formed by the primary particles (sand, silt and clay) which are grouped into units defined by limits. Marcos, in [53] cited that the morphological evaluation of the soil structure is qualitative, while the physical evaluation is functional.

It is known that soil macrostructure is strongly affected by climatic changes, biological activity, as well as the land use and soil tillage, being vulnerable to mechanical and physical-chemical forces, by according to Hillel, in [41]. In another words, composite structural units or aggregates are formed by aggregation of primary mineral particles in association with organic particles, especially the humidified ones, in [91], originating the soil structure, which influences the porosity. Thus, the aggregates have their own genesis reflected in their size, shape, composition and stability, in [9, 98].

According to with this soil structure model, there is a strong influence of the mineralogical components of the clay fraction upon formation of a particular structure type. It is reported, for instance, that oxides (mainly gibbsite, followed by iron oxides-goethite and hematite) jointly with organic matter, in this order of importance, tend to disorganize the particles at microscopic scale, in [30, 77].

Therefore, the higher content of these components has a greater degree of disorganization and, consequently, the structure tends to become of the granular type. So, gibbsite, iron oxides and organic matter are precursors and maintainers of the granular structure, which is typical of oxidic Latosols in the Cerrado Biome, and it results in high permeability values, in [30, 31].

In Latosols, the granular structure type is responsible for a lower bulk density and a higher porosity values compared to the blocky structure (kaolinitic Latosols), in [10, 22]. The developments of structural- or among micro-pores (> 50 μm diameter) are more expressive in oxidic Latosols, followed by textural- or inside -micro-pores (< 50 μm diameter), in [14, 50, 63]. In oxidic Latosols, the structural pores exhibit a relationship with clay content reflecting on their hydro-physical attributes such as water retention. This feature can be considered a special characteristic of oxidic Latosols, in [31, 86-89].

Therefore, the presence of this type of structure formed by stable micro-aggregates, especially in the Bw horizons of oxidic Latosols, consequently determines the dominance of structural porosity over textural porosity, giving to these soils excellent permeability and moderate to low water retention, in [14].

The voids of the soil are formed by various processes that result in different pore shapes and sizes that affect the soil functions. For instance, the water and gases transportation occurs through the interconnected pores. The soil structure is considered to have various hierarchical levels, namely: a) groups of primary particles which comprise micro-aggregates; b) groups of micro-aggregates comprising aggregates; and c) groups of aggregates comprising much larger aggregates or soil clumps, in [19].

The pore distribution by size affects the soil hydro-physical dynamics. In the literature there are several classification schemes for pore diameter, highlighting the most simplified ones that separate two classes of pores: macro-pores, when the pores have a diameter > 50 μm, and micro-pores, when the pores have a diameter < 50 μm, as proposed by Kiehl, in [49] and Richards, in [79]. Pores of intermediate size, meso-pores, have lower expression in the Latosols from the Cerrado Biome, in [10].

There are equations that aim to quantify the pore size. Bouma, in [6] proposed the following equation: $D = 4\sigma \cos \theta / \Psi_m$, being: D = pore diameter (mm), σ = water surface tension (73.43 kPa at 20 °C); θ = contact angle between the meniscus and the wall of the capillary tube (assumed to be 0) e Ψ_m = matric potential (kPa). However, there are simple and straightforward methods to determine the pore size distribution, for example, using mathematical models to describe the water retention curve, because it is known that the shape and slope of the curve correspond to the homogeneity of the distribution of pore diameter, in [2, 19, 36].

Thus, the bimodal pore distribution of oxidic Latosols can be represented from the water retention curve, in [10]. When using the shape of the curve, the first inflection point occurs at low matric potentials (between 1 and 3 kPa, in absolute value) identifying structural pores, while the textural pores are represented by the second inflection point that occurs at extremely high matric potentials (between 10.000 and 20.000 kPa). Between these maximum points it can be observed that the asymptotes, related to the presence of intermediate pores, have low expression in oxidic Latosols in [10, 21, 22, 75]. In soils of temperate regions, the bimodality has been observed within the range of the standard curve of water retention, in the range from 1 to 1.500 kPa in [21, 22] due to the more uniform pore distribution, compared to soils of tropical regions.

It is noteworthy to remember that the soil water retention depends on pore distribution, and this is influenced by various factors such as structure, particle size distribution, organic matter, clay mineralogy, as well as biological activity. There are two possible reasons for the influence of mineralogy on the soil water retention: a) specific surface area; and b) presence of electrical charge of clay minerals. The larger the specific surface area and the higher the electrical charge is, the more water can be bound to the clay minerals, in [62, 34].

Thus, there is a substantial process of water being adsorbed on the surface of clay minerals by electrostatic forces and, hence, the water retention. Gaiser et al. [35] observed significant differences in soil water retention with different mineralogy, noting that soils with low activity clays (1:1 clay minerals and iron- and aluminum-oxides) retain less water when compared to soils that have high activity clays (2:1 clay minerals), using pedotransfer functions. Several studies have indicated a strong influence of clay fraction on water retention in Latosols, in [1, 4, 10, 11, 70, 73, 89]. A few authors claim that clayey Latosols having oxidic mineralogy favor higher water content and more gradual decrease of soil water content with increasing matric potential (in absolute value). The study of water retention developed by van den Berg et al., in [96] in Latosols from different regions showed that the increased release of water occurs at low potentials (between 5 and 10 kPa) similar to what happens with very sandy soils. The spatial variability of water retention in clayey Latosols was studied by Cichota & van Lier, in [12]. These authors observed that the water retained at matric potentials ranging from 1 to 100 kPa is not strictly related to the content of clay, which confirms the theory of Raws et al., in [70] that at low matric potentials the retention curve is directly influenced by structure stability and consequently by the formation of pores in addition to the indirect effects of organic matter.

Many advances have been made in order to better characterize soil water retention. More sophisticated devices as the WP4-T, in [18] should be highlighted, which allows the quantification of the residual water retained at high matric potentials. The residual water retained in the textural pores of oxidic Latosols, although considered unavailable to crops, in [50, 80], may reflect significant water content (up to 0.25 g g^{-1}) in more clayey soils, in [10]. So, it becomes of great interest in studies involving regulation of microbial and biochemical processes in the soil, in [60], re-induction of desiccation tolerance of germinating seeds and seedlings when subjected to high matric potentials ($\Psi_m > 1.500$ kPa), in [81] and it can act as a lubricant between aggregates, when the soil undergoes external pressure during mechanized operations in [23].

4. Modeling the water retention curve of oxidic Latosols

Water retention curve has been used to describe the dynamics of the soil water, in [20, 36]. This curve graphically represents the relationship between the energy of water retention (matric potential, in logarithmic scale) and water content, which is dependent on the intrinsic characteristics of each soil, the result of joint action of soil attributes such as texture, structure, mineralogy and organic carbon, in [4, 19, 37, 40].

Several types of adjustments to the water retention curve have been used, in [25, 36] for describing the soil hydro-physical performance. However, in order to identify the bimodal distribution of pores in oxidic Latosols the double van Genuchten model was recently proposed by Carducci et al., in [10]. Based upon the shape of the curve, the first inflection point usually occurs at low matric potentials, representing the structural pores, while the textural pores are represented by the second inflection point that occurs at higher matric potentials. For soils from temperate regions, the bimodality of pore distribution has been observed within the range of the standard curve of soil water retention, in other words, in the range from 1 to 1.500 kPa (in absolute value), in [22] because there is a more uniform distribution of pores, when compared to soils from tropical regions. This mathematical model allows to identify, with high predictive power, the bimodal density function for the pore size distribution of tropical soils in a more superior range than to the one of the standard curve: $1 < \Psi_m < 300.000$ kPa).

One purpose of science is to find, describe and predict the possible relationships between events occurring in the environment. A common practice is to develop models that relate these events. For this purpose, statistical modeling is widely used, mainly by the use of linear and nonlinear regression models, in [56]. The two classes of regression models differ mainly in aspects related to their application and the characteristics linked to the mathematical form. The choice of which model to consider in fitting a certain set of data can be made intuitively, or through a graphical which expresses the function of the variables or prior knowledge of the phenomenon in question.

The linear models are widely used for presenting analytical solution for estimating parameters and statistical properties. The interpretation of these parameters is purely mathematical, based on rate of variation of the dependent variable in relation to the independent variables, in [94]. Furthermore, the use of a linear model for predicting values outside the range of observed values of independent variables is not advisable. Although the linear model is very flexible, since many models can be formulated by the combination of independent variables, in [26], there are several types of models which are based on theoretical considerations inherent to the phenomenon which one has interest in knowing, i.e., the called mechanistic models, in [56, 84]. Generally in these models the parameters have practical interpretation and the prediction of values is allowed, since when considering the mechanistic model the restrictions which ensure the model utility are imposed, in [3].

A model is considered nonlinear when the mathematical expectation of a dependent variable "Y" cannot be written as a linear function of parameters in a regression model. Historically, nonlinear regression models date from early 1920's, in [33]. However, the application and a detailed investigation of these models had to wait for the advancements allowed by computational calculations after 1970, in [24].

The rise up of nonlinear models often accompanies the forecasts involving physical and/or biological dynamics about the phenomenon under study in [102]. Such expectations are based upon models in which the parameters have practical significance in describing the phenomenon that is observed.

The function of statistics in this scenario is to evaluate, select, and provide models and tools for better understanding of these phenomena. An overview of a nonlinear model considers a set of p columns of a matrix X and a vector of parameters $\theta = (\theta_1,...,\theta_k)^T$ such that the average related to a response Y is given by:

$$E(Y|X = x) = f(x, \theta) \tag{1}$$

Where f is the function average or expectation of Y. Unlike linear models, the numbers of columns of the matrix X does not necessarily need to be equal to the number of parameters in the vector θ. Many of the functions impose restrictions on parameters (eg: $\theta_i > 0$, i = 1,...,k) due to both practical interpretations of the compatibility of mathematical relationships. The variance of Y in turn is given by:

$$Var\ (Y|X = x_i) = \sigma^2 \tag{2}$$

The above equations, including the presupposition of independence between observations, define the classic nonlinear model. The only difference between the classes of models is the form of the expectation function. The function is nonlinear regarding the parameters, and therefore, many parallels can be drawn regarding the procedures for estimation of parameters and statistical inference. The fitting of nonlinear models can be obtained by minimizing the residual sum of squares, $RSS(\theta)$, where:

$$RSS(\theta) = \sum_{i=1}^{n}\left(y_i - f\left(x_i, \theta\right)\right)^2 \tag{3}$$

By inspection of all values from the parameter space of $\theta \in \theta$.

For linear models there is an analytical solution for the estimating $\hat{\theta}$ that minimizes $RSS(\theta)$. For nonlinear models the search for the minimum point of equation (3) is usually a problem with the numerical solution. Such problem uses a linear approximation of nonlinear function that converges to the minimum point at each iteration, in [48]. This procedure, as expected, also provides rough estimates for standard errors and hypothesis tests, and such approach is a function of how strong the nonlinearity of the model.

The Taylor series approximation of the function of expectation around a value θ^* considering expanding to the second term, can be written as:

$$f(\theta) \cong f(\theta^*) + F(\theta^*)(\theta - \theta^*) + \frac{1}{2}(\theta - \theta^*)^T H(\theta^*)(\theta - \theta^*) \tag{4}$$

Where F (θ^*) and H (θ^*) are the score matrix and hessian arrangement, respectively. The j-th column vector of the score matrix is given by $\partial f\ (x, \theta) = \partial\theta_j$ and the jl-th column vector of the hessian matrix is given by $\partial^2 f(x, \theta) = \partial\theta_j\partial\theta_l$, both evaluated at $\theta = \theta^*$.

Omitting the second term of the expansion (4), we can rewrite (3) as:

$$RSS(\theta) \cong \sum_{i=1}^{n}\left(yi - f(xi, \theta^*) - F(\theta^*)(\theta - \theta^*)\right)^2$$
$$\cong \sum_{i=1}^{n}(\hat{e}_i^* - F(\theta^*)(\theta) - (\theta^*))^2 \tag{5}$$

$\hat{e}_i^*$ is the current residuum that depends on the current value of θ^* in the iterative process. In the re-writing of the matrix form, the minimization process can be written as:

$$\widehat{\theta - \theta^*} \cong [F(\theta^*)^T F(\theta^*)]^{-1} F(\theta^*)^T \hat{e}^*$$
$$\hat{\theta} \cong \theta^* + [F(\theta^*)^T F(\theta^*)]^{-1} F(\theta^*)^T \hat{e}^* \tag{6}$$

The equations (6), below, are applied in two forms: first to support the algorithm for the estimation of θ and the second as the basis for statistical inference on the parameter estimates, in [48]. The majority of statistical packages use the Gauss-Newton algorithm to find the parameter estimates in nonlinear models. Other packages also present derivative forms or algorithms based on other optimization processes. Practically algorithms differ at execution time. However, the efficiency of any one of them is very dependent on the value $\theta^{(0)}$ and $\hat{\theta}$ given at the beginning of iterative process. Depending on the numerical distance between $\theta^{(0)}$ and the algorithm can converge to a local minimum, or even not converge, therefore, suitable choices for $\theta^{(0)}$ in this sense are more important than the iterative method.

At each iteration the algorithms gets closer to the θ value which minimizes the sum of squared residuals, and hence $\hat{e}^*$ increasingly approaches the final residue. In this process one can think that $\hat{\theta}$ is equal to the parametric value plus a linear combination of random variables (e), so by the limit central theorem and satisfying certain regularity conditions, $\hat{\theta}$ will present approximately normal distribution, in [84]:

$$\hat{\theta} \sim N(\theta, \sigma^2 [F(\theta)^T F(\theta)]^{-1}) \tag{7}$$

An estimate of the variance is obtained by replacing in θ by $\hat{\theta}$ in equation (7),

$$\widehat{Var}(\hat{\theta}) = \hat{\sigma}^2 \left[F(\hat{\theta})^T F(\hat{\theta}) \right]^{-1} \tag{8}$$

In which the second estimate σ^2 is:

$$\hat{\sigma}^2 = \frac{RSS(\hat{\theta})}{n-k} \tag{9}$$

Where k is the number of estimated parameters of the expectation function and n is the sample size.

These results are generalizations of those obtained in linear models, and hence, the inferential methods such as F test for comparison of corresponding models, t for testing hypotheses about the parameters, can be applied to nonlinear models. These tests are simple extensions of the applied ones to linear models that are submitted to an appropriate linear approximation. Due to this, in contrast to the linear case where the same hypothesis is inspected similarly by different procedures with the same descriptive level, in nonlinear models equivalent tests may lead to differing conclusions. For instance, the Wald test for H_0: $A\theta = d$ may not produce the same result by the F test of model reduction, in [26]. The properties of these tests depend both on the sample size and the intensity of nonlinearity of the model.

Once obtained the estimate of the parameters it is possible to establish the asymptotic standard error for expectation E(Y) at a given point x_i:

$$ase\left(f(x_i,\hat{\theta})\right) = \sqrt{\hat{\sigma}^2\,\hat{F}_i\left(\hat{F}^T\hat{F}\right)^{-1}\hat{F}_i^T} \tag{10}$$

Where is $\hat{F}_i$ is an abbreviated representation of $\hat{F}$ $(x_i; \hat{\theta})$.

An confidence interval of $(1 - \alpha)$ to covering E (Y) in a given association x_i can be obtained by:

$$f(x_i,\hat{\theta}) \pm t_{\alpha/2,n-k}ase\left(f(x_i,\hat{\theta})\right) \tag{11}$$

As discussed above, one can see that all inference procedures for nonlinear models admit supposition of adequate linear approximation and make use of asymptotic arguments.

With the method of generalized nonlinear least squares it is possible to model the heterogeneity of variance in a specification similar to that used to model the response variable average. Davidian & Giltinan, in [17] presented the following expressions for the general definition of the variance function:

$$Var(y_i) = \sigma^2 g^2(\mu_i, x_i, \delta), \mu_i = f(x_i, \theta) \tag{12}$$

The variance of the response in equation (12) is a function $g(\cdot)$, which in turn may be a function of the response average (μ) of the fixed effect of independent variables (x) and of the parameter vector (δ) associated with the variance function $g(\cdot)$. Not necessarily $g(\cdot)$ must be specified as a function of all arguments. The variance function can be represented by any continuous positive function, being the most common are the exponential function and power function:

$$g(\mu_i, x_i, \delta) = exp(\delta\mu_i)$$

$$g(\mu_i, x_i, \delta) = |\mu_i|^\delta \tag{13}$$

The process of estimating the variance function is based on generalized least squares. Following the estimative parameter δ and choice of initial values for θ, an iterative process to generates definitive values for the parameters by minimization of the pseudo-verisimilitude function with respect to θ:

$$PV\left(\theta^{(0)}, \sigma, \delta\right) = \sum_{i=1}^{n}\left(\frac{\{y_i-f(x_i,\hat{\theta}^{(0)})\}^2}{\sigma^2 g^2(f(x_i,\hat{\theta}^{(0)}),x_i,\delta)} + log\left(\sigma^2 g^2\left(f(x_i,\hat{\theta}^{(0)}),x_i,\delta\right)\right)\right) \tag{14}$$

Technically, the above minimization means the verisimilitude maximization in relation to θ. For minimization of the above expression, by iteration, it is necessary the knowledge of θ. Regardless of the variance and $g(\cdot)$ function, minimization of the equation implicates on minimization of the sum of squares errors ($\{y_i\text{-}f (x_i; \hat{\theta})\}^2$). However, the most suitable the estimated variance values and of the $g(\cdot)$ function, smaller the sum squared errors. Computationally, the algorithm employed provides the joint estimation of the θ, σ^2 and δ parameters.

The heterogeneity of variance is corrected by specifying the variance function and estimates the associated parameters. Therefore, those observations that have larger deviation have their influence on the estimation of parameters ponderated by its variance. The standard errors of the parameter estimatives at the end of the estimation procedure are considering only the variance due to residual error, free of the difference in dispersion observed for the response variable.

Based on concepts of nonlinear models mentioned above, there are applications of these in various areas of Soil Science. As plausible examples it can be mentioned nonlinear regression models to predict soil nitrogen mineralization, in [67, 68] models of potassium release from various sources of organic residue in Latosol, in [101] extraction of zinc from sewage sludge, in [95] as well as the nonlinear model of Genuchten, in [36] which is the most used worldwide to describe the soil water retention.

The soil water retention curve is a nonlinear theoretical model which relates to water content with the matric potential. This feature is specific for each soil, in [4] being that the water content held in a given Ψ_m depends on the structure, the pore distribution and bulk density in which capillary phenomena are of greater importance. However, when the adsorption phenomenon governs, it is dependent on the texture and specific surface area of the mineral particles of clay fraction, in [1, 4, 41, 70].

Its graphic representation is based on the survey of a certain number of points, usually selected arbitrarily, by plotting the abscissa axis the logarithm of matric potential (log Ψ_m) and on the ordinates axis the soil water content (U, g g^{-1}, θ, dm^3 dm^{-3}). Based on these points, a curve is delineated to represent the soil water retention characteristics.

The knowledge of the water retention curve has practical and scientific applications, including: determining the inflection point as being the field capacity, in [20, 32, 57,58] the slope of retention curve at the inflection point, in another words, obtaining the physical parameter "S", in [19] total water availability and drainable porosity, in [57], water content and pore size distribution, in [63, 50] non saturated hydraulic conductivity, in [36, 103] among others.

Several nonlinear models are used to describe the relation between water content and matric potential, in [2, 7, 8, 19, 36, 42, 45, 57, 58, 61, 82]. These empirical models continue to be used in order to adjust the soil water retention curves because it has not been developed theoretical mathematical expressions capable of adequately represent this physical-hydrical relationship. In adjusting the water retention curve is expected that the greater the number of points, the better representation of the soil water retention in [90].

At low matric potentials, the retention curve is directly influenced by the stability of the structure and, consequently, by the formation of structural pores in addition to the indirect effects of organic matter, in [64, 72]. In high matric potentials, the water retention is influenced by textural pores associated with particle size distribution and soil mineralogy, becoming the more important due to the available surface for water adsorption, in [51]. This relation between the factors mentioned above characterizes the non-increasing monotonic function, which is common to all mathematical models of the water retention curve.

For the soil physical-hydrical description, the theoretical model proposed by Genuchten, in [36] has been universally adopted and allows to relate, with high predictive power, the retention energy and the water availability, in [19]. This model is characterized by two asymptotes, related to soil water content corresponding to saturation and the residual content, and an inflection point between the plateaus, which is dependent on soil properties, being its shape and its slope regulated by empirical parameters of adjusting of the model ("α", "n" and "m"). The estimative of the water retention curve is given by fitting the tested model to the data from the undisturbed soil samples, submitted to the interval of the standard matric potential (1 at 1.500 kPa).

Despite its extensive use in relation to other available models, in [25] it does not adequately fit to soils with bimodal distribution of pores, i.e., soils with two contrasting classes of pores, classified into structural and textural pores, in [22]. As a result, modelings have been proposed which employ equations capable of identifying this distribution, in which these pore classes are quantified by means of two maximum points, obtained by derivation of the water retention curve, in [2, 19] and consequently, two inflection points.

The double exponential model proposed by Dexter et al, in [19] allows identifying the bimodal pore distribution in soils from temperate region in the matric potential interval related to the saturation water content (U_{sat}) up to the residual water content (U_{res}). On the other hand, the Alfaro Soto et al. model, in [2] identifies the bimodal pores distribution in tropical soils in a matric potential interval upper to the standard determination (1 < Ψ_m <100.000kPa) of the water retention curve.

The application of theoretical models, both in unimodal- and bimodal-pore distribution soils, provides only the description of the water content average value as a function of the matric potential and does not consider the possible correlation attributable to observed measurements in the sample at different matric potentials. In addition, these models do not consider the heterogeneity of variance, which was studied by Moraes et al, in [59] which found reduction of dispersion of the water content by increasing the matric potential.

A new model of adjustment for the water retention curve was proposed, in [10], denominated double van Genuchten (Figure 2). So, as well as other models, in [2, 19] the derivative of this model presents the bimodal density function for the pore size distribution of soil tropical, which stratifies the porosity of these soils into structural and textural pores, obtained by two inflection points which are evident from the nonlinear relation among the variables, expressed by this model, considering, however, the different matric potential interval for establishment of water retention curve (1< Ψ_m < 300.000 kPa). However, due to the higher number of parameters, the template double van Genuchten becomes more flexible.

The equation below (Figure 2) shows m=1-1/n restriction, in [61] for both curve segments, structural (m$_{str}$) and textural (m$_{tex}$). The gravimetric water content and matric potential are represented by U and Ψ, respectively. The parameters U$_{res}$, U$_{pwp}$, U$_{sat}$ represent the inferior asymptotic plateau ($\Psi \rightarrow \infty$) or asymptotic residual water content, the intermediate plateau, represent the value of water content which is slightly constant around the permanent wilting point and the upper asymptotic plateau ($\Psi \rightarrow 0$), indicates the saturation water content, respectively. The α and n parameters are associated with the scale and shape of the curve between top, middle and bottom asymptotes; α_{str} and n$_{str}$ (structural) correspond to the first segment and α_{tex} and n$_{tex}$ (texture) to the second segment of the curve. This procedure of adjusting of nonlinear models can be obtained by employing the 2.14.1software R, in [69].

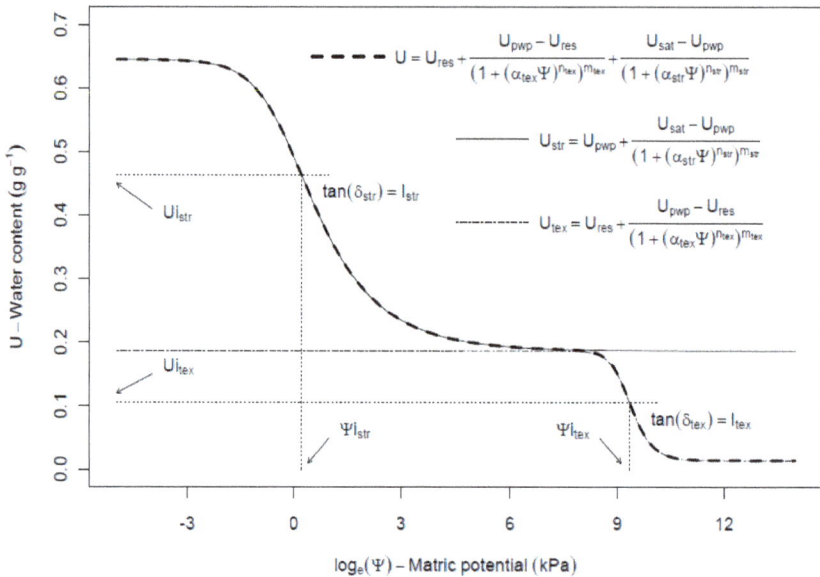

Figure 2. Proposed model for adjustment of the double van Genuchten function for retention curve, with the locations of parameters associated with the model, being matric potential (Ψ) and water content (U) estimated from the first inflection point (I$_{str}$) and the second inflection point (I$_{tex}$), in [10]

On the other hand, the water retention curve represents a cumulative distribution, thus, its derivative is proportional to the probability density function, and this function represents the distribution density of pores by size. The slopes represent the class of pore diameter that occurs most frequently. It explains why a larger quantity of water is removed when it is applied a tension corresponding to that diameter of the pores and therefore there is a great loss of water around this matric potential. The double van Genuchten model generalizes this assumption to accommodate the bimodal pore distribution, and therefore, the function has two inflection points.

5. Final remarks

A higher content of iron- and aluminum- oxides in the clay fraction of clayey Latosols (Oxisols), widely dominant soils in the plateaus of the Brazilian Cerrado Biome, currently the most demanded for sustainable grain production, is associated with the soil granular structure of these oxidic soils. This structure, when well expressed as in the B horizon of these very old Latosols, favors in the soil the existence of two distinct populations of pores: the bigger pores or structural pores (among aggregates) and the smaller pores or textural pores formed between the mineral particles (inside aggregates). This means that in these soils practically there are no pores between these two limits. This condition is also valid for sandy soils.

In this context, the model recently proposed by Carducci et al, in [10], and much detailed in this chapter adequately contemplates this bimodal distribution of pore size and functionality with respect to soil water retention in these peculiar soils in this important Brazilian Biome (one of the last agricultural frontiers in the world).

This represents a conceptual and methodological advance and an adapted modeling to the mineralogical and structural characteristics of these soils, in a region characterized by well-defined wet and dry season, with direct consequences on the water dynamics in these soils and in the environment in general.

Author details

Carla Eloize Carducci, Geraldo César de Oliveira and Nilton Curi
Department of Soil Science, Federal University of Lavras, Brazil

Eduardo da Costa Severiano
Federal Institute of Education, Science and Technology of Goiás State, Brazil

Walmes Marques Zeviani
Department of Statistics, Federal University of Paraná, Brazil

6. References

[1] Ajayi A. E, Dias Junior M. S, Curi N, Gontijo I, Araujo Junior C. F, Inda Junior A. V (2009) Relation of strength and mineralogical attributes in Brazilian Latosols. Soil and Tillage Research, Amsterdam, 102: 14-18.

[2] Alfaro-Soto M. A, Kiang C. H, Vilar O. M (2008) Avaliação do escalonamento fractal de alguns solos brasileiros. Revista Brasileira de Geociências, São Paulo, 38: 253-262.

[3] Berthouex P. M, Brown L. C (2002) Statistics for environmental engineers. 2. ed. Boca Raton: Lewis Publishers/CRC Press.

[4] Beutler A. N, Centurion J. F, Souza Z. M, Andrioli I, Roque C. G (2002). Retenção de água em dois tipos de Latossolos sob diferentes usos. Revista Brasileira de Ciência do Solo, Viçosa, MG, 26: 829-834.

[5] Bioma Cerrado/Ibge (2012) Available:
http://www.ibge.gov.br/home/presidencia/noticias/noticia_visualiza.php?id_noticia=16
9. Accessed 2012 Fev 12.

[6] Bouma J (1991) Influence of soil macroporosity on environmental quality. Adv. Agron.,
46: 2-37.

[7] Brooks R. H, Corey A. T (1966) Properties of porous media affecting fluid flow. Journal
of Irrigation and Drainage Division, Reston, 92: 61-88.

[8] Burdine N. T (1953) Relative permeability calculations from pore-size distribution data:
petroleum transportation. Bulletin of the American Institute of Mining and
Metallurgical Engineers, New York, 198: 71-77.

[9] Camargo L. A, Marques Júnior J, Pereira G. T, Horvat R. A (2008) Variabilidade espacial
de atributos mineralógicos de um Latossolo sob diferentes formas do relevo. Ii -
correlação espacial entre mineralogia e agregados. Revista Brasileira de Ciência do Solo,
Viçosa, MG, 32: 2279-2288.

[10] Carducci C. E, Oliveria G. C, Severiano E. C, Zeviani W. M (2011) Modelagem da curva
de retenção de água de Latossolos utilizando a equação duplo van Genuchten. Revista
Brasileira de Ciência do Solo,Viçosa, MG, 35: 77-86.

[11] Centurion J. F, Freddi O. S, Aratani R. G, Metzner A. F. M, Beutler A. N, Andrioli I
(2007) Influência do cultivo da cana-de-açúcar e da mineralogia da fração argila nas
propriedades físicas de latossolos vermelhos. Revista Brasileira de Ciência do Solo,
Viçosa, MG, 31: 199-209.

[12] Cichota R, Jong Van Lier Q (2004) Análise da variabilidade de pontos amostrais da
curva de retenção da água no solo. Revista Brasileira de Ciência do Solo, Viçosa, MG,
28: 585-596.

[13] Cline M. G, Buol S. W (1973) Soils of the Central Plateau of Brazil and extension of
results of field research conducted near Planaltina, Federal District, to them. Ithaca,
Cornell University. 43 p. (Agronomy Mimeo 73-13).

[14] Cooper M, Vidal-Torrado P (2005) Caracterização morfológica, micromorfológica e
físico-hídrica de solos com horizonte B Nítico. Revista Brasileira de Ciência do Solo,
Viçosa, MG, 29: 581-595.

[15] Curi N (1983) Lithosequence and toposequence of Oxisols from Goiás and Minas Gerais
states, Brazil. 158f. thesis (Ph.D.)-Purdue University, West Lafayette, Indiana, USA.

[16] Curi N, Franzmeier D (1984) Toposequence of oxisols from the central plateau of Brazil.
Soil Science of American Journal, Madison, 48: 341-346.

[17] Davidian M, Giltinan D. M (1995) Nonlinear models for repeated measurement data.
London: Chapman and Hall, 359 p.

[18] Decagon Devices (2000) Operator's manual version 1.3 WP4-T dewpointmeter.
Pullman, 70p.

[19] Dexter A. R (2004) Soil physical quality part I: theory, effects of soil texture, density, and
organic matter, and effects on root growth. Geoderma, Amsterdam, 120: 201-214.

[20] Dexter A. R, Bird N. R. A (2001) Methods for predicting the optimum and the range of
soil water contents for tillage based on the water retention curve. Soil and Tillage
Research, Amsterdam, 57: 203-212.

[21] Dexter A. R, Czyż E. A, Richard G, Reszkowska A (2008) A user-friendly water retention function that takes account of the textural and structural pore spaces in soil. Geoderma, Amsterdam, 143: 243-253.

[22] Dexter A. R, Richard G (2009) Tillage of soils in relation to their bi-modal pore size distributions. Soil and Tillage Research, Amsterdam, 103: 113-118.

[23] Dias Junior M. S (2000) Compactação do solo. In: Novais R. F, Alvarez V. V. H, Schaefer C. E Tópicos em ciência do solo. Viçosa, MG: SBCS, pp. 55-94.

[24] Dodge Y (2008) The concise encyclopedia of statistics. 2. ed. London: Springer, 616 p.

[25] Dourado Neto D, Nielsen D. R, Hopmans J. W, Reichardt K, Bacchi O. O. S (2000) Software to model soil water retention curves: SWRC, version 2.0. Science Agricola, Piracicaba, 57: 191-192.

[26] Draper N. R, Smith H (1998) Applied regression analysis. New York: J. Wiley, 706 p.

[27] Eberhardt D. N, Vendrame P. R. S, Becquer T, Guimarães M. F (2008) influência da granulometria e da mineralogia sobre a retenção do fósforo em Latossolos sob pastagens no cerrado. Revista Brasileira de Ciência do Solo, Viçosa, MG, 32: 1009-1016.

[28] Empresa Brasileira De Pesquisa Agropecuária (2006) Sistema brasileiro de classificação de solos. 2. ed. Rio de Janeiro: Ministério da Agricultura e Abastecimento, 306 p.

[29] Ferreira I. M (2003) Bioma Cerrado: um estudo das paisagens do cerrado. 81f. Tese (Doutorado em Geografia, área de concentração em Organização do Espaço), Universidade Estadual Paulista/UNESP-Campus de Rio Claro, São Paulo.

[30] Ferreira M. M, Fernades B, Curi N (1999) Mineralogia da fração argila e estrutura de Latossolos da região sudeste do brasil. Revista brasileira de Ciência do Solo, Viçosa, MG, 23: 507-514.

[31] Ferreira M. M, Fernades B, Curi N (1999) Influência da mineralogia da fração argila nas propriedades físicas de Latossolos da região sudeste do Brasil. Revista Brasileira de Ciência do Solo, Viçosa, MG, 23: 515-524.

[32] Ferreira M. M, Marcos Z. Z (1983) Estimativa da capacidade de campo de Latossolo Vermelho Roxo distrófico e Regossolo através do ponto de inflexão da curva característica de umidade. Ciência e Prática, Lavras, 7: 96–101.

[33] Fisher R. A, Mackenzie W. A (1923) Studies in crop veriation II. The manurial response of different potato varieties. Journal of Agricultural Science, Cambridge, 13: 311-320.

[34] Fontes M. P. F, Camargo O. A, Sposito G (2001) Eletroquímica das partículas coloidais e sua relação com a mineralogia de solos altamente intemperizados. Scientia Agricola, Piracicaba, 58: 627-646.

[35] Gaiser T, Graef F, Cordeiro J. C (2000) Water retention characteristics of soils with contrasting clay mineral composition in semi-arid tropical regions. Australian Journal of Soil Research, Australia, 38: 523-36.

[36] Genuchten M. T. van (1980) A closed-form equation for predicting the hydraulic conductivity of unsaturated soils. Soil Science Society American Journal, Madison, 44: 892-898.

[37] Giarola N. F. B, Silva A. P, Imhoff S (2002) Relações entre propriedades físicas e características de solos da região sul do Brasil. Revista Brasileira de Ciência do Solo, Viçosa, MG, 26: 885-893.

[38] Gomes J. B. V, Curi N, Motta P. E. F, Ker J. C, Marques J. J. G. S. M, Schulze D. G (2004) Análise de componentes principais de atributos físicos, químicos e mineralógicos de solos do bioma cerrado. Revista Brasileira de Ciência do Solo, Viçosa, MG, 28: 137-153.

[39] Gomes J. B. V, Curi N, Schulze D. G, Marques J. J. G. S. M, Ker J. C, Motta P. E. F (2004) Mineralogia, morfologia e análise microscópica de solos do bioma cerrado. Revista Brasileira de Ciência do Solo, Viçosa, MG, 28: 679-694.

[40] Gupta S. C, Larson W. E (1979) Estimating soil water retention characteristics from particle size distribution, organic matter percent, and bulk density. Water Resources Research, New York, 15: 1633-1635.

[41] Hillel D (1982) Introduction to soil physics. San Diego: Academic. 365 p.

[42] Hutson J. L, Cass A (1987) A retentivity functions for use in soil-water simulation models. Soil Science Journal, Baltimore, 38: 105-113.

[43] Iuss Working Group Wrb (2006) World Reference Base for Soil Resources. World Soil Resources Reports No. 103. FAO, Rome. 128 p.

[44] Kampf N, Curi N (2003) Argilominerais em solos brasileiros.In: Curi, N.; Marques, J. J.. G. S. M.; Guilherme, L. R. G.; Lima, J. M.; Lopes, A. S.; Venegas, V. H. A. (Org.).Tópicos em Ciência do Solo. Viçosa, MG:Sociedade Brasileira de Ciência do Solo, 3: 1-54.

[45] Kastanek F. J, Nielsen D. R (2001) Description of soil water characteristics using cubic spline interpolation. Soil Science Society of America Journal, Madison, 65: 279-283.

[46] Ker J. C (1997) Latossolos do Brasil: uma revisão. Geonomos, Belo Horizonte, 5: 17-40.

[47] Ker J. C, Resende M. (1996) Recursos edáficos dos Cerrados: ocorrência e potencial. In: Simpósio Sobre O Cerrado: biodiversidade e produção sustentável de alimentos e fibras no cerrado, 8., Planaltina, 1996. Anais. Planaltina, Empresa Brasileira de Pesquisa Agropecuária- CPAC, pp.15-19.

[48] Khuri A. I (2003) Advanced calculus with applications in statistics. 2. ed. Hoboken: J. Wiley.

[49] Kiehl E. J (1979) Manual de edafologia. São Paulo, Agronômica Ceres, 262 p.

[50] Klein V. A, Libardi P. L (2002) Densidade e distribuição do diâmetro dos poros de um Latossolo Vermelho sob diferentes sistemas de uso e manejo. Revista Brasileira de Ciência do Solo, Viçosa, MG, 26: 857-867.

[51] Larson W.E, Gupta S. C (1980) Estimating critical stress in unsaturated soils from changes in pore water pressureduring confined compression. Soil Science society of American Journal, Madison, J. 44: 1127-1132.

[52] Macedo J (1996) Os solos da região dos cerrados.In:Alvares, V. H.; Fontes, L. E. F.; Fontes, M. P. F. (Ed.). O solo nos grandes domínios morfoclimáticos do Brasil e o desenvolvimento sustentado. Viçosa, MG: SBCS/UFV, pp.135-155.

[53] Marcos Z. Z (1968) Estrutura, agregação e água do solo. 55f. (Doutorado em Solos e Nutrição de Plantas) - Escola Superior de Agricultura Luiz de Queiróz, Piracicaba.

[54] Marques J. J, Schulze D. G, Curi N, Mertzman S. A (2004) Major element geochemistry and geomorphic relationships in Brazilian Cerrado soils. Geoderma, Amsterdam, 119: 179-195.

[55] Marshall T. L (1962) The nature, development and significance of soil structure. In: Neale, G.J., ed. Trans. Of Joint Meeting Of Commissions, 4 & 5. (ISSS). Palmerston North, New Zealand, pp.243-257.

[56] Mazucheli J, Achcar J. A (2002) Algumas considerações em regressão não linear. Acta scientiarum, Maringá, 24: 1761–1770.

[57] Mello C. R, Oliveira G. C, Ferreira D. F, Lima J. M (2002) Predição da porosidade drenável e disponibilidade de água para Cambissolos da Microrregião Campos das Vertentes, MG. Pesquisa Agropecuária Brasileira, Brasília, 37: 1319-1324.

[58] Mello C. R, Oliveira G. C, Resck D. V. S, Lima J. M, Dias Junior M. S (2002) Estimativa da capacidade de campo baseada no ponto de inflexão da curva característica. Ciência e Agrotecnologia, Lavras, 26: 836-841.

[59] Moraes S. O, Libardi P. L, Reichardt K, Bacchi O. O. S (1993) Heterogeneidade dos pontos experimentais de curvas de retenção da água no solo. Scientia Agricolae, 50: 393–403.

[60] Moreira F. M. S, Siqueira J. O (2007) Microbiologia e bioquímica do solo. 2. ed. Lavras: UFLA. 626 p.

[61] Mualem Y (1976) A new model for predicting the hydraulic conductivity of unsatured porous media. Water Resource Research, New York, 12: 513-522.

[62] Ochiston H.D (1954) Adsorption of water vapor: II. clays at 25°C. Soil Science. 78: 463–480.

[63] Oliveira G. C, Dias Junior M. S, Resck D. V. S, Curi N (2004) Caracterização química e físico-hídrica de um Latossolo Vermelho após vinte anos de Manejo e cultivo do solo. Revista Brasileira de Ciência do Solo, Viçosa, MG, 28: 327-336.

[64] Oliveira G. C, Dias Junior M. S, Resck D. V. S, Curi N (2003) Alterações estruturais e comportamento compressivo de um Latossolo Vermelho Distrófico argiloso sob diferentes sistemas de uso e manejo. Pesquisa Agropecuária Brasileira, Brasília, 38: 291-299.

[65] Pedrotti A, Ferreira M. M, Curi N, Silva M. L. N, Lima J. M, Carvalho R (2003) Relação entre atributos físicos, mineralogia da fração argila e formas de alumínio no solo. Revista Brasileira de Ciência do Solo, 27: 1-9.

[66] Pereira T. T. C, Ker J. C, Schaefer C. E. G. R, Barros N. F, Neves J. C. L, Almeida C. C (2010) Gênese de Latossolos e Cambissolos desenvolvidos de Rochas Pelíticas do Grupo Bambuí – Minas Gerais. Revista Brasileira de Ciência do Solo, Viçosa, MG, 34: 1283-1295.

[67] Pereira J. M, Muniz J. A, Silva C. A (2005) Nonlinear models to predict nitrogen mineralization in an Oxisol. Scientia Agricola, 62: 548–554.

[68] Pereira J. M, Muniz J. A, Sáfadi T, Silva, C. A (2009) Comparação entre modelos para predição do nitrogênio mineralizado: uma abordagem bayesiana. Ciência e Agrotecnologia, 33: 1792–1797.

[69] R Development Core Team (2011) R: a language and environment for statistical computing. Vienna: R Foundation for Statistical Computing. Avaiable: http://www.R-project.org. Acessed: 2011 Fev. 2.

[70] Rawls W. J, Gish T. J, Brakensiek D. L (1991) Estimating soil water retention from soil physical properties and characteristics. Advencend Soil Society, New York, 16: 213-234.

[71] Reatto A (2009) Nature et propriétés de l'horizon diagnostic de Latosols du Plateau Central brésilien. 220f. Thesis (Docteur de l'Université d'Orléans: Sciences du Sol)- Université D'Orléans, Orléans, France.

[72] Reatto A, Bruand A, Martins E. S, Muller F, Silva E. M, Carvalho Junior O. A, Brossard M, Richard G (2009) Development and origin of the microgranular structure in Latosols of the Brazilian central plateau: significance of texture, mineralogy, and biological activity. Catena, Cremlingen, 76: 122-134.

[73] Reatto A, Bruand A, Silva E. M, Martins E. S, Brossard M (2007) Hydraulic properties of the diagnostic horizon of Latosols of a regional toposequencia across the Brazilian central platea. Geoderma, Amsterdam, 139: 251-259.

[74] Reatto A, Martins E. S, Spera S. T, Correia J. R (2000) Variabilidade mineralógica de Latossolos da área da Embrapa Cerrados em relação aos do bioma cerrado. Planaltina, DF: Embrapa Cerrados, 39p. (Embrapa Cerrados. Boletim de Pesquisa, 19).

[75] Resck D. V. S, Pereira J, Silva J. E (1991) Dinâmica da Matéria Orgânica na Região dos Cerrados. Planaltina: Empresa Brasileira de Pesquisas Agropecuárias. 1991. 22p. (Série documentos, 36).

[76] Resende M, Curi N, Ker J. C, Rezende S. B (2011) Mineralogia de solos Brasileiros: Interpretação e aplicações. Lavras, Editora UFLA. 201p. Il. (2ª edição revisada e ampliada).

[77] Resende M, Curi N, Rezende S. B, Corrêa G. F (2007) Pedologia: base para distinção de ambientes. 5. ed. Lavras: UFLA, 322 p.

[78] Resende M, Ker J. C, Bahia Filho A. F. C (1996) Desenvolvido sustentado do cerrado. In: Alvarez V. V. H, Fontes L. E. F, Fontes M. P. F, eds. Os solos nos grandes domínios morfoclimáticos do Brasil e o desenvolvimento sustentado. Viçosa, Sociedade Brasileira de Ciência do Solo/UFV-DPS, pp.169-199

[79] Richards L. A (1965) Physical conditions of water in soil. In: Black, C.A., ed. Methods of soil analysis. Part 1. Madison, American Society for Testing and Materials, (Agronomy, 9) 770 p.

[80] Richards L. A, Weaver L. R (1943) Fitten-atmosphere percentage as related to the permanent wilting percentage. Soil Science, Baltimore, 56: 331-339.

[81] Rodrigues A. C (2010) Germinação e reindução da tolerância à dessecação em sementes de Bauhinia forficata Link (Fabaceae). 64 f. Dissertação (Mestrado em Fisiologia Vegetal) - Universidade Federal de Lavras, Lavras.

[82] Rossi C, Nimmo J. R (1994) Modeling of soil water retention from saturation to oven dryness. Water Resources Research, New York, 30: 701-708.

[83] Schaefer C. E. G. R, Fabris J. D, Ker J. C (2008) Minerals in the clay fraction of brazilian Latosols (Oxisols): A Review. Clay Minerals, 43: 1-18.

[84] Seber G. A. F, Wild C. J (1989) Nonlinear regression. New York: J. Wiley.

[85] Severiano E. C (2010) Alterações estruturais de latossolos representativos da região do cerrado e potencial de uso de solos cultivados com cana-de-açúcar. 148f. Tese (Doutorado em Ciência do Solo) - Universidade Federal de Lavras, Lavras, Brasil.

[86] Severiano E. C, Oliveira G. C, Dias Júnior M. S, Castro M. B, Oliveira L. F. C, Costa K. A. P (2010) Compactação de solos cultivados com cana-de-açúcar: I - modelagem e quantificação da compactação adicional após as operações de colheita. Engenharia Agrícola, Jaboticabal, 30: 404-413.

[87] Severiano E. C, Oliveira G. C, Dias Júnior M. S, Castro M. B, Oliveira L. F. C, Costa K. A. P (2010) Compactação de solos cultivados com cana-de-açúcar: II - quantificação das restrições às funções edáficas do solo em decorrência da compactação prejudicial. Engenharia Agrícola, Jaboticabal, 30: 414-423.

[88] Severiano E. C, Oliveira G. C, Dias Júnior M. S, Costa K. A. P, Silva F. G, Ferreira Filho S. M (2011). Structural changes in latosols of the cerrado region: i – relationships between soil physical properties and least limiting water range. Revista Brasileira de Ciência do Solo, Viçosa, MG, 35: 773-782.

[89] Severiano E. C, Oliveira G. C, Dias Júnior M. S, Costa K. A. P, Benites V. M, Ferreira Filho S. M (2011) Structural changes in latosols of the cerrado Region: ii – soil compressive behavior and Modeling of additional compaction. Revista Brasileira de Ciência do Solo, Viçosa, MG, 35: 783-791.

[90] Silva E. M, Lima J. E. F. W, Azevedo J. A, Rodrigues L. N (2006) Valores de tensão na determinação da curva de retenção de água de solos do Cerrado. Pesquisa Agropecuária Brasileira, Brasília, 41: 323-330.

[91] Soares J. L. N, Espíndola C. R, Castro S. S (2005) Alteração física e morfológica em solos cultivados sob sistema tradicional de manejo. Revista Brasileira de Ciência do Solo, Viçosa, MG, 29: 1005-1014.

[92] Soil Survey Staff 1998. Keys to soil taxonomy, 8 ed. United States Department of Agriculture, Natural Resources Conservation Service, Washington.326 p.

[93] Sousa D. M. G, Lobato E (2004) (Ed.) Cerrado: correção do solo e adubação. Planaltina: Embrapa Cerrados, 416 p.

[94] Souza G. S (1998) Introdução aos modelos de regressão linear e não linear. Brasília: Embrapa- SPI/Emprapa-SEA. 489 p.

[95] Souza E. M, Muniz J. A, Marchi G, Guilherme L. R. G (2010) Modelagem não-linear da extração de zinco de um lodo de esgoto. Acta Scientiarum. Technology (Impresso), 32: 193–199.

[96] Van Den Berg M, Klamt E, Van Reeuwijk L. P, Sombroek W.G (1997) Pedotransfer functions for the estimation of moisture retention characteristics of Ferralsols and related soils. Geoderma, 78: 161–180.

[97] Vendrame P. R. S, Eberhardt D. N, Brito O. R, Marchão R. L, Quantin C, Becquer T (2011) Formas de ferro e alumínio e suas relações com textura, mineralogia e carbono orgânico em Latossolos do Cerrado. Semina: Ciências Agrárias, Londrina, 32: 1657-1666.

[98] Viana J. H. M, Fernandes Filho E. I, Schaefer C. E. G. R, (2004) Efeitos de ciclos de umedecimento e secagem na reorganização da estrutura microgranular de latossolos. Revista Brasileira de Ciências do Solo, Viçosa, MG, 28: 11-19.

[99] Vitorino A. C. T, Ferreira M. M, Curi N, Lima J. M, Silva M. L. N, Motta P. E. F (2003) Mineralogia, química e estabilidade de agregados do tamanho de silte de solos da região Sudeste do Brasil. Pesquisa Agropecuária Brasileira, 38: 133-141.

[100] Vollant-Tuduri N, Bruand A, Brossard M, Balbino L.C, Oliveira M. I. L, Martins E. S (2005) Mass proportion of microaggregates and bulk density in a Brazilian clayey Oxisol. Soil Science Society American Journal, Amsterdam, 69: 1559-1564.

[101] Zeviani W. M (2009) Avaliação de modelos de regressão não linear na cinética de liberação de potássio de resíduos orgânicos. 85f. (Mestrado em Estatística e Experimentação Agropecuária)-Universidade Federal de Lavras, Lavras.

[102] Weisberg S (2005) *Applied Linear Regression*. 3. Ed. Nova Iorque: John Wiley & Sons, 310 p.

[103] Wösten J. H. M, Schuren C. H. J. E, Bouma J, Stein A (1990) Functional sensitivity analysis of four methods to generate soil hydraulic functions. Soil Science Society of America Journal, Madison, 55: 832-836.

Clay Minerals Characterization, Modification and Application

Chapter 10

Vermiculite:
Structural Properties and Examples of the Use

Marta Valášková and Gražyna Simha Martynková

Additional information is available at the end of the chapter

1. Introduction

The effort to clarify the meanings of the terms 'clay', 'clays', and 'clay minerals' was the subject of the joint nomenclature committees (JNCs) of the Association Internationale pour l'Etude des Argiles (AIPEA) and the Clay Minerals Society (CMS). The JNCs have proposed the term a class of hydrated phyllosilicates forming the fine-grained fraction of rocks, sediments, and soils and have defined 'clay' as ''a naturally occurring material composed primarily of fine-grained minerals, which is generally plastic at appropriate water contents and will harden when dried or fired'' [1]. According to this definition synthetic clays and clay-like materials are not regarded as clay even though they may be fine grained, and display the attributes of plasticity and hardening on drying and firing.

For phyllosilicates, the terms ''planes'', ''sheet'' and ''layer'' refer to specific parts of the structure, with atomic arrangements that increase in thickness and they cannot be used interchangeably [2,3].

A '*plane*' can occur consisting of one or more types of atoms (e.g. a plane of Si and Al atoms, a plane of basal oxygen atoms).

A '*tetrahedral sheet*' is composed of continuous corner-sharing tetrahedra like 'octahedral sheet' is composed of the edge-sharing octahedra.

A '*layer*' contains one or more tetrahedral sheets and an octahedral sheet.

'*Interlayer material*' separates the layers and generally may consist of cations, hydrated cations, organic materials, and/or hydroxide octahedral sheets. In certain cases (e.g. talc, pyrophyllite) there is no interlayer material, and thus an empty interlayer separates the layers.

Layer charge (x) per half unit cell (p.f.u.), is the *net negative charge per layer,* expressed as a positive number (eq/(Si,Al)$_4$O$_{10}$). The net negative layer charge arises from substitution of Al^{3+} for Si^{4+} in tetrahedra and substitution for lower charge cations in octahedra and from the presence of vacancies. The negative layer charge is balanced by the positively charged interlayer material.

A *'unit structure'* is the total assembly and includes the layer and interlayer material. Therefore it is inappropriate to refer to a 'tetrahedral layer' or an 'octahedral layer', although these incorrect terms are commonly found in the literature.

Guggenheim et al [4] give the revised classification scheme for planar hydrous phyllosilicates [5].

The phyllosilicates are divided by layer type, and within the layer type, and by groups based on charge p.f.u. Further subdivisions by subgroups is based on dioctahedral or trioctahedral character, and finally by species based on chemical composition. Two types of layers, depending on the component sheets are a '1:1 layer' consisting of one tetrahedral sheet and one octahedral sheet, and a '2:1 layer' containing an octahedral sheet between two opposing tetrahedral sheets.

The structure of 2:1 phyllosilicates is composed of tetrahedral (T) and octahedral (O) sheets (Fig. 1). According to the AIPEA Nomenclature Committee [4], tetrahedral sheet is composed of continuous two-dimensional corner-sharing tetrahedra [TO$_4$]$^+$ involving three basal oxygens and the apical oxygen. The tetrahedral sheet has a composition of [T$_4$O$_{10}$]$^+$ where T = Si^{4+}, Al^{3+}, Fe^{3+}. The apical oxygens form a corner of the octahedral coordination unit around larger octahedral cations. The octahedral sheet consists of two planes of closely packed O^{2-}, OH$^-$ anions of octahedra with the central cations Mg^{2+} or Al^{3+}. The smallest structural unit contains three octahedral sites. The *trioctahedral* structures of phyllosilicates have all three sites occupied with cations (e.g. hydroxide sheet Mg$_6$(OH)$_{12}^{12-}$). The *dioctahedral* phyllosilicates have two octahedral sites occupied with cations (e.g. hydroxide sheet, Al$_4$(OH)$_{12}^{12-}$) and one site is vacant.

Smectites are 2:1 phyllosilicates with a total (negative) layer charge between 0.2 and 0.6 p.f.u.. The octahedral sheet may either be dominantly occupied by trivalent cations (dioctahedral smectites) or divalent cations (trioctahedral smectites).

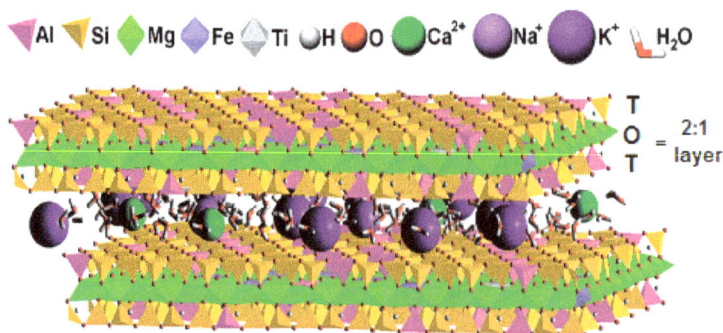

Figure 1. Model of the structure of vermiculite from Brazil; (with courtesy of dr. Jonáš Tokarský).

Vermiculites are generally trioctahedral and are termed (according to the joint nomenclature committees AIPEA and CMS) on the basis of a negative layer charge, which is between 0.6 to 0.9 p.f.u. [4,7,8]. The negative layer charge of vermiculites results from the substitution of Si^{4+} by trivalent cations in tetrahedral positions [9,10]. Vermiculites were mostly formed by removal of potassium from biotite, phlogopite or muscovite [5,6] and therefore vermiculite chemistry from this perspective is closely linked to that of mica. The thickness of the structural unit (2:1 layer and interlayer space) is about 1.4 nm, depending on the water interlamellar layers and the interlayer cations. Bailey [9] and Lagaly [10] described a method to determine the layer charge based on the measurement of basal spacings after exchange with alkylammonium cations of varying chain lengths. Alkylammonium ions in the interlayer spaces of vermiculite acquire distinct arrangements: monolayer and two-layer (bilayer) structures with the alkyl chains parallel to the surface, pseudotrimolecular layers and paraffin type structures. Vermiculites have paraffin-type interlayers if the layer charge is at least 0.75 p.f.u.

The identification of trioctahedral and dioctahedral vermiculites is based on the position of reflection (060) on their XRD patterns. Dioctahedral vermiculites show an interlayer space value $d(060)$ between 0.149 nm and 0.150 nm, while trioctahedral vermiculites have $d(060)$ between 0.151 nm and 0.153 nm [11].

The cation exchange capacity (CEC) of clay minerals is defined as "the quantity of cations available for exchange at a given pH expressed in meq/100g which is equivalent to cmol(+)/kg [13]. The CEC varies between the 120 and 200 cmol+/kg (air dried vermiculites) or 140 and 240 cmol(+)/kg (dehydrated vermiculites) [14,15]. Water molecules associated with internal surfaces evokes hydration of interlayer cations. The total amount of alkylammonium ions bound by 2:1 clay minerals is often slightly higher than the total CEC determined by other methods [16]. This is a consequence of the charge regulation at the edges. Density of the charges at the crystal edges depend on the pH of the dispersion and arises from adsorption or dissociation of protons. In an acidic medium an excess of protons creates positive edge charges, the density of which decreases with rising pH. Negative charges are produced by the dissociation of silanol and aluminol groups [17,18].

Vermiculite structures contain water interlamellar layers which are subjected to the hydration and dehydration processes [19-26]. The hydration properties are controlled by the interlayer cations Mg^{2+} and minor amounts of Ca^{2+}, Na^+, and K^+. The cation radius and charge influence the degree of hydration state in the interlayer and the stacking layer sequences [9,15,19,22,28]. The hydration state of vermiculite was defined by the number of water layers in the interlayer space. The basal space of Mg-vermiculite was declared 0.902 nm for zero-water layer, 1.150 nm for one-water layer and 1.440 nm for two water layer hydration state [29].

Ordering in vermiculite layered structure occurs when the pseudohexagonal cavities (made of six-membered tetrahedra rings) are facing each other in adjacent tetrahedral layers. In most cases the vermiculites have various possible layer-stacking sequences and therefore the regular arrangement of the layers in vermiculites occurs rarely [30-33].

Criteria for defining *the degree of regularity of alternation of different layer types* (e.g. 1:1 or 2:1 layers) were recommended in criteria for interstratification nomenclatures [4,12].

The structure of vermiculite is called semi-ordered when the transition from a layer to the next layer can be obtained in two or more different ways. For semi-ordered stacks the reciprocal space cannot be described by a set of hkl indexes but rather by modulated reciprocal hk with the variable intensity along them [15]. The regular alterations of 1.150 and 0.902 nm domains can create the regular interstratification with $d = 2.060$ nm. Collins et al. [21] studied the variability in d values of the basal spacings of vermiculite from Llano (Texas). Authors assigned interlayer values $d = 1.04$–1.03 nm to the interstratified one-zero layer hydrate and the value $d = 1.28$ nm as random interstratified phase of two- and one-layer hydrates. The X-ray diffraction patterns of the South African raw vermiculite with high potassium content in the interlayer space ($K_{0.53}$) showed the value $d = 1.45$ nm corresponding to the vermiculite and additional peaks with $d = 2.52$ nm, $d = 1.26$ and 1.205 nm, which were attributed to a biotite-vermiculite mixed-layer mineral [34]. Reichenbach and Beyer [22] evaluated two superstructures formed by a regular 1:1 interstratification. One of them with $d = 2.541$ nm as a result of the altering layers with $d = 1.376$ and 1.165 nm. Other superstructure with the value $d = 2.153$ nm was assigned to the altering layers 1.151nm and 1.002 nm. The hydrated states and interstratified phases in the vermiculites from Sta. Olalla (Spain), Paulistana (Brazil), Palabora (South Africa) and West China described Marcos et al. [25, 26]. Authors came to the founding that vermiculite from China consists of alternating mica layers ($d = 1.02$ nm, zero-water layer hydration) with vermiculite layers ($d = 1.47$ nm, two-water layer hydration) and two–one layer hydration interstratification with $d = 1.21$ nm.

The structural formula of vermiculites is often reported on the basis of the structure unit (half unit-cell content). The general formula can be written as:

$X_4 (Y_{2-3}) O_{10} (OH)_2 M \cdot n \; H_2O$,

where M is exchangeable (Mg^{2+}, Ca^{2+}, Ba^{2+}, Na^+, K^+) cations positioned in the interlayer space, that compensate negative layer charge, Y is octahedral Mg^{2+},Fe^{2+} or Fe^{3+}, Al^{3+}, and X is tetrahedral Si, Al.

The half unit cell compositions of vermiculites given in literature are listed below for comparison.

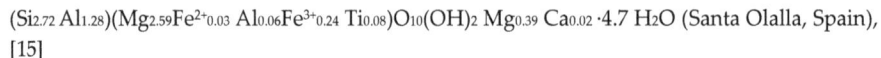

$(Si_{2.86} Al_{1.14})(Mg_{2.83}Al_{0.15}Fe^{3+}_{0.02})O_{10}(OH)_2 \; Mg_{0.41} \cdot 3.72 \; H_2O$ (Llano), [9]

$(Si_{2.72} Al_{1.28})(Mg_{2.36}Al_{0.16}Fe_{0.58})O_{10}(OH)_2 \; Mg_{0.32} \cdot 4.32 \; H_2O$ (Kenya), [19]

$(Si_{2.72} Al_{1.28})(Mg_{2.59}Fe^{2+}_{0.03} Al_{0.06}Fe^{3+}_{0.24} Ti_{0.08})O_{10}(OH)_2 \; Mg_{0.39} \; Ca_{0.02} \cdot 4.7 \; H_2O$ (Santa Olalla, Spain), [15]

$(Si_{2.64}Al_{1.36})(Mg_{2.48} Fe^{2+}_{0.04} Al_{0.14} Fe^{3+}_{0.32}Ti_{0.01}Mn_{0.01})O_{10}(OH)_2 \; Mg_{0.44}$ (Santa Olalla, Spain), [35]

$(Si_{2.69}Al_{1.31})(Mg_{2.48}Fe^{3+}_{0.324}Fe^{2+}_{0.036}Al_{0.14} Ti_{0.01} Mn_{0.01})O_{10}(OH)_2Mg_{0.39}$ (Santa Olalla, Spain), [57]

$(Si_{2.83}Al_{1.17})$ $(Mg_{2.01}Al_{0.2}Fe^{2+}_{0.16} Fe^{3+}_{0.40} Ti_{0.14})O_{10}(OH)_2Mg_{0.235}$ (Ojen, Spain), [35]

$(Si_{2.83}Al_{1.17})(Mg_{2.01}Fe^{3+}_{0.4}Fe^{2+}_{0.16}Al_{0.2}Ti_{0.12})$ $O_{10}(OH)_2Mg_{0.275}$ (Ojen, Spain), [57]

$(Si_{2.64} Al_{1.36})(Mg_{2.38}Fe^{2+}_{0.02}Al_{0.06}Fe^{3+}_{0.51}Ti_{0.03})O_{10}(OH)_2$ $Mg_{0.35} Ca_{0.01}Na_{0.01} \cdot$ 4.9 H_2O (Letovice, Czech Republic), [28].

$(Si_{3.02}Al_{0.98})$ $(Mg_{2.27}Al_{0.12}Fe^{3+}_{0.28} Fe^{2+}_{0.05}Ti_{0.07})O_{10}(OH)_2Ca_{0.09}Na_{0.21}K_{0.50}$ (West China), [36].

$(Si_{3.02}Al_{0.79}Ti_{0.05}Fe^{3+}_{0.14})(Mg_{2.50}Fe^{2+}_{0.38}Fe^{3+}_{0.09})O_{10}(OH,F)_2Ba_{0.29}K_{0.14}Ca_{0.08}$ (Palabora, South Africa), [37].

$(Si_{3.43} Al_{0.57})(Al_{0.26} Fe_{0.32} Mg_{2.34}) O_{10}(OH)_2 Ca_{0.064}Na_{0.016}K_{0.047}$ (Brazil), [38].

2. Obtaining vermiculite small particles

Natural vermiculite flakes are characterized by high values of aspect ratio. In clay science, *exfoliation* involves a degree of separation of the layers of a host structure where units, either individual layers or stacking of several layers, are dispersed (freely oriented and independent) in a solvent or polymer matrix [39]. This may be achieved by intercalation, by mechanical procedures, or by other methods. Exfoliation implies that the orientation between the layers of the host structure is lost, and that interlayer cohesive forces are overcome. If delamination or exfoliation cannot be distinguished, the terms "intercalation" or "delamination/exfoliation" should be use [40].

Delamination is a term used to describe a layer-separation process between the planar faces of adjacent layers of a particle. Delamination describes a process where intercalation occurs: guest material introduces between the layers while the stacking of layers remains. When delamination cannot be distinguished from exfoliation, the terms *"intercalation"* or *"delamination/exfoliation"* should be use to describe the process [40].

Intercalation is a general term to describe the movement of atoms, ions or molecules into a layered host structure, often a swelling clay mineral. The resulting structure is an *"intercalated structure"* [40].

Exfoliated vermiculite can be produced either by a thermal or a chemical treatment. Heating a particle of vermiculite rapidly above about 200 °C results in the transformation of the interlayer water into steam. *The thermal shock* is the best-known procedure for dissociating the macroscopic packet of vermiculite [41]. The pressure of the steam separates silicate layers to the several orders thicker than were the fundamental layers. Such exfoliated particles exhibit the "accordion" type *morphology*. The large thermal expansion of the vermiculite after explosive dehydration of interlayer water causes cracking of vermiculite flakes. This exfoliation generates a 10 to 20 times volume expansion but the basal dimensions of the particles are unchanged [42]. This exfoliation is associated with a sudden release of water molecules between the silicate layers and also the hydroxyl water gradually releasing on heating from about 500 to 850°C [44]. The water rapidly vaporized cause a disruptive effect upon the particles, in industry called exfoliation. Several authors have found that the presence of the mixed-layer vermiculite–mica or mixed-layer minerals phases containing vermiculite in different hydration states contributed to the exfoliation of vermiculites [26,45,46].

Preparation of submicron-sized vermiculites is usually accomplished by applying *wet and dry grinding* [47].

Progressive amorphization and agglomeration of vermiculite particles takes place when grinding time increases [48].

Vermiculite can be also delaminated using the mechanical shearing force [49,50].

Balek et al.[51] found that vermiculite from Santa Olalla after grinding for 2 min increased specific surface area from 1 m^2g^{-1} to 39 m^2g^{-1}. Prolongation of grinding time to 10 min led to the formation of amorphous phase and the surface area decrease to 20 m^2g^{-1}. The thermogravimetry measurement recorded different thermal behavior of original and milled vermiculite samples. Original vermiculite showed two dehydration steps after heating up to 250°C and one dehydroxylation step at 900°C. The grounded sample showed no such steps but dehydration was observed as continuous mass loss from 50 to 350°C and dehydroxylation was completed at 800°C.

Ultrasound is an alternative method for reduction of the particle size using treatment by high power. The ultrasound caused delamination and particle size reduction not only along the basal planes but also in different directions [52-55]. Péréz-Maqueda et al. [56] compared vermiculites after the grinding and ultrasonic treatments. According to authors, the sonication resulted in particle size reduction and particles retained the plate-like morphology of the original vermiculite. The prolonged grinding caused the loss of long-range order, crystallinity and agglomeration of the particles. The edges of vermiculite particles were changed and allowed easier access nitrogen during measurements and therefore yielding surface areas larger than those expected only from the particle size reduction. The ultrasound on the high-and low-charge vermiculites studied also Wiewiora et al. [35]. A high-charge vermiculite from Santa Olalla (0.88 p.f.u.) preserved the flake shape, whereas the low-charge vermiculite from Ojén (0.47 p.f.u.) displayed scrolling of the flakes into tubes. Sonication differently influenced the specific surface area and increased from the 1 m^2g^{-1} at the raw samples to 36 and 54 m^2 g^{-1} at high-charged and low-charged vermiculites, respectively.

The particle size reduction by sonication may be accompanied by a change of the redox state and the layer charge of the material. Sonication in a 1:1 mixture (volume ratio) of water and hydrogen peroxide (30% H_2O_2) is a soft method for particle size reduction of phyllosilicate minerals like vermiculites [57]. The oxidation state of the iron in the high-charge vermiculite from Santa Olalla and the low-charge vermiculite from Ojén was found different. In spite that the chemical composition of vermiculites is similar (see unit half cell composition formulas), the layer charge is different 0.78 and 0.52 p.f.u. for the Santa Olalla and Ojén vermiculite, respectively. Both of the original vermiculites had the specific surface area 1 m^2g^{-1}. The area in Ojén-vermiculite increased to 51 m^2g^{-1} after eight cycles of sonication in water and after the same numer of sonication cycles in an aqueous 15% H_2O_2 to 54 m^2g^{-1}. The surface area of the Santa Olalla vermiculite after six cycles of sonication was smaller and decreased to the 36 m^2g^{-1} after sonication in hydrogen peroxide and 38 m^2g^{-1} in water. In both vermiculites, one cycle of sonication provided by half greater value increase in the surface area in water in comparison with sonication in hydrogen peroxide. This difference disappeared after three cycles of sonication. Authors stated that delamination by sonication plays a significant role in particle size reduction and in the increase of the surface areas only after a short time. The pH during sonication in water 6.5 increased after three cycles to 7.5–8 and in hydrogen peroxide from 3.7 to 6.5. After sonication no structural change in Santa Olalla vermiculite was observed in comparison with the Ojén-vermiculite. Whereas the ratio Fe^{3+}/Fe_{total} was found almost constant in the Santa Olalla vermiculite, it increased from 0.79 to 0.85 in case of the Ojén sample together with a decrease of the layer charge.

A strong hydrogen peroxide concentration (up to 50 %) for exfoliation vermiculite was used by Obut and Girgin [58] and Üçgül and Girgin [59]. Authors interpreted the exfoliation mechanism by a hydrogen peroxide penetration into the interlayer spaces and its decomposition with evolution of atomic oxygen. These interactions disrupt the electrostatic equilibrium between the layers and the interlayer cations. Weiss et al. [60] studied exfoliation of Mg-vermiculite after thermal, microwave and/or hydrogen peroxide treatments. A stronger than hydrogen peroxide oxidation agent potassium persulfate ($K_2S_2O_8$) used Matějka et al. [61] to delaminate/exfoliate Mg-vermiculite. The vermiculite treatment with potassium persulfate solution caused a collapse of the layered structure at considerably lower molar concentration in comparison to treatment using hydrogen peroxide solution.

Kehal et al. [62] modified vermiculite from Palabora by the combination of the thermal shock (700°C), chemical exfoliation (80°C in the presence of H_2O_2) and ultrasonic treatments (20 kHz, H_2O or H_2O_2) to improve the adsorption of boron. Authors found that only 1 h treatment of vermiculite in ultrasound and 35 wt.% H_2O_2 produced small particles of drastically decreased density from 1.026 g cm^{-3} for raw vermiculite to about 0.23 g cm^{-3}. Furthermore, it was found by acido–basic titrations, that breaking of the particles by 20 kHz sonication induced the generation of OH groups on the edges of the layers which acted as active adsorption sites.

3. Vermiculite particles as carrier of silver nanoparticles

The negative surface charge that results from the ion substitution or from the site vacancies at the tetrahedral and/or octahedral sheets predetermined the use of layered 2:1 phyllosilicates as substrates for the growth of metallic nanoparticles. Uncompensated charges occur at the broken edges of the clay mineral layers, predominantly at the hydroxyl groups. The cations like Al^{3+} or Fe^{3+}, which typically occupy the octahedral positions, remain at the crystal edges, i.e. at the Lewis acid sites, where they coordinate water molecules. The exchangeable cations between the layers (at the Brønsted acid sites) compensate the negative charge and may be easily exchanged by other metal cations [63,64]. The reduction of the metal cations on the clay minerals matrix initiates the growth of nanoparticles preferably on the surface, because the interlamellar space of clay minerals limits the particle growth [65-70].

The metal nanoparticles are investigated in the function of biosensors, label for cell and cancer therapeutics. The silver nanoparticles have shown to be a promising antiviral material. The strong toxicity of several silver compounds to a wide range of microorganisms is well known. Silver acts as catalyst of the oxidation of the microorganisms, which leads to the disruption of electron transfer in bacteria [71,72]. Silver offers sufficiently small repulsion to oxygen, thus only a small amount of thermal energy is required to move the atomic oxygen readily through its crystal lattice [73]. Monovalent silver ions have a high affinity to sulfhydryl (–S–H) groups in bacteria cells. The resulting stable –S–Ag bonds inhibit the hydrogen transfer and prevent the respiration and the electron transfer. After exposing silver containing molecular oxygen to aqueous media, the molecular oxygen reacts quickly with sulfhydryl (–S–H) groups on the surface of bacteria and replaces hydrogen [74]. Silver nanoparticles exhibit cytoprotectivity towards HIV-1 infected cells [75]. The biological distribution as well as the potential toxicity of silver nanoparticles on the montmorillonite substrate was studied using Swiss mice [76]. The animal study demonstrated that the Ag-montmorillonite was nontoxic, showed no immune response, exhibited increased blood half-life and neurotransmission. Based on the study authors noted that Ag-montmorillonite enables diverse applications in life sciences such as drug development, protein detection and gene delivery for any organs, lungs and brain in particular.

The catalytic activity of silver particles depends on their size, shape and the size distribution as well as chemical–physical environment [77, 78].

Silver nanoparticles, size about 2 nm, were usually prepared by chemical reduction from silver nitrate under ultraviolet irradiation [79]. Another method employed reduction with sodium borohydride [80]. During this reaction, irregular silver nanoparticles with the size between 1.5 and 2.5 nm nucleate first. This rapid nucleation was followed by the growth of these nuclei into aggregates of circular particles with the size of approximately 12–16 nm. Several approaches have been investigated to prepare silver nanoparticles with well-defined size and morphology. The available free-network spaces between hydrogel networks helped to grow and to stabilize the nanoparticles [81-86].

The antibacterial activity of silver-montmorillonite and copper-montmorillonite studied Magaňa et al. [87]. The authors stated that the overall antibacterial effect was related to the surface characteristics of the sample and to the quantity of silver. Similarly, the antibacterial behaviour of silver grown attached to copper-palygorskite characterized Zhao et al. [88]. Valášková et al. [89] have prepared and characterized Ag nanoparticles on vermiculite and compared them with Ag nanoparticles on montmorillonite. The antibacterial activity of both nanocomposite types was tested on two bacteria strains. The Gram negative (G-) strain was represented with two bacteria *Klebsiella pneumoniae* (*K. pneumoniae*, 9CCM 4415) and *Pseudomonas aeruginosa* (*P. aeruginosa*, CCM 1960), and the Gram positive (G+) strain using two bacteria *Staphylococcus aureus* (*S. aureus*, CCM 3953) and *Enterococcus faecalis* (*E. faecalis*,

CCM 4224). *K. pneumoniae* resides in the normal flora of the mouth, skin and intestines. *P. aeruginosa* is the most common pathogen isolated from patients, who have been hospitalized longer than one week. *S. aureus* is a constituent of the skin flora that is frequently found in the nose and on the skin. *E. faecalis* inhabits the gastrointestinal tracts of humans and other mammals.

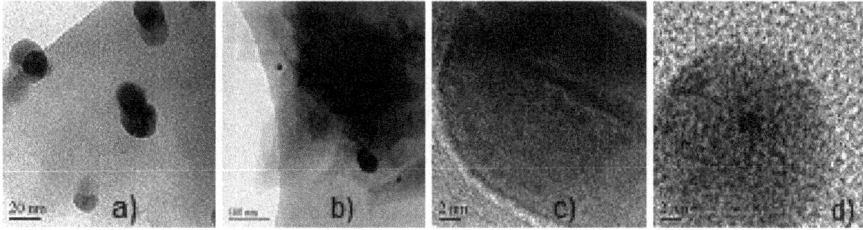

Figure 2. TEM images of silver nanoparticles on montmorillonite (a) and vermiculite substrates (b). HRTEM images of nanoparticles grew together (c) and microstructure of the Ag nanoparticles (d); (with courtesy of dr. Valter Klemm)

Nanocrystalline particles of silver reduced from the solution of silver nitrate on the clay mineral substrates were characterized using TEM and HRTEM. Transmission electron microscopy (TEM) revealed silver nanoparticles grown on the surface of both clay minerals of the mean size between 40 and 50 nm. Small Ag particles were substantially smaller than 20 nm. Furthermore, TEM/HRTEM found essential differences in the size distribution of the Ag particles grown on the surface of the montmorillonite and on the surface of the vermiculite. On montmorillonite, all Ag particles grew with a similar size and were well distributed on the surface (Fig. 2a). Huge Ag particles with the size much larger than 50 nm were observed only on the edges of the montmorillonite flakes. The size of the Ag particles grown on the surface of vermiculite was very heterogeneous (Fig. 2b). Some of nanoparticles were agglomerated or grew together (Fig. 2c). A lot of small particles contained smaller domains with high defect density. Microstructure of the Ag nanoparticles showed crystallites disorientation with a lot of defects, especially different planar defects (Fig. 2d).

Based on the results of the microstructure analysis of silver nanoparticles and clay mineral substrates authors [36, 70] assumed the manner of growth of nanoparticles on the surface of vermiculite and montmorillonite.

The particle growth process began after the previous docking of silver (precipitated from the solution of AgNO3) on the clay mineral surface. During their further growth, the Ag nanoparticles separated the clay mineral crystallites into several domains. Single nanocrystalline particles of Ag grow next to each other and then growing to larger nanoparticles with many microstructure defects. Still, the atomic ordering of the Ag nanoparticles at the silver/clay mineral interface is controlled by the local orientation of the clay mineral matrix. During the further growth, larger Ag nanoparticles re-crystallize by reducing the energy of their internal defects through the rearrangement and through the conservation or formation of the low-energy boundaries (as it was observed by HRTEM).

Differences between silver nanoparticles on the vermiculite and montmorillonite substrates authors confronted with the different sources of negative layer charge. When at dioctahedral montmorillonite negative layer charge results from an octahedral substitution of magnesium for iron and the substitution of Si^{4+} in tetrahedra is negligible [90] the trioctahedral vermiculite has a negative charge resulting mainly from the substitution of trivalent cations for Si^{4+} in tetrahedra [9]. Further it was found that montmorillonite and vermiculite after treatment with the silver nitrate solution released the Na^+, K^+, Ca^{2+} and Mg^{2+} cations from the interlayer space of both clay minerals and from the octahedral positions in the montmorillonite. Such structural change allowed the Ag^+ to be incorporated in the host structure of the clay minerals through their surface and edges. The total amount of silver on vermiculite was higher than on the montmorillonite. The size distribution of the Ag nanoparticles was much more homogeneous on the montmorillonite substrate than on the vermiculite. The lower negative layer charge of montmorillonite results from the substitution of cations, which are located at the octahedral positions, is responsible for a uniform size of the Ag nanoparticles. The Ag particles with the size much larger than 50 nm were only on the edges. The negative layer charge of vermiculite as a consequence of the charge on tetrahedra can hold higher content of silver than montmorillonite.

Both sample series of silver/vermiculites and silver/montmorillonites, showed good inhibitory action against the Gram negative bacteria strains. The higher content of silver reduced on vermiculite predestines silver/vermiculite to be stronger antibacterial agent than silver/montmorillonite.

4. Vermiculite nanofillers to polymer/clay nanocomposites

Nanocomposites have at least one ultrafine phase dimension, typically in the range of 1–100 nm, and exhibit other properties with comparison to the micro- and macro-composites. The clay–polymer nanocomposite could be considered as "one-nano-dimensional" because the clay filler has one dimension at the nanometer scale [91]. The high aspect ratio of layered silicate nanoparticles is ideal to modify the properties of the polymer, but the hydrophilic nature of silicate surfaces impedes their homogeneous dispersion in the organic polymer phase [92]. On the other hand it was found that polymers can interact with the external surface of clay minerals and also penetrate into the nano- structural spaces, holes and tunnels [93].

The polymer nanocomposites exhibit new and sometimes improved properties that are not displayed by the individual phases or by their conventional composite counterparts. Significant improvement of mechanical properties, thermal stability, resistance to solvent swelling and suppression of flammability have been achieved with the only up to 5 wt.% exfoliated layered silicates nanoparticles in polymer matrix [94-97]. The development of clay–polymer nanocomposites at low clay loadings (3–7 wt %) as advanced structural materials brings significant improvements in mechanical strength and stiffness, enhanced gas barrier behavior, reduced linear thermal expansion coefficients, and increased solvent resistance in comparison with pristine polymers.

The hydrophilic clay minerals are commonly treated with ammonium cations with long alkyl chains to improve the compatibility between the silicate layers and the polymer matrix. The methods used for the preparation of clay–polymer nanocomposites, include solvent intercalation, in situ polymerization, and melt-compounding [98-102]. Both solvent intercalation and in situ polymerization allow polymer chains to enter into the galleries of silicate clays. Melt blending is more attractive because the nanocomposites can be processed by conventional methods such as extrusion and injection molding. The influences of the addition of various types of clays on the non-isothermal crystallization process of thermoplastics have been studied by several researchers. Fornes and Paul [103] found that depending on the dispersion of clays in thermoplastic matrices the clay nanofillers can either promote or retard the crystallization of polymers and observed that the degree of crystallinity of nanocomposites showed a strong dependence on the cooling rates.

Polyethylene (PE) is one of the most widely used polyolefins. The molecular structure is generally simply written by the formula $(CH_2)n$, where n is very large. The PE specimen may contain chains of different lengths [104]. Low-density PE (LDPE) contains many statistically placed paraffinic branches. The defects in the arrangement of chains reduce the structure crystallinity [105]. When mixing PE with hydrophilic clay minerals the additives can play a role of a polymeric surfactant, or they may act as a compatibilizer when mixed with organophilized clay minerals. The polar additives or charge of carriers introduce dipole moments in PE [106-109].

Clay mineral vermiculite (VER) was used as the clay mineral nanofiller into polymer matrix and was intercalated with the maleic anhydride into polyamide [110], polyethylene [109], and polypropylene [89, 111]. It should be noted that nanocomposites of polypropylene with nanofiller of VER were prepared without any compatibilizer by solid-state shear compounding (S3C) using pan-mill equipment [112]. Vermiculite particles were only partly intercalated with polypropylene and exfoliated in PP matrix (Shao et al., 2006).

Organovermiculite nanofiller octadecylamine/vermiculite (ODA/VER) was exfoliated in polypropylene (PP) [89]. The organovermiculite nanofillers were prepared in three ways. Sample 1 was original VER (Fig. 3a) intercalated using melt intercalation with octadecylamine (ODA). Sample 2 was VER milled in jet mill and intercalated with ODA (Fig. 3b). Sample 3 was VER exfoliated using an oxidizing agent potassium persulfate [61] and subsequently intercalated with ODA.

The thermal compounding of the organovermiculites (4 wt. %) with maleated PP caused the partial deintercalation of organovermiculites. As the jet milled original VER flakes have corrugated edges [113] in PP remained intercalated only in their central parts. The frayed edges and small particles showed somewhat higher interlayer distances and the penetration of polymeric segments from the softened polymer matrix was possible. The sample 2 was very well dispersed within PP matrix. The layered vermiculite structure after exfoliation with potassium persulfate was destroyed [61] and therefore intercalation of ODA molecules into the interlayer was limited (sample 3). On the X-ray diffraction patterns of composite sample 2 in PP authors identified new reflection of the orthorhombic γ-PP in addition to the α-PP. According to the literature it is known that the formation of γ-PP in α-PP takes place when the VER particles reduce the PP chain mobility within the narrow space surrounded by the dispersed clay mineral particles [114].

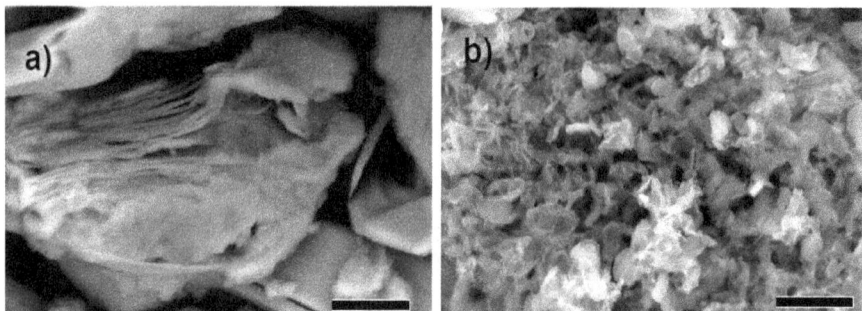

Figure 3. SEM images of original particle VER (a) and jet milled VER intercalated with ODA (sample 2). Bar shows 10 μm.

Thermal stability of PP and PP/VER nanocomposites was evaluated by thermogravimetric analysis (Fig. 4).

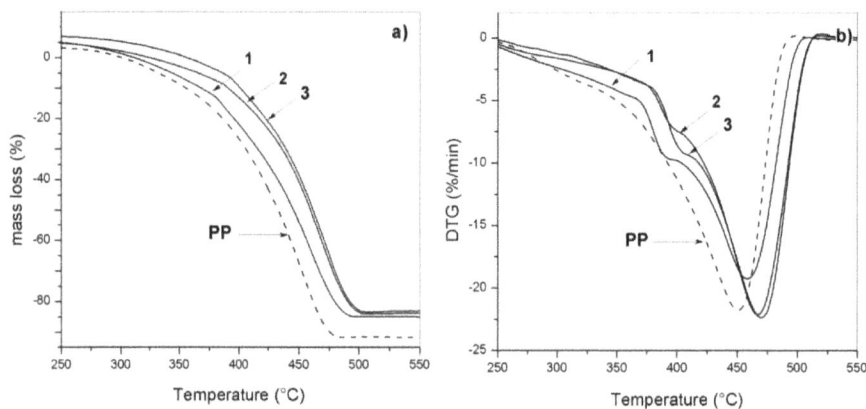

Figure 4. TG and DTA of ODA/VER nanofillers in PP.

The thermal resistance of the nanocomposites was evaluated by comparing temperatures at the certain weight loss points. Decomposition of pure PP started at 247°C. The degradation point characterized by onset temperature followed the sequence: sample 3 (286°C) → sample 2 (273°C) → sample 1 (267°C).

At 5 % mass loss, the dispersed exfoliated organovermiculite nanofillers sample 2 showed about 51°C, and sample 3 (weekly exfoliated in PP) even about 65°C higher thermal resistance than pure PP.

This agrees with observations of Gilman [112], who reported that intercalated clay mineral particles in polymer matrix make composite a much more resistant to temperatures than exfoliated particles.

Nanoclays are clay minerals optimized for use in clay nanocomposites– multi-functional material systems with several property enhancements targeted for a particular application. The polymer–clay nanocomposites have been reviewed in several literature sources [112-118].

Depending on the structure of dispersed clay-filler in the polymeric matrix, the composites can be classified as intercalated or exfoliated nanocomposites. Polymer nanocomposites can be prepared by two main processes, that is, in situ polymerization and melt compounding.

Use of nano-sized filler particles to form polymer composites has attracted much attention in recent years because of the potential performance advantages that could create new technological opportunities. Potential benefits include increased mechanical strength, decreased gas permeability, superior flame-resistance, and even enhanced transparency when dispersed nanoclay plates suppress polymer crystallization [119-123].

The key issue is to obtain an effective dispersion and exfoliation of the platelets into the polymer matrix to yield well-aligned, high-aspect ratio particles for mechanical reinforcement or a tortuous diffusion pathway for improved barrier properties [124]. In situ polymerization process consists of intercalation of monomer as precursor species, followed by their polymerization inside the interlayer of clay mineral. Numerous nanocomposite materials were prepared by method with clay having inorganic metallic interlayer cation, which can promote formation of monomer radical inducing its polymerization, or modifying agent for nonpolar polymers. Both thermosets and thermoplastics have been incorporated into nanocomposites.

Simulation techniques become an integral part of experimental techniques, since the information about the spatial arrangement of molecules within the interlayer is hard to obtain without the aid of computer simulation. It offers a range of modeling and simulation methods, covering the length range from the subatomic quantum scale, through the molecular level, to the micrometer scale. Methods to model the behavior of systems on each of these scales are combined with analytical instrument simulation and statistical correlation techniques, allowing detailed study of structure, properties, and processes. Molecular modeling using empirical force field represents a way to preliminary estimation of the host-guest complementarity and prediction of structure and properties for nanomaterial design in molecular nanotechnology.

5. Industrial vermiculite nanocomposites

The group of vermiculite–polymeric nanocomposites, where vermiculite or its silicate layers in the final composite stage is altering properties of full system, represents economically and technologically unpretentious materials. Vermiculite here enters the composite system unmodified; it means that it keeps its original characteristics. Usually, polymer and/or layered silicate nanocomposites are synthesized from polymeric macromolecules directly or from monomers followed by in situ polymerization to intercalate into and then enlarge the interlayer of the used clay.

The struggle to prepare nanocomposites is affected by many factors such as the blending or reaction time, temperature and the affinity between the host and the matrix. Moreover, the intercalated nanocomposites instead of the exfoliated nanocomposites are commonly gained in most cases. Generally, dispersion as well as delamination of vermiculite in polymer matrices is observed using electron microscopy and XRD techniques.

Three main approaches have been used in the preparation of clay–polymer composites: (1) melt blending, (2) solution blending, and (3) in situ polymerization. However, homogeneity at nanoscale level may not be fully achieved using these methods. One way of approaching this problem is complete delamination of the clay particles to give colloidal dispersions of single layers in a suitable solvent and restacking the layers in presence of guest species.

Utilization of vermiculite in nanocomposite is not as extensive as for montmorillonite. Most of the applications limits low swell ability of vermiculite compare to montmorillonite. However, several main streams of applications were introduced.

Superabsorbents are a type of loosely crosslinked hydrophilic polymer that can swell, absorb, and retain a large volume of aqueous or other biological fluid. The superabsorbents have found extensive applications in many fields such as agriculture, hygienic products, wastewater treatment, drug-delivery systems, etc.

The conventional superabsorbents are based on expensive fully petroleum-based polymers. Their production consumes lots of petroleum and their usage can also cause a nonnegligible environment problem. New types of superabsorbents by introducing naturally available raw materials as additives were desired. The incorporation of clays reduces production cost, and also improves the properties (such as swelling ability, gel strength, mechanical, and thermal stability) of superabsorbents and accelerates the generation of new materials for special applications. The properties of traditional superabsorbent could be enhanced by incorporating vermiculite [125]. It is expected that organomodification of vermiculite can further improve dispersion and performance of the resultant nanocomposite. The superabsorbent nanocomposites were prepared from natural guar gum and organovermiculite by solution polymerization and analyses indicated that organovermiculite was exfoliated during polymerization and uniformly dispersed in the polymeric matrix [126].

In the superabsorbent field recently, much attention has been paid especially to layered silicate as favorable compound of absorbent composites. *Polyacrylate* (PAA)–*vermiculite superabsorbent* composites were tested for their absorption abilities. Acrylamide and acrylic acid mixed with vermiculite were polymerized. Final nanopowders had the particle size in range 40–80 mesh. Nanocomposite was immersed in examined medium and let to reach swelling equilibrium for 4 h, which resulted in absorption of water into network of composite and the formation of hydrogel. Swelling behavior of the superabsorbent composites of various cations from salt solutions (NaCl, $CaCl_2$, and $FeCl_3$), anions salt solutions (NaCl, Na₂SO₄, and Na₃PO₄), and pH solutions were observed. Equilibrium water absorbency for the PAA–vermiculite superabsorbent composite was significantly affected by the content of vermiculite. The highest water absorbency was obtained when 20 wt.% of vermiculite was incorporated. Water absorbency of these composite materials was also significantly dependent on properties of external saline solutions, including valence of ions, ionic strength, and pH values. FTIR indicated that the reactions occur between –COO⁻ groups and –OH groups on the surface of vermiculite. SEM and TEM studies illustrated more finely dispersion of the clay particle in the polymer matrix. In addition, XRD analysis showed that the polymerization reaction is performed on the surface of vermiculite and the d-space is not changed. TGA indicated that introduction of vermiculite into the polymer network leads to an increase in thermal stability of the composites. The reaction between organics and clay is related to the structure and properties of clay. The equilibrium water absorbency decreases with increasing clay content owing to the increasing crosslinking points and the decreasing percentage of hydrophilic groups in polymeric network. The vermiculite based nanocomposite acquired high- water absorbency in CaCl₂ aqueous solution [133].

Simple mixing technique of vermiculite into polymer matrix could be applied in case of preparation of *membranes*. Several minerals, including vermiculite 10–100 μm particle size, were tested for having desirable features to be used in membrane. Many membranes are designed to be selectively permeable, commonly intended for use in membrane separations or to be impermeable barriers to protect a surface or a product. Paint, food wrap, and electronic packaging are examples. For these barrier membranes, we seek polymers through which solutes such as water, oxygen, and chloride permeate slowly, rather than rapidly. Developing barrier membranes containing aligned mineral flakes could be solution for those requirements. Since mineral flakes are generally crystalline, they have very low permeability, and can reduce the permeability of the composite film. The permeability of polymer films can be reduced dramatically with many layers of thin mineral flakes aligned parallel to the film's surface.

As the *gas barrier membrane* could be utilized of *butyl rubber–vermiculite nanocomposite* coating in form of thing layer. The coating formulation consisting of a butyl rubber (polyisobutylene (C₄H₈)ₙ) latex, which is well known among elastomers for its superior gas barrier characteristics, to which dispersion of exfoliated vermiculite was added. The gas permeability and diffusion coefficients of the nanocoating compare to pure polymer were altered remarkably by the presence of the high loadings of vermiculite. The gas permeability was reduced by 20–30-fold by the vermiculite. Diffusion coefficients computed from time lag data were reduced by two orders of magnitude. Solubility coefficient obtained from the time lag observation increased significantly with vermiculite content in contradiction to that expected by theory (the solubility coefficient should decrease with filler loading since the volume of available polymer is decreased). The excess sorption appears to be the effect of gas adsorption by the vermiculite [127].

Mixing dilution of *polyvinyl alcohol (PVA) (C₂H₄O)ₙ with vermiculite suspension* both in water was prepared membrane – precursor, which was later cast on block heated to 75°C. For a film with a volume fraction ϕ of flakes of aspect ratio α, the permeability reduction is expected to be proportional to $\alpha\phi$ for the dilute limit ($\phi \ll 1$ and $\alpha\phi < 1$) but proportional to $(\alpha\phi)^2$ in the semidilute limit ($\phi \ll 1$ but $\alpha\phi > 1$). Permeability of hydrochloric acid and sodium hydroxide across films containing mica or vermiculite in PVA agree with the second, semidilute prediction. These improvements in barrier properties are independent on the flake size, permeate, and polymer chemistry [128].

However, it is very difficult to prepare an entirely exfoliated nanocomposite using natural vermiculite as nanofiller by conventional techniques. Therefore, vermiculite has been successfully delaminated with acid treatment and used to mix with engineering polymer directly to synthesize polymer–vermiculite nanocomposites [129, 130]. Acid delaminated vermiculite was successfully utilized for preparation of PVA nanocomposite. The properties of composite significantly depend on the preparation procedure for the reason that chemical reactions and physical interactions involved. Two steps preparation, delamination with hydrochloric acid and then addition to the PVA solution using various mixing times. The positive effect of the vermiculite content on the thermal behavior of the PVA–vermiculite blends was observed [131].

The utility of the *PVP–vermiculite nanocomposite materials* can be found in area of *wastewater cleaning* as the remover of color from dye wastewater, since vermiculite is an excellent adsorbent for basic blue dye, requiring moderately short contact times. The phenols and iodine form a molecular complex with pyridine, as well as with PVP. Thus, common impurities from the industrial waste stream such as iodine, weak acids, and phenol can be removed using the PVP–vermiculite nanocomposite. It can also be used as a reusable mild acid scavenger, which can be easily removed from the reaction medium by filtration. The cost of vermiculite is 10% of activated carbon and it is possible to regenerate it by simple heating, so it offers an attractive alternative for waste removal and recovery.

Nanocomposites poly(4-vinylpyridine(VP)) (PVP) and poly(N-vinyl-2-pyrrolidinone(NVP)) (PNVP) both combined with vermiculite have been synthesized by the intercalative redox polymerization of monomer in the gallery of Cu^{2+} ion-exchanged vermiculite (Cu^{2+} serves as polymerization agent). The formation of a single filament of the PVP polymer in the vermiculite gallery is confirmed by the increase in gallery spacing of 0.47 nm as indicated by XRD analysis. XRD analysis following intercalative polymerization of PNVP indicates the presence of two prominent peaks with the corresponding basal spacing $d_{002} = 1.43$ nm (intercalated) and 0.99 nm (not intercalated), suggesting the formation of a partially intercalated hybrid material. The amount of polymers present in the gallery is found to be around 20 wt.% by TG analysis. Presence of the polymer in vermiculite gallery results in enhanced thermal stability that is evident from the increase of initial decomposition temperature by 300°C. Differential scanning calorimetry of the nanocomposite indicates that the polymer is confined to a restricted geometry because of the absence of a glass-transition temperature, which confirms the XRD finding. The IR absorption peaks corresponding to PVP and the expected PVP of UV $\pi-\pi^*$ transition at 275 nm, along with the XRD, and thermal data confirms that the gallery expansion is owing to the PVP filament [132].

The nanocomposites consisting of the emeraldine salt of polyaniline (PANI) and layered vermiculite were synthesized to improve *thermal stability of nanocomposite* [134]. Emeraldine salt (base) is regarded as the most useful form of PANI owing to its high stability at room temperature. *PANI–vermiculite nanocomposites* were prepared via in situ polymerization of monomer compounded with vermiculite. Vermiculite silicate layers could be highly dispersed within PANI bulk thanks to pretreatment of vermiculite with hydrochloric acid. The introduction of nanolayers of vermiculite improves greatly the thermal stability of nanocomposite, but its electrical conductivity decrease slightly only, as demonstrated by TGA and electrical conductivity measurements.

An increased understanding, on the micro- and nanolevel, of polymers at surfaces and in confined geometries assists in the development and improvement of new technologies. Of special interest is to understand the nature of the glass transition – not fully solved problem of condensed matter physics. By confining molecules in very small spaces (e.g., clay gallery), the existence or not of a length scale associated with molecular motions responsible for the glass transition can be established. The increment of its relaxation strength in the clay has been related to the 2D geometry of the confinement. These polymers have a dipole moment component parallel to the chain and, therefore, the total dipole vector is proportional to the end-to-end vector and so the overall chain dynamics can be measured by dielectric spectroscopy.

Vermiculite was selected for study the molecular dynamics of oligomeric poly(propylene glycol (PG)) (PPG) liquids (M_w = 1200, 2000, and 4000 g mol^{-1}). The thickness of the liquid layers was 0.55 nm in the case of PG and 0.37 nm for 7-PG and PPG. The PG oligomers form a flat monolayer in vermiculite gallery with their methyl groups pointing in the direction of the clay surfaces. In contrast, the monomers are probable orientated either perpendicular to the clay layers or parallel to the layers, but with the methyl groups pointing toward a clay surface. The dynamics generally become slower with increasing chain length but the dynamics of the 7-mers was established to be faster than for the single monomers. A possible explanation for the fact may be that the OH end groups of the monomers are linked together forming a network, and, as a result, slow down the diffusion at temperatures low enough to keep the network structure intact [135]. The PPG confined in Na-vermiculite has been studied by broadband dielectric spectroscopy. In addition to the temperature dependence of the main (α-) relaxation process and the related high-temperature translational or segmental diffusion, the normal mode relaxation process was studied for all samples in both bulk and confinement. For the normal mode process the relaxation rate and the temperature dependence of the relaxation time in the clay is drastically shifted to lower frequencies compared to that of the bulk material. The α-process relaxation time is only slightly affected by the confinement.

The fact that interactions of PPG with the clay surfaces are very weak was implied based on similar temperature dependence of the relaxation time the α-relaxation in both bulk and confinement [136]. Relaxation process corresponds to the molecular motions of translational character and that it is almost unaffected by the present true 2D confinement, in contrast to the dielectrically active normal mode of PPG which is substantially slower in the confinement. Thus, there is no indication, for none of the confined liquids, that the OH-end groups should form strong hydrogen bonds to the clay surfaces. In fact, the rather small effect of the present confinement on the diffusive dynamics and the main relaxation time suggest that the surface interactions are considerably weaker than in many other model systems [135].

Nanocomposites based on magnetic nanoparticles supported within thermally-expanded vermiculite are extremely interesting for many applications, including its use in removing oil from water after oil spills. Two distinct magnetic nanocomposites based on thermally-expanded vermiculite pellets were treated with two different magnetic fluid samples. One of them was an ionic magnetic fluid (IMF), and the second a surfaced magnetic fluid (SMF) was consisting of oleic acid-coated nanoparticles suspended in organic medium. Magnetic fluids (MFs) are highly stable colloids consisting of nanosized magnetic particles (mainly iron oxides) suspended in a hydrophilic or hydrophobic liquid carrier. Besides numerous industrial and biomedical applications MFs can be successfully used to introduce nanosized magnetic particles into a variety of hosting structures and templates [137].

Recently, *layered silicates have been used to modify bitumen*. It has been found that physical properties, rheological behaviors and aging resistance of bitumen and polymer modified bitumen could be obviously improved due to barrier properties of MMT. Expanded vermiculite (EVMT) clays modified bitumen was prepared by using EVMT, cetyltrimethyl ammonium bromide (CTAB)–EVMT and octadecyl dimethyl benzyl ammonium chloride (ODBA)–EVMT as modifiers. The morphology characteristic of the modified bitumen was investigated by X-ray diffraction (XRD). The binders were aged by pressure aging vessel (PAV), in situ thermal aging and ultraviolet (UV) radiation. The phase-separated structure was observed in EVMT modified bitumen, while CTAB–EVMT and ODBA–EVMT modified bitumen formed the intercalated and exfoliated nanostructures, respectively [138].

In situ ring-opening polymerization of the *cyclic oligomers in the gallery of vermiculite* can form high performance polymer nanocomposites, in the absence of any catalyst and without any volatile by-products formed. The new approach allows to intercalate the silicate layers with cyclic(arylene disulfide) oligomers via direct melt compounding, and in turn to prepare in situ the corresponding nanocomposites. The cyclic arylene disulfide oligomer acts both as a precursor of polymeric matrix and as a swelling agent of the vermiculite layers. High molecular weight polymer can be formed in a few minutes [139].

A conducting polymer/inorganic host hybrid nanocomposite was created from the conductive polypyrrole (PPy) self-assembled monolayer (SAM) coated on expanded vermiculite (VMT) [140]. The conducting polymer/inorganic host hybrid composites provide the new synergistic properties, which cannot attain from individual materials. Their conductivity is more easily controlled, and the mechanical or thermal stability is improved through the synthesis of the composites. resulting in PPy/VMT nanocomposites after VMT particles surface modification.

X-ray diffraction (XRD) analysis confirmed that the main peaks of PPy/VMT nanocomposites are similar to the SAM–VMT particles, which reveal that the crystal structure of SAM–VMT is well maintained after the coating process under polymerization conditions and exhibit semi-crystalline behavior. Thermogravimetric analysis showed that the thermal stability of PPy/VMT nanocomposites was enhanced and these can be attributed to the retardation effect of amine-functionalized VMT as barriers for the degradation of PPy. The morphology of PPy/VMT nanocomposites showed the layered structure and encapsulated morphology. The composites possess high electrical conductivity at room temperature, weakly temperature dependence of the conductivity.

Intercalated nanocomposites comprised of poly(propylene carbonate) (PPC) and organo-vermiculite (OVMT) was *first prepared* via *direct melt compounding of the alkali-vermiculite intercalated host with PPC in a twin rotary mixer.* The dispersion and morphologies of OVMT within PPC were investigated by X-ray diffraction and transmission electron microscopic techniques. The results revealed the formation of intercalated-exfoliated vermiculite sheets in the PPC matrix. Because of the thermally sensitive nature of PPC, thermal degradation occurred during the melt compounding. The degradation led to a deterioration of the mechanical properties of the nanocomposites. Tensile test showed that the yield strength and modulus of the nanocomposites decrease with increasing vermiculite content [141].

6. Polymeric bio-related nanocomposites

Polymeric bio-related nanocomposites can be classified into two classes: (1) natural-based materials, including polysaccharides (starch, alginate, chitin/chitosan, hyaluronic acid derivatives) or proteins (soy, collagen, fibrin gels, silk); (2) synthetic polymers, such as poly(lactic acid) (PLA), poly(glycolic acid) (PGA), poly(3-caprolactone) (PCL), poly(hydroxyl butyrate) (PHB) [142].

Polymer	Type	Utilization
Poly(lactic acid) (PLA)	aliphatic polyester	Fracture fixation, interference screws, suture anchors, meniscus repair
poly(glycolic acid) (PLG)	aliphatic polyester	non-woven fibrous fabrics;suture anchors, meniscus repair, medical devices, drug delivery
poly(ε-caprolactone) (PCL)	aliphatic polyester	Suture coating, dental orthopaedic implants, bone tissue implantant
poly(hydroxylbutyrate) (PHB)	Polyester	material for waste management strategies; biocompatibility in the medical devices

Table 1. Biopolymers for composites

Many advantages and disadvantages characterize these two different classes of biomaterials. Synthetic polymers have relatively good mechanical strength and their shape and degradation rate can be easily modified, but their surfaces are hydrophobic and lack of cell-recognition signals. Naturally derived polymers have the potential advantage of biological recognition that may positively support cell adhesion and function, but they have poor mechanical properties. Many of them are also limited in supply and can therefore be costly [142].

During the last decade, significant attention has been focused on biodegradable polymers. Among all these polymers, poly(lactic acid) (PLA) is one of the most promising because it is thermoplastic, biocompatible and has a high strength, a high modulus, and good processability [143]. PLA also has been revealed an inefficient crystallization process for both the lower crystallization rate and crystallinity as compared with other polymers [144-147].

To maximize therapeutic activity while minimizing negative side effects is the main driven force for the continuous development of new controlled drug delivery systems. Because the release of drugs in drug-intercalated layered materials is potentially controllable (Fig.5), these new materials have a great potential as a delivery host in the pharmaceutical field. Calcium clay has been used extensively in the treatment of pain, open wounds, colitis, diarrhea, hemorrhoids, stomach ulcers, intestinal problems, acne, anemia, and a variety of other health issues.

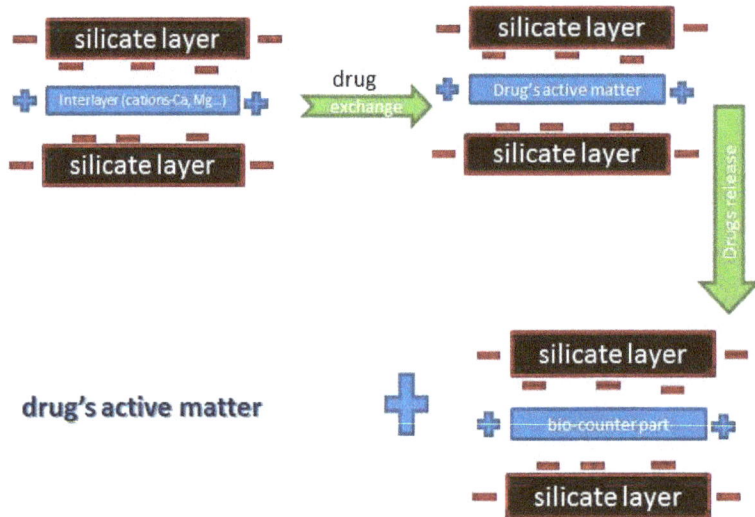

Figure 5. Schema of drugs delivery.

Bionanocomposites belong to group of materials, being the result of the combination of biopolymers and inorganic solids at the nanometer scale. These hybrid organic–inorganic materials are extraordinarily versatile as they could be formed from a large variety of biopolymers and also from different inorganic solid particles such as layered silicates, hydroxyapatite, cellulose and other metal oxides [148]. Among them, chitosan/layered silicate nanocomposites have received much more attention. These nanocomposites are of interest for advanced biomedical materials, as for instance tissue engineering, artificial bones or gene therapy. Other possible fields of applications are related to their mechanical, thermal and barrier properties, making this class of materials attractive for potential uses in controlled drug and pesticides delivery, membranes for food processing, drinking water purification, oxygen barrier films and food package.

Chitosan is the deacetylated product of chitin, a natural polymer found in the cell wall of fungi and microorganisms. The active groups in the chitosan structure are the free amine groups, located in the C2 position of the glucose residue in the polysaccharide chain, and the hydroxyl groups [148]. *The chitosan/vermiculite nanocomposites* have been successfully prepared with different modified vermiculite, which was treated by acid, sodium and CTAB cations. The modification and the nano-scale dispersion of the modified vermiculites were confirmed.

Author details

Marta Valášková[1,2,*] and Gražyna Simha Martynková[1,2]

[1]*Nanotechnology Centre, VŠB – Technical University of Ostrava, Ostrava-Poruba, Czech Republic*
[2]*IT4Innovations Centre of Excellence, VŠB-Technical University of Ostrava, Ostrava-Poruba, Czech Republic*

Acknowledgement

Financial support of the Czech Grant Agency (projects GA ČR 210/11/2215) and the IT4Innovations Centre of Excellence (project reg. no. cz.1.05/1.1.00/02.0070) are gratefully acknowledged.

7. References

[1] Guggenheim S, Martin RT (1995) Definition of Clay and Clay Mineral. Joint Report of the AIPEA Nomenclature and CMS Nomenclature Committees. Clays clay miner. 43: 255-256.

[2] Brindley GW, Bailey SW, Faust GT, Forman SA, Rich CI (1968) Report of the Nomenclature Committee (1966-67) of The Clay Minerals Society. Clays clay miner. 16: 322-324.

[3] Brindley GW, Pedro G (1972) Report of the AIPEA Nomenclature Committee. AIPEA Newsletter 7: 8-13.

---

* Corresponding Author

[4] Guggenheim S, Adams JM, Bain DC, Bergaya F, Bigatti MF, Drits VA, Formoso MLL, Galán E, Kogue T, Stanjek H (2006) Summary of Recommendations of Nomenclature Committees Relevant to Clay Mineralogy: Report of the Association Internationale pour L'Étude des Argiles (AIPEA) Nomenclature Committee for 2006. Clays clay miner. 54: 761-772; Clay miner. 41: 863-877.

[5] Martin RT, Bailey SW, Eberl DD, Fanning DS, Guggenheim S, Kodama H, Pevear DR, Srodon J, Wicks FJ (1991) Report of the Clay Minerals Society Nomenclature Committee: Revised Classification of Clay Materials. Clays clay miner. 39: 333-335.

[6] Norrish K (1973) Factors in the Weathering of Mica to Vermiculite. Proc. int. clay. conf. 1972. Madrid, 417–432.

[7] Brindley GW, Pedro G (1976) Meeting of the Nomenclature Committee of AIPEA, Mexico City, July 21, 1975, AIPEA Newsletter 12: 5-6.

[8] Guggenheim S, Alietti A, Drits VA, Formoso MLL, Galán E, Köster HM, Paquet H, Watanabe T, Bain DC, Hudnall WH (1997) Report of the Association Internationale pour L'Étude des Argiles (AIPEA) Nomenclature Committee for 1996 Clays clay miner. 45: 298-300; Clays clay miner. 32: 493-496.

[9] Shirozu H, Bailey SW (1966) Crystal Structure of a Two-Layer Mg–vermiculite. Am. miner. 51: 1124–1143.

[10] Lagaly G (1982) Layer Charge Heterogeneity in Vermiculites. Clays clay miner. 30: 215-222.

[11] Bailey SW (1980) Structures of Layer Silicates. In: Brindley GW, Brown G, editors. Crystal Structures of Clay Minerals and their X-ray Identification. Monograph 5, Mineralogical Society, London: pp. 1-124.

[12] Bailey SW, Brindley GW, Fanning DS, Kodama H, Martin RT (1984) Report of the Clay Minerals Society Nomenclature Committee for 1982 and 1983. Clays clay miner. 32: 239.

[13] Bergaya F, Vayer M. (1997) CEC of Clays: Measurement by Adsorption of a Copper Ethylene-Diamine Complex. Appl. clay sci. 12: 275-280.

[14] Rausell-Colom JA, Fernández M, Serratosa JMJ, Alcover F, Gatineau L (1980) Organisation de L'espace Interlamellaire dans les Vermiculites Monocouches et Anhydres. Clay miner. 15: 37-58.

[15] de la Calle C, Suquet H, Pons CH (1988) Stacking Order in 14.30 Å Mg–vermiculite. Clays clay miner. 36: 481-490.

[16] Jasmund K, Lagaly G, editors (1993) Tonminerale und Tone, Struktur, Eigenschaften, Anwendung und Einsatz in Industrie und Umwelt. Steinkopff Verlag, Darmstadt. 490p.

[17] Tournassat C, Greneche JM, Tisserant D, Charlet L (2003a). The Titration of Clay Minerals. I. Discontinuous Backtitration Technique Combined with CEC Measurements. J. colloid interface sci. 273: 224–233.

[18] Tournassat C, Ferrage E, Poinsignon C, Charlet L (2003b). The Titration of Clay Minerals. II. Structure-Based Model and Implications for Clay Reactivity. J. colloid interface sci. 273: 234–246.

[19] Mathieson AM, Walker GF (1954) Crystal Structure of Magnesium–Vermiculite. Am. miner. 39: 231–255.

[20] Vali H, Hesse R (1992) Identification of Vermiculite by Transmission Electron Microscopy and X-ray Diffraction. Clay miner. 27: 185–192.

[21] Collins DR, Fitch AN, Catlow RA (1992) Dehydration of Vermiculites and Montmorillonites: a Time-Resolved Powder Neutron Diffraction Study. J. mat. chem. 8: 865–873.

[22] Reichenbach HG, Beyer J (1994) Dehydration and Rehydration of Vermiculites: IV. Arrangement of Interlayer Components in the 1.43 nm and 1.38 nm Hydrates of Mg-vermiculite. Clay miner. 29: 327–340.

[23] Reichenbach HG, Beyer J (1995) Dehydration and Rehydration of Vermiculites: II. Phlogopitic Ca-vermiculite. Clay miner. 30: 273–286.

[24] Ruiz-Conde A, Ruiz-Amil A, Pérez-Rodríguez JL, Sánchez-Soto PJ (1996) Dehydration–Rehydration in Magnesium Vermiculite: Conversion from Two-One and One-Two Water Hydration States through the Formation of Interstratified Phases. J. mat. chem. 6: 1557–1566.

[25] Marcos C, Argüelles A, Ruíz-Conde A, Sánchez-Soto PJ, Blanco JA (2003) Study of the Dehydration Process of Vermiculites by Applying a Vacuum Pressure: Formation of Interstratified Phases. Mineral. mag. 67: 1253–1268.

[26] Marcos C, Arango YC, Rodriguez I (2009) X-ray Diffraction Studies of the Thermal Behaviour of Commercial Vermiculites. Appl. clay sci. 42: 368-378.

[27] de la Calle C, Suquet H (1988). Vermiculite. In: Bailey SW, editor. Reviews in Mineralogy, Vol. 19, Hydrous Phyllosilicates. Mineralogical Society of America, Washington: pp. 455-496.

[28] Weiss Z, Valvoda V, Chmielová M (1994) Dehydration and Rehydration of Natural Mg-vermiculite. Geol. carpath. 45: 33-39.

[29] Walker GF (1956) Mechanism of Dehydration of Mg-vermiculite. Clays clay miner. 4: 101–115.

[30] de la Calle C, Dubernat J, Suquet H, Pezerat H, Gaulthier J, Mamy J (1975) Crystal Structure of Two-Layer Mg-vermiculites and Na-, Ca-vermiculites. In: Bailey SW, editor. Proc. Internat. Clay Conf. Mexico City: Applied Publishing, Wilmette, Illinois: pp. 201–209.

[31] de la Calle C, Suquet H, Pezerat H (1975) Glissement de Feuillets Accompagnant Certains Echanges Cationiques dans les Vermiculites. Bull. du groupe français des argiles 27: 31–49.

[32] de la Calle C, Suquet H, Dubernat J, Pezerat H (1978) Mode d'Empilement des Feuillets dans les Vermiculites Hydratees a 'deux Couches.' Clay miner. 13: 275–297.

[33] de la Calle C, Suquet H, Pezerat H (1985) Vermiculites Hydratees a Une Couche. Clay miner. 20: 221–230.

[34] Temuujin J, Okada K, Mackenzie KJD (2003) Preparation of Porous Silica from Vermiculite by Selective Leaching. Appl. clay sci. 22: 187-195.

[35] Wiewóra A, Pérez-Rodríguez JL, Pérez-Maqueda LA, Drapala J (2003) Particle Size Distribution in Sonicated High- and Low-Charge Vermiculites. Appl. clay sci. 24: 51-58.

[36] Valášková M, Hundáková M, Kutláková Mamulová K, Seidlerová J, Čapková P, Pazdziora E, Matějová K, Heřmánek M, Klemm V, Rafaja D (2010) Preparation and

Characterization of Antibacterial Silver/Vermiculites and Silver/Montmorillonites. Geochim. cosmochim. ac. 74: 6287-6300.

[37] del Rey-Perez-Caballero FJ, Poncelet G (2000) Microporous 18 Å Al-pillared Vermiculites: Preparation and Characterization. Micropor. mesopor. mat. 37: 313-327.

[38] da Fonseca MG, Wanderley AF, Sousa K, Arakaki LNH, Espinola JGP (2006) Interaction of Aliphatic Diamines with Vermiculite in Aqueous Solution. Appl. clay sci. 32: 94–98.

[39] Bergaya F, Jaber M, Lambert JF (2011) Clays and Clay Minerals. In: Galimberti M, editor. Rubber Clay Nanocomposites: Science, Technology and Applications, Wiley: pp 3-44.

[40] Guggenheim S (Chair), Bergaya F, Brigatti MF, Galán E, Drits V, Formoso MLL, Kogure T, Stanjek H, ex officio: Adams J, Stuck J (2012) 9. Nomenclature Committee - 2010-2012, Report of the AIPEA Nomenclature Committee for AIPEA Newsletter. Aipea Newsletter 44: 53-54.

[41] Hindman JR (1994) Vermiculite. In: Carr DD editor. Industrial Minerals and Rocks. Littleton, CO: Society for Mining, Metalurgy and Exploration, Inc.: pp. 1103-1111.

[42] Walker GF (1961) Vermiculites. In: Brown G (editor), The X-ray Identification and Crystal Structures of Clay Minerals. Mineralogical Society, London: pp. 297–324.

[43] Marcos C, Arango YC, Rodríguez I (2009) X-ray Diffraction Studies of the Thermal Behaviour of Commercial Vermiculites. Appl. clay sci. 42: 368–378.

[44] Walker GF (1951) Vermiculites and Some Related Mixed-Layer Minerals. In: Brindley GW, editor. X-ray Identification and Crystal Structures of Clay Minerals. Mineralogical Society, London: pp. 199–223.

[45] Midgley HG, Midgley CM (1960) The Mineralogy of Some Commercial Vermiculites. Clay miner. bull. 4: 142–150.

[46] Justo A, Maqueda C, Perez-Rodriguez JL, Morillo E (1989) Expansibility of Some Vermiculites. Appl. clay sci. 4: 509–519.

[47] Potter MJ (1997) Vermiculite. Am. ceram. soc. bull. 76: 135.

[48] Sanchez-Soto PJ, Ruiz Conde A, Aviles MA, Justo A, Pérez-Rodríguez JL (1995) Mechanochemical Effects on Vermiculite and its Influence on the Synthesis of Nitrogen Ceramics. In: Vicenzine P, editor. Ceramic Charting the Future. Techna Srl, Italia: pp. 1383- 1390.

[49] Ou CCY, Yang JC (1978). European Patent Application Eppl EP212930.

[50] Nelson LL (1988). European Patent Application EP 282928 A2.

[51] Balek V, Pérez-Rodríguez JL, Pérez-Maqueda LA, Šubrt J, Poyato J (2007) Thermal Behaviour of Ground Vermiculite. J. therm. anal. calorim. 88: 819–823.

[52] Peters D (1996) Ultrasound in Material Chemistry. J. mat. chem. 6: 1605–1618.

[53] Pérez-Maqueda LA, Caneo OB, Poyato J, Pérez-Rodríguez JL (2001) Preparation and Characterization of Micron and Submicron-Sized Vermiculite. Phys. chem. miner. 28: 61-66.

[54] Pérez-Rodríguez JL, Carrera F, Poyato J, Perez-Maqueda LA (2002) Sonication as a Tool for Preparing Nanometric Vermiculite Particles. Nanotechnology 13: 382– 387.

[55] Jiménez de Haro MC, Martínez Blanes JM, Poyato J, Pérez-Maqueda LA, Lerf A, Pérez-Rodríguez JL (2004) Effects of Mechanical Treatment and Exchange Cation on the Microporozity of Vermiculite. J. phys. chem. solids 65: 435-439.

[56] Pérez-Maqueda LA, Jiménez de Haro MC, Poyato J, Pérez-Rodríguez JL (2004) Comparative Study of Ground and Sonicated Vermiculite. J. mat. sci. 39: 5347- 5351.

[57] Poyato J, Pérez-Rodríguez JL,Ramírez-Valle V, Lerf A, Wagner FE (2009) Sonication Induced Redox Reactions of the Ojén (Andalucía, Spain) Vermiculite. Ultrason. sonochem. 16: 570–576.

[58] Obut A, Girgin I (2002) Hydrogen Peroxide Exfoliation of Vermiculite and Phlogopite. Miner. eng. 15: 683-687.

[59] Üçgül E, Girgin I (2002) Chemical Exfoliation Characteristics of Karakoç Phlogopite in Hydrogen Peroxide Solution. Turk. j. chem. 26: 431-439.

[60] Weiss Z, Valášková M, Seidlerová J, Šupová-Křístková M, Šustai O, Matějka V, Čapková P (2006) Preparation of Vermiculite Nanoparticles using Thermal Hydrogen Peroxide Treatment. J. nanosci. nanotechnol. 6: 726-730.

[61] Matějka V, Šupová-Křístková M, Kratošová G, Valášková M (2006) Preparation of Mg-vermiculite Nanoparticles using Potassium Persulfate Treatment. J. nanosci. nanotech. 6: 2484-2488.

[62] Kehal M, Laurence R, Duclaux L (2010) Characterization and Boron Adsorption Capacity of Vermiculite Modified by Thermal Shock or H_2O_2 Reaction and/or Sonication. Appl. clay. sci. 48: 561–568.

[63] Brown DR, Rhodes CN (1997) Brønsted and Lewis Acid Catalysis with Ion-Exchanged Clays. Catal. let. 45: 35–40.

[64] Schoonheydt RA, Johnston CT (2006) Surface and Interface Chemistry of Clay Minerals. In: Bergaya F, Theng BKG, Lagaly G, editors. Handbook of Clay Science Developments in Clay Science, Vol. 1, Elsevier Ltd.: pp. 87–113.

[65] Ayyappan S, Subbanna GN, Goplan RS, Rao CNR (1996) Nanoparticles of Nickel and Silver Produced by the Polyol Reduction of the Metal Salts Intercalated in Montmorillonite. Solid state ion. 84: 271–281.

[66] Aihara N, Torigoe E, Esumi K (1998) Preparation and Characterization of Gold and Silver Nanoparticles in Layered Laponite Suspensions. Langmuir 14: 4945–4949.

[67] Patakfalvi R, Oszkó A, Dékány I (2003) Synthesis and Characterization of Silver Nanoparticle/Kaolinite Composites. Colloid surf. A-physicochem. eng. asp. 220: 45–54.

[68] Patakfalvi R, Dékány I (2004) Synthesis and Intercalation of Silver Nanoparticles in Kaolinite/DMSO Complexes. Appl. clay sci. 25: 149–159.

[69] Praus P, Turicová M, Valášková M (2008) Study of Silver Adsorption on Montmorillonite. J. braz. chem. soc. 19: 549–556.

[70] Valášková M, Simha Martynková G, Lešková J, Čapková P, Klemm V, Rafaja D (2008) Silver Nanoparticles /Montmorillonite Composites Prepared using Nitrating Reagent at Water and Glycerol. J. nanosci. nanotechnol. 8: 3050–3058.

[71] Davies RL, Etris SF (1997) The Development and Functions of Silver in Water Purification and Disease Control. Catal. today 36: 107–114.

[72] Feng QL, Wu J, Chen GQ, Cui FZ, Kim TN, Kim JO (2000) A Mechanic Study of the Antibacterial Effect of Silver Ions on Escherichia Coli and Staphylococcus Aureus. J. biomed. mater. res. part A 52: 662–668.

[73] Eberhart ME, Donovan MJ, Outlaw RA (1992) Ab Inito Calculations of Oxygen Diffusivity in Group-IB Transition Metals. Phys. rev. B 46: 12744–12747.

[74] Outlaw RA, Davidson MR (1994) Small Ultrahigh Vacuum Compatible Hyperthermal Oxygen Atom Generator. J. vac. sci. technol. A 12: 854–860.

[75] Elechiguerra JL, Burt JL, Morones JR, Camacho Bragado A, Gao X, Lara HH, Yaca Man MJ (2005) Interaction of Silver Nanoparticles with HIV-I. J. nanobiotechnol. 3: 6–16.

[76] Kiruba Daniel SCG, Tharmaraj V, Anitha Sironmani T, Pitchumani K (2010) Toxicity and Immunological Activity of Silver Nanoparticles. Appl. clay sci. 48: 547–551.

[77] Chimentão RJ, Kirm I, Medina F, Rodríguez X, Cesteros Y, Salagre P, Sueiras JE (2004) Different Morphologies of Silver Nanoparticles as Catalysts for the Selective Oxidation of Styrene in the Gas Phase. Chem. commun. 4: 846–847.

[78] Lofton C, Sigmund W (2005) Mechanisms Controlling Crystal Habits of Gold and Silver Colloids. Adv. funct. mater. 15: 1197–1208.

[79] Shao K, Yao J (2006) Preparation of Silver Nanoparticles via a Non-Template Method. Mater. lett. 60: 3826–3829.

[80] Zhang Z, Han M (2003) Template-Directed Growth from Small Clusters into Uniform Silver Nanoparticles. Chem. phys. lett. 374: 91–94.

[81] Esumi K, Suzuki A, Yamahira A, Torigoe K (2000) Role of Poly(amidoamine) Dendrimers for Preparing Nanoparticles of Gold Platinum and Silver. Langmuir 16: 2640–2680.

[82] Zhang Z, Zhang L, Wang S, Chen W, Lei Y (2001) A Convenient Route to Polyacrylonitrile/Silver Nanoparticle Composite by Simultaneous Polymerization–Reduction Approach. Polymer 42: 8315–8318.

[83] Bajpai SK, Mohan YM, Bajpai M, Tankhiwale R, Thomas V (2007) Synthesis of Polymer Stabilized Silver and Gold Nanostructures. J. nanosci. nanotechnol. 7: 2994–3010.

[84] Mohan YM, Lee K, Premkumar T, Geckeler KE (2007) Hydrogel Networks as Nanoreactors: a Novel Approach to Silver Nanoparticles for Antibacterial Applications. Polymer 48: 158–164.

[85] Lee WF, Huang YC (2007) Swelling and Antibacterial Properties for the Superabsorbent Hydrogels Containing Silver Nanoparticles. J. appl. polym. sci. 106:1992–1999.

[86] Vimala K, Sivudu SK, Mohan MY, Sreedhar B, Raju MK (2009) Controlled Silver Nanoparticles Synthesis in Semihydrogel Networks of Poly(acrylamide) and Carbohydrates: a Rational Methodology for Antibacterial Application. Carbohydr. polym. 75: 463–471.

[87] Magaña SM, Quintana P, Aguilar DH, Toledo JA, Ángeles-Chávez C, Cortés MA, León L, Freile-Pelegrín Y, López T, Torres Sánchez RM (2008) Antibacterial Activity of Montmorillonites Modified with Silver. J. mol. catal. a-chem. 281: 192–199.

[88] Zhao D, Zhou J, Liu N (2006) Preparation and Characterization of Mingguang Palygorskite Supported with Silver and Copper for Antibacterial Behavior. Appl. clay sci. 33: 161-170.

[89] Valášková M, Simha Martynková G, Matějka V, Barabaszová K, Plevová E, Měřínská D (2009) Organovermiculite Nanofillers in Polypropylene. Appl. clay sci. 43: 108–112.

[90] Grim RE, Kulbicky G (1961) Montmorillonite: High Temperature Reactions and Classification. Am. miner. 46: 1329-1369.

[91] Ruitz-Hitzky E, van Meerbeek A (2006) Clay Mineral and Organoclay–Polymer Nanocomposites. In: Bergaya F, Theng BKG, Lagaly G, editors. Handbook of Clay Science Developments in Clay Science, Vol. 1, Elsevier Ltd. pp. 583-621.

[92] Lagaly G, Ogawa M, Dékány I (2006) Clay Mineral Organic Interactions. In: Bergaya F, Theng BKG, Lagaly G, editors. Handbook of Clay Science Developments in Clay Science, Vol. 1, Elsevier Ltd.: pp. 309-377.

[93] Tuney JJ, Detellier C (1996) Aluminosilicate Nanocomposites Materials. Poly(ethyleneglycol)-Kaolinite Intercalates. Chem. mat. 8: 927-35.

[94] Gilman JW, Jackson CL, Morgan AB, Harris R, Manias E, Giannelis EP, Wuthenow M, Hilton D, Phillips SH (2000) Flammability Properties of Polymer–Layered-Silicate Nanocomposites. Polypropylene and Polystyrene Nanocomposites. Chem. mat.12: 1866–1873.

[95] Osman MA, Rupp JEP, Suter UW (2005) Tensile Properties of Polyethylene/Layered Silicate Nanocomposites. Polymer 46: 1653-1660.

[96] Yuan Q, Misra RDK (2006) Impact Fracture Behavior of Clay–Reinforced Polypropylene Nanocomposites. Polymer 47: 4421–4433.

[97] Peneva Y, Tashev E, Minkova L (2006) Flammability, Microhardness and Transparency of Nanocomposites Based on Functionalized Polyethylenes. Eur. polym. j. 42: 2228–2235.

[98] Balazs AC, Singh Ch, Zhulina E (1998) Modeling the Interactions between Polymers and Clay Surfaces through Self-Consistent Field Theory. Macromolecules 31: 8370-8381.

[99] Rong M, Zhang M, Liu H, Zeng H (1999) Synthesis of Silver Nanoparticles and Their Self-Organization Behavior in Epoxy Resin. Polymer 40: 6169-6178.

[100] Liu X, Wu Q, Berglund L A (2002) Polymorphism in Polyamide 66/Clay Nanocomposites. Polymer 43: 4967-4972.

[101] Xu WB, Bao SP, Shen SJ, Hang GP, He, PS (2003) Curing Kinetics of Epoxy Resin–Imidazole–Organic Montmorillonite Nanocomposites Determined by Differential Scanning Calorimetry. J. appl. polym. sci. 88: 2932-2941.

[102] Yu ZZ, Yang M, Zhang Q, Zhao Ch, Mai YW (2003) Dispersion and Distribution of Organically Modified Montmorillonite in Nylon-66 Matrix. J. polym. sci. pt. b-polym. phys. 41: 1234-1243.

[103] Fornes TD, Paul DR (2003) Crystallization Behavior of Nylon 6 Nanocomposites. Polymer 44: 3945-3961.

[104] Das-Gupta DK (1994) Polyethylene: Structure, Morphology, Molecular Motion and Dielectric Behavior. IEEE Electr. insul. mag. 10: 5–15.

[105] Androsch R, Di Lorenzo ML, Schick Ch, Wunderlich B (2010) Mesophases in Polyethylene, Polypropylene, and Poly(1-butene). Polymer 51: 4639–4662.

[106] Tjong SC, Meng,YZ (2003) Preparation and Characterization of Melt-Compounded Polyethylene/ Vermiculite Nanocomposites. J. polym. sci. pt. b-polym. phys. 41: 1476-1484.

[107] Hotta S, Paul DR (2004) Nanocomposites Formed from Linear Low Density Polyethylene and Organoclays. Polymer 45: 7639–7654.

[108] Chrissopoulou K, Altintzi I, Anastasiadis SH, Giannelis EP, Pitsikalis M, Hadjichristidis N, Theophilou N (2005) Controlling the Miscibility of Polyethylene/ Layered Silicate Nanocomposites by Altering the Polymer/Surface Interactions. Polymer 46: 12440–12451.

[109] Tanniru M, Yuan Q, Misra RDK (2006) On Significant Retention of Impact Strength in Clay–Reinforced High-Density Polyethylene (HDPE) Nanocomposites. Polymer 47: 2133–2146.

[110] Tjong SC, Meng YZ, Xu J (2002) Preparation and Properties of Polyamide 6/Polypropylene–Vermiculite Nanocomposite/Polyamide 6 Alloys. J. appl. polym. sci. 86: 2330–2337.

[111] Tjong SC, Meng YZ (2003). Impact-Modified Polypropylene/Vermiculite Nanocomposites. J. polym. sci. pt. b-polym. phys. 41: 2332–2341.

[112] Shao W, Wang Q, Chen Y, Gu Y (2006) Preparation and Properties of Polypropylene/ Vermiculite Nanocomposite through Solid-State Shear Compounding (S3C) Method using Pan-Mill Equipment. Mater. manuf. process. 21: 173–179.

[113] Simha Martynková G., Barabaszová K, Valášková M (2007) Effect of Size and Preparation Method of Particles on Intercalation Ability of Vermiculite. Acta metall. slovaca 6: 275–279.

[114] Nam PH, Maiti P, Okamoto M, Kotaka T, Hasegawa N, Usuki A (2001) A Hierarchical Structure and Properties of Intercalated Polypropylene/Clay Nanocomposites. Polymer 42: 9633–9640.

[115] Gilman JW (1999) Flammability and Thermal Stability Studies of Polymer Layered Silicate (Clay) Nanocomposites. Appl. clay sci. 15: 31–49.

[116] Wang Z, Massam J, Pinnavaia TJ (2000) Epoxy - Clay Nanocomposites. In: Pinnavaia TJ, Beall GW, editors. Polymer–Clay Nanocomposites. New York: Wiley, New York: pp.127-148.

[117] Sinha Ray S, Bousmina M (2005) Biodegradable Polymers and their Layered Silicate Nanocomposites. In: Greening the 21st Century Materials World, Prog. mater. sci. 50: 962-1079.

[118] D'Souza NA (2004) Epoxy+clay nanocomposites. In: Nalwa HS, editor. Encyclopedia of Nanoscience and Nanotechnology, Vol.3 American Scientific Publishers, California: pp. 253-265.

[119] Giannelis EP (1996) Polymer Layered Silicate Nanocomposites. Adv. mater. 8: 29–35.

[120] Vaia RA, Giannelis EP (1997) Polymer Melt Intercalation in Organically-Modified Layered Silicates: Model Predications and Experiment. Macromolecules 30: 8000–8009.

[121] Vaia RA, Jandt KD, Kramer EJ, Giannelis EP (1995) Kinetics of Polymer Melt Intercalation. Macromolecules 28: 8080–8085.

[122] Vaia RA, Jandt KD, Kramer EJ, Giannelis EP (1996) Microstructural Evolution of Melt Intercalated Polymer–Organically Modified Layered Silicates Nanocomposites. Chem. mater. 8: 2628–2635.

[123] LeBaron PC, Wang Z, Pinnavaia TJ (1999) Polymer-Layered Silicate Nanocomposites: An Overview. Appl. clay sci. 15: 11–29.

[124] Šupová M, Martynková GS, Barabaszová K (2011) Effect of Nanofillers Dispersion in Polymer Matrices: A Review. Sci. adv. mater. 3: 1–25.

[125] Zheng YA, Li P, Zhang JP,Wang AQ (2007) Study on Superabsorbent Composite XVI. Synthesis, Characterization and Swelling Behaviors of Poly(sodium acrylate)/ Vermiculite Superabsorbent Composites. Eur. polym. j. 43: 1691–1698.

[126] Wang W, Zhang J, Wang A (2009) Preparation and Swelling Properties of Superabsorbent Nanocomposites based on Natural Guar Gum and Organo-vermiculite. Appl.clay sci. 46: 21–26.

[127] Takahashi S, Goldberg HA, Feeney CA, Karim DP, Farrell M, O'Leary K, Paul DR (2003) Gas Barrier Properties of Butyl Rubber/Vermiculite Nanocomposite Coatings. Polymer 47: 3083-3093.

[128] De Rocher J, Gettelfinger BT, Wang J, Nuxoll EE, Cussler EL (2005) Barrier Membranes with Different Sizes of Aligned Flakes. J. memb. sci. 254: 21-30.

[129] Ravichandran J, Sivasankar B 1997 Properties and Catalytic Acivity of Acid Modified Montmorillonite and Vermiculite. Clays clay miner. 45: 854-858.

[130] Okada K, Arimitsu N, Kameshima Y, Nakajima A, Kenneth JD, MacKenzie KJD (2006) Solid Acidity of 2:1 type Clay Minerals Activated by Selective Leaching. Appl. clay sci. 31: 185-193.

[131] Xu S, Meng YZ, Li RKY, Xu Y, Rajulu A V (2003) Preparation and Properties of Poly(vinyl alcohol)-Vermiculite Nanocomposites. J. polym. sci. pt. b-polym. phys. 41: 749 -755.

[132] Dhamodharan R, Jeyaprakash JD, Samuel V, Rajeswari MK (2001) Intercalative Redox Polymerization and Characterization of Poly(4-vinylpyridine)–Vermiculite Nanocomposite. J. appl. polym. sci. 82: 555-561.

[133] Zhang J, Wang A (2007) Study on Superabsorbent Composites. IX: Synthesis, Characterization and Swelling Behaviors of Polyacrylamide/Clay Composites Based on Various Clays. React. funct. polym. 67: 737-745.

[134] Liu D, Du X, Meng Y (2006) Facile Synthesis of Exfoliated Polyaniline/Vermiculite Nanocomposites. Mater. lett. 60: 1847-1850.

[135] Swenson J, Schwartz GA, Bergman R, Howells WS (2003) Dynamics of Propylene Glycol and its Oligomers Confined in Clay. Eur. phys. j. e 12: 179-183.

[136] Schwartz GA, Bergman R, Mattsson J, and Swenson J (2003) Eur. phys. j. e 12: 113-116.

[137] da Silva DC, Skeff Neto K, Coaquira JAH, Araujo PP, Cintra DOS, Lima ECD, Guilherme LR, Mosiniewicz-Szablewska E, Morais PC (2010) Magnetic Characterization of Vermiculite-based Magnetic Nanocomposites. J.non-cryst. solids 356: 2574–2577.

[138] Zhang H, Yu J, Kuang D (2012) Effect of Expanded Vermiculite on Aging Properties of Bitumen. Constr. build. mater. 26: 244–248.

[139] Du XS, Xiao M, Meng YZ, Hung TF, Rajulu AV, Tjong SC (2003) Synthesis of Poly(arylene disulfide)-Vermiculite Nanocomposites via in situ Ring-Opening Reaction of Cyclic Oligomers. Eur. polym. j. 39: 1735-1739.

[140] Yanga C, Liua P, Guoa J, Wang Y (2010) Polypyrrole/Vermiculite Nanocomposites via Self-Assembling and in Situ Chemical Oxidative Polymerization. Synth. met. 160: 592–598.

[141] Xu J, Li RKY, Xu Y, Li L, Meng YZ (2005) Preparation of Poly(propylene carbonate)/Organo-Vermiculite Nanocomposites via Direct Melt Intercalation. Eur. polym. j. 41: 881–888.

[142] Armentano I, Dottori M, Fortunati M, Mattioli S, Kenny JM (2010) Biodegradable Polymer Matrix Nanocomposites for Tissue Engineering: A Review. Polym. degrad. stabil. 95: 2126-2146.

[143]   Drumright RE, Gruber PR, Henton DE (2000) Polylactic Acid Technology. Adv. mater. 12: 1841-1846.

[144] Nam JY, Sinha Ray S, Okamoto M (2003) Crystallization Behavior and Morphology of Biodegradable Polylactide/Layered Silicate Nanocomposite. Macromolecules 36:7126 - 7131.

[145] Paul MA, Alexandre M, Degee P, Calberg C, Jerome R, Drbois P (2003) Exfoliated Polylactide/Clay Nanocomposites by in situ Coordinationinsertion Polymerization. Macromol. rapid commun. 24: 561-566.

[146] Ogata N, Jimenez G, Kawai H, Ogihara T (1997) Structure and Thermal/Mechanical Properties of Poly(L-lactide)-Clay Blend. J. polym. sci. pt. b-polym. phys. 35: 389-396.

[147] Xu H, Bai Y, Mao Z (2011) Preparation of PLLA-co-bis A ER/VMT Nanocomposites by In-Situ Polymerization Process. The open mat. sci. j. 5: 56-60.

[148] Zhang K, Xu J, Wang KY, Cheng V, Wang V, Liu V (2009) Preparation and Characterization of Chitosan Nanocomposites with Vermiculite of Different Modification. Polym. degrad. stabil. 94: 2121–2127.

Chapter 11

Synthesis and Characterization of Fe-Imogolite as an Oxidation Catalyst

Masashi Ookawa

Additional information is available at the end of the chapter

1. Introduction

Imogolite is a hydrated aluminosilicate with a unique tubular structure, which is found in volcanic ash soil and is shown in Figure 1. It was first discovered in glassy volcanic ash soil in Japan and was named after the soil in Hitoyoshi, Kumamoto Prefecture [1]. Its chemical composition is $(OH)_3Al_2O_3Si(OH)$.

Figure 1. Photograph of natural imogolite films in soil.

The tubular structure of imogolite was proposed by Cradwick et al. [2] based on results from electron diffraction observation and is shown in figure 2. The tube wall consists of a single continuous Al(OH)₃ (gibbsite) sheet and orthosilicate anions (O₃SiOH groups) associated with each vacant octahedral site of the gibbsite sheet. The imogolite has an outer diameter of ca. 2 nm and an inner diameter of ca. 1 nm.

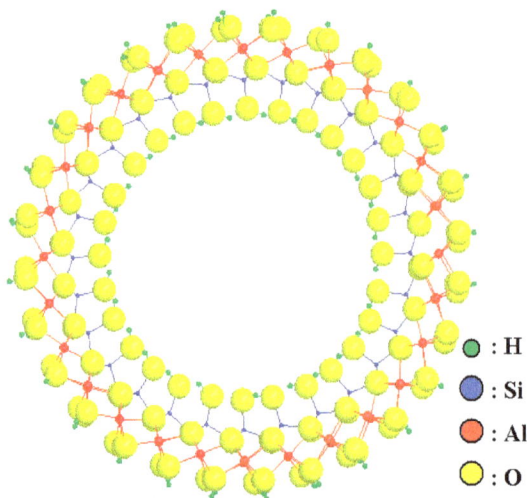

●	: H
●	: Si
●	: Al
○	: O

Figure 2. A cross-section of the structural model of an imogolite tube.

Recently, imogolite has drawn a new attention as a new nano-material because of its unique nano-scale tubular structure similar to that of a single-walled carbon nanotube [3]. Studies on synthesis [4-9], mechanisms of formation [10], structural evolution [11, 12], stability [13, 14], electronic states [14, 15] and application have been carried out. Proposed applications, such as a polymer composite [16,17], a fuel gas storage [18], an absorbent [19], an exchange material for heat pump system [20], a humidity-controlling material [6] and an anti-deweling material [6] have been discussed.

Application as a catalyst or a catalyst support also attracts attention because it is expected to have a shape-selective characteristic property as molecular sieving zeolites due to its unique tubular structure. However, few investigations [21, 22] have been reported using natural imogolite as a catalyst because the extraction of pure imogolite from the soil is difficult and time consuming [21]. Therefore, synthesis of imogolite has become necessary in order to utilize it as a functional material.

We synthesized imogolite containing Fe^{3+} ions (Fe-imogolite) using $NaSiO_4$, $FeCl_3$ and $AlCl_3$ to investigate its catalytic properties [23]. Because of its chemical stability, an incorporation of another element is necessary to generate a chemical function. We found that it served as a catalyst of liquid-phase oxidation reactions of some hydrocarbons.

In this chapter, the synthetic methods, characterization and the general properties of imogolite is described at first. And then the characterization and catalytic properties of Fe-imogolite obtained from our researches is described.

2. Synthetic imogolite

In the latter half of the 1970's, synthesis of imogolite was succeeded from a dilute solution. Recently, various synthesis methods have been investigated and many characterizations have been carried out. In this section, synthesis methods, structural characterization and catalytic properties of synthetic imogolite are described.

2.1. Synthesis method of imogolite

Farmer et al. reported [24] the synthesis method of imogolite in 1977. Imogolite was synthesized from a dilute solution containing of hydroxyaluminum cations (2.4 mmol L^{-1}) and orthosilicic acid (1.4 mmol L^{-1}). Afterward, the solution was adjusted to pH 5 with sodium hydroxide, 1 mmol L^{-1} of hydrochloric acid solution and 2 mmol L^{-1} of acetic acid. Imogolite was obtained in this solution by heating it near the boiling point. Wada et al. [25] later also investigated the effects of Al-to-OH^{-} (as sodium hydroxide) ratio to synthesizing imogolite and allophane using a dilute inorganic solution.

A diluted inorganic solution is necessary to form imogolite in these methods, because preventing the condensation of orthosilicic acid and the formation inhibition of it by anions is important to form the nanotube structure. However, it is difficult to obtain a large sample. Suzuki et al. [6] developed a synthetic method of producing imogolite using a concentrated inorganic solution. Sodium orthosilicate was used as a starting material to prevent the condensation of orthosilicic acid. Furthermore, a desalination process is carried out by centrifugation. We have synthesized Fe containing imogolite based on improving Suzuki's method which will be described below in detail.

More recently, new synthesis methods of imogolite have been reported. Levard et al. [8] synthesized it from a decimolar concentration solution at 95 °C for 60 days. Abidin et al. [9] proposed a new supplying method of a silicon source using colloidal silica for the synthesis of imogolite.

2.2. Characterization of imogolite

2.2.1. Morphology

Transmission electron microscopy (TEM) or scanning electron microscopy (SEM) is used mostly in order to observe the morphology and structure of imogolite nanotubes. Figure 3 shows an SEM image of synthetic imogolite.

Bursill et al. [2] have observed the various aggregations of them such as randomly oriented single tube, close-packed arrays or fiber bundles by high resolution TEM.

On the other hand, atomic force microscopy (AFM) is a powerful tool to investigate them under ambient conditions. The tapping-mode AFM especially shows the morphological features of the synthetic imogolite clearly [26, 27]. Figure 4 shows the tapping mode AFM image of synthetic imogolite. Fibrous materials with a length of 100 - 1000 nm can be observed in this image.

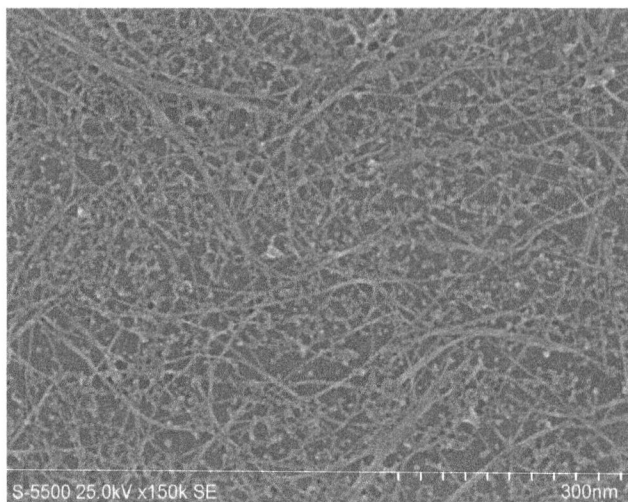

Figure 3. The FE-SEM image of synthetic imogolite.

2.2.2. Structural characterization

The analytical methods, such as X-ray diffraction (XRD), infrared (IR) spectroscopy and solid state nuclear magnetic resonance (NMR) are used in order to identify or characterize imogolite. In this section, the features of imogolite obtained using these analytical methods are mentioned.

The XRD profiles which were obtained by Cu Kα irradiation are given in Figure 5. Imogolite is characterized by three broad peaks in the low angle region. There are three peaks at 2θ = 5.1°, 11° and 15.6° in the XRD profile of natural imogolite and at 2θ = 4.6°, 9.6° and 14.3° in that of synthetic imogolite. The difference in peak position is due to the difference in the diameter of a tube of a natural and synthetic imogolite [7, 24]. The diameter of 1.8 - 2.2 nm in natural imogolite [1] and 2.7 - 3.2 nm in synthetic imogolite [25] were reported.

FT-IR spectra of Natural and synthetic imogolite, which are shown in figure 6, have a characteristic absorption that appears as a doublet at around 1000 cm^{-1}. These absorptions are attributed to Si-O (higher frequency) and Si-O-Al (lower frequency) stretching [28]. These samples also have some absorption bands in the region from 400 cm^{-1} to 750 cm^{-1}. The absorption bands at 685 cm^{-1}, 563 cm^{-1} and 427 cm^{-1} in natural imogolite arise from various Al-O vibrations [29].

Imogolite is also characterized by ^{29}Si and ^{27}Al solid state NMR [30-34]. Magic angle spinning (MAS) techniques are used in general in order to obtain a high resolution NMR spectrum of sold state. Goodman et al. [34] have carried out ^{29}Si and ^{27}Al MAS NMR measurements of synthetic imogolite. The peak is observed at -78.8 ppm in ^{29}Si MAS NMR spectrum of it as well as natural imogolite. Barron et al. [30] have shown that the observed

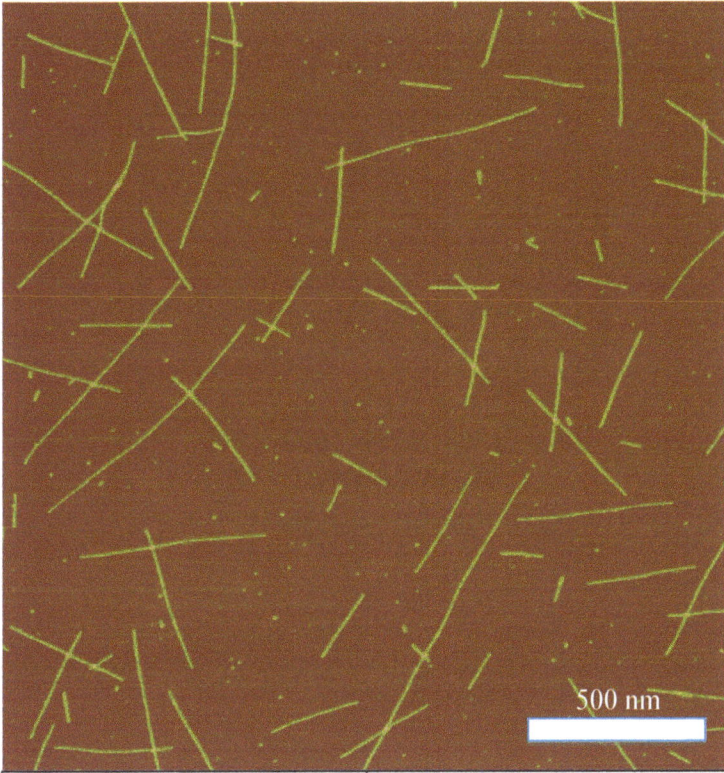

Figure 4. The tapping mode AFM image of synthetic imogolite.

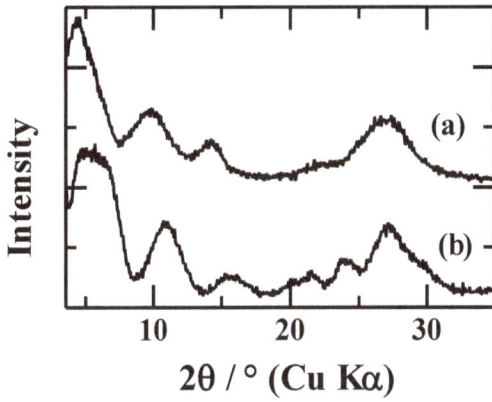

Figure 5. XRD profiles of synthetic imogolite (a) and natural imogolite (b).

chemical shift of natural imogolite is consistent with silicon tetrahedra which are isolated by coordination through oxygen with three aluminum atoms and one proton. On the other hand, ^{27}Al MAS NMR spectrum of imogolite has one peak at 0 ppm and it is attributed to the six coordinated octahedral Al^{3+} species [31-34].

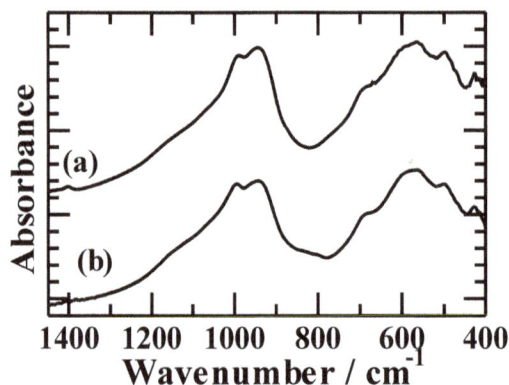

Figure 6. FT-IR spectra of synthetic imogolite (a) and natural imogolite (b).

2.2.3. Thermal transformation of synthetic imogolite

The differential thermal analysis (DTA) and thermo gravimetric analysis (TGA) traces of synthetic imogolite are shown in Figure 7. Two broad endothermic peaks with a weight loss and an exothermic peak without a weight change were observed.

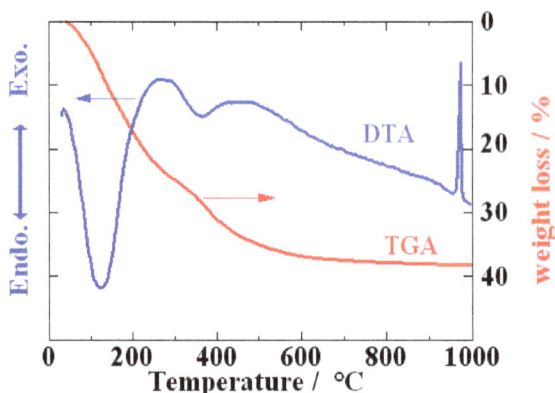

Figure 7. DTA-TGA curves of synthetic imogolite.

MacKenzie et al. [31] investigated thermal transformation of natural imogolite by using DTA-TGA, ^{29}Si MAS NMR and ^{27}Al MAS NMR and proposed the structural models which were changed by heating. The results of DTA-TGA showing thermal transformation of imogolite are explained based on their literature. The endothermic peaks at 110 °C and 400 °C were attributed to the loss of adsorbed water and dehydroxylation, respectively. Amorphism of imogolite occurred by these dehydroxylations. The exothermic peak at 950 °C was attributed to crystallization to mullite ($Al_6Si_2O_{13}$). Donkai *et al.* [11] investigated thermal transformation of natural imogolite up to 1600 °C using XRD and IR and reported the formation of tridymite (SiO_2) with mullite crystals above 1200 °C. Hatakeyama et al. [34] investigated the transformation heat-treated synthetic imogolite by using ^{27}Al MAS NMR and ^{27}Al multiple-quantum magic-angle-spinning (MQMAS) NMR. These results show five- and four-coordinated Al is formed above 350 °C in amorphous materials clearly.

2.2.4. Catalytic properties

It has been known that imogolite has surface acidity [35]. The acid strength of it is increased by heat treatment. Natural imogolite calcined at various temperatures were used as catalysts for the isomerization of 1-butene [21]. This reaction proceeded effectively over samples calcined at 400 °C. The decomposition reaction of organic peroxides also was investigated using Cu^{2+} loaded imogolite calcined at 500 °C. Furthermore, it was mentioned that imogolite calcined at temperature up to 750 °C exhibited the shape selective adsorption. Bonelli et al. [36] have studied *in situ* IR spectroscopy of synthetic imogolite which adsorbed CO, ammonia, methanol or phenol and catalytic tests using gas-phase phenol reactions with methanol. They showed that a small amount of Al^{3+} Lewis acid sites to adsorb CO existed, the adsorbed ammonia on imogolite evacuated at 150 °C was observed as NH_4^+ species and, in addition, a probe such as CO, ammonia, methanol or phenol could interact with inner silanols. The reaction of phenol with methanol was performed over imogolite after thermal treatment at 300 °C and at 500 °C. The activity was shown over samples heated at 500 °C and o-cresol and anisole were obtained as products.

We have investigated the acidic property of imogolite without heat treatment [37]. At first, we attempted an isomerization reaction of α-pinene on synthetic imogolite. It is known that the isomerization products of α-pinene depend on the acid or base property of the catalyst. In the case of acid catalysts, α-pinene isomerize to limonene, camphene and tricyclene [38]. Prior to the reaction imogolite was dried at 120 °C for 12 hours. The isomerization reaction of α-pinene was carried out at 80 °C for 3 hours or 24 hours using an evacuated batch reactor. In a typical experiment, the reactor was loaded with 12.6 mmol of α-pinene and 50 mg of catalyst. The α-pinene did not react on it. It was reported that imogolite had a weak acid property, however it was not detected in this reaction.

Although isomerization of α-pinene did not occur, some oxidation products of it were detected slightly. So the oxidation reaction of cyclohexene using hydrogen peroxide was carried out to test the possibility as an oxidation catalyst.

In a typical catalytic experiment for oxidation [38], the reaction was carried out by using 25 mmol of cyclohexene, 25 mmol of H_2O_2 (30 wt%) and 30 ml of acetic acid or acetonitrile as a solvent with 100 mg of catalyst under being stirred at 50 °C. The results of the oxidation reaction of cyclohexene with H_2O_2 are listed in Table 1. Without a catalyst the oxidation products such as trans-1,2-cyclohexanediol, cis-1,2-cyclohexanediol and 2-cyclohexene-1-ol were observed using acetic acid as solvent. With imogolite the yield of these oxidation products was increased, but the selectivity was almost the same as the control experiment. This result indicates that imogolite has the potential of concentration and acts as a field of reaction to increase collision frequency.

Product		1-ol	trans-diol	cis-diol
Yield	imogolite	32.3	21.5	2.9
/ %	no catalyst	14.6	12.0	1.3
Selectivity	imogolite	56.9	38.0	5.1
/ %	no catalyst	52.3	43.2	4.5

Table 1. Oxidation of cyclohexene with H_2O_2 over imogolite in acetic acid as a solvent. [38]
Reaction condition: temperature 50 °C, time 6h, imogolite 100 mg
1-ol : 2-cyclohexene-1-ol, trans-diol : trans-1,2-cyclohexanediol, cis-diol: 1,2-cyclohexanediol

In the case of using acetonitrile as the solvent, the oxidation products were not detected for 27 hours of reaction time. In the case of imogolite, 2-cyclohexene-1-ol and 1,2-epoxycyclohexene were produced with 0.5 % and 0.8 % yields, respectively. This results shows that imogolite has the possibility as an oxidation catalyst.

3. Fe-imogolite

Although we found the new possibility of synthetic imogolite as an oxidation catalyst, a chemical modification of imogolite like an introduction of Fe^{3+} ion was necessary to promote the reaction because imogolite is chemically stable. In this section, our results of the synthetic method, characterization and catalytic test of Fe-imogolite are described.

3.1. Synthesis of Fe-imogolite

Fe-imogolite was synthesized based on improving Suzuki's method [6, 7]. The typical method [23] is described here and its flowchart is shown in Scheme 1.

1. We prepared 100 mL of 0.15 mol L^{-1} aqueous solutions consist of $FeCl_3$ and $AlCl_3$ with x = 0.05 (x = Fe/Al+Fe, atomic ratio). Another way of saying, 0.00075 mol $FeCl_3 \cdot 6H_2O$ and 0.01425 mol $AlCl_3 \cdot 6H_2O$ were dissolved in 100 mL of water.
2. 100 mL of 0.1 mol L^{-1} Na_4SiO_4 aqueous solution was prepared.
3. Na_4SiO_4 aqueous solution was added to $AlCl_3$ and $FeCl_3$ mixture and the solution was stirred for 90 minutes.
4. 0.1 mol L^{-1} NaOH aqueous solution was added to the solution consist of Na_4SiO_4, $FeCl_3$ and $AlCl_3$ at the rate of 1.0 mL min^{-1} under stirring until the pH of the mixture become 5.5.

5. The salt-free precursor was obtained from thick solution by centrifugal separation three times.
6. It was dispersed in 2 L of water and 40 mL of 0.1 mol L^{-1} HCl was added.
7. The solution was stirred for two hours at room temperature.
8. It was aged at 100 °C for 40 hours.
9. This aging solution was dialyzed and then dried at 100 °C for two days.
10. The film-like Fe-imogolite was obtained.

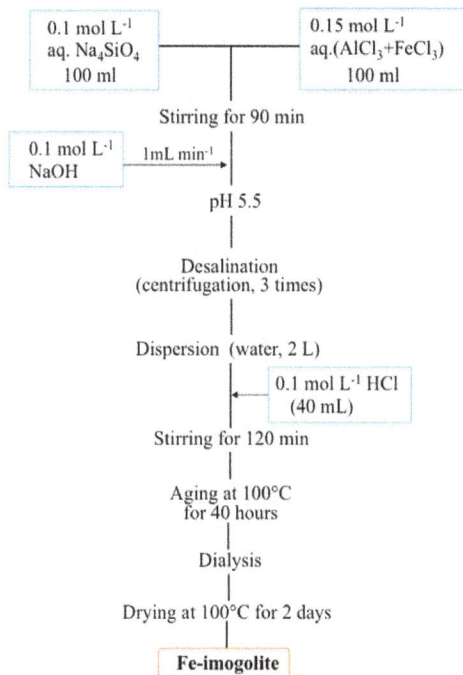

Scheme 1. Flowchart of a synthetic method of Fe-imogolite.

3.2. Characterization of Fe-imogolite

Fe containing samples with x = 0, 0.05, 0.1 were prepared in order to survey the effect of Fe^{3+} contents in starting solutions to formation of imogolite nanotubes,. The XRD profiles and IR spectra of these samples are shown in Figure 8.

As mentioned above, imogolite (x = 0) was characterized by three broad peaks in the low angle region of the XRD profile and a doublet at around 1000 cm^{-1} in the FT-IR spectrum. In the case of x = 0.1, it lacks the three broad peaks in XRD and a doublet in FT-IR, meaning the sample does not have the imogolite structure. In the case of x=0.05, the XRD profile and FT-IR spectrum were similar to the imogolite. The intensity of the three broad peaks is weaker than that of imogolite because XRD measurements were carried out using Cu Kα irradiation. The color and the profiles of XRD and FT-IR indicate the sample with x=0.05 is considered Fe containing imogolite and is thus called Fe-imogolite.

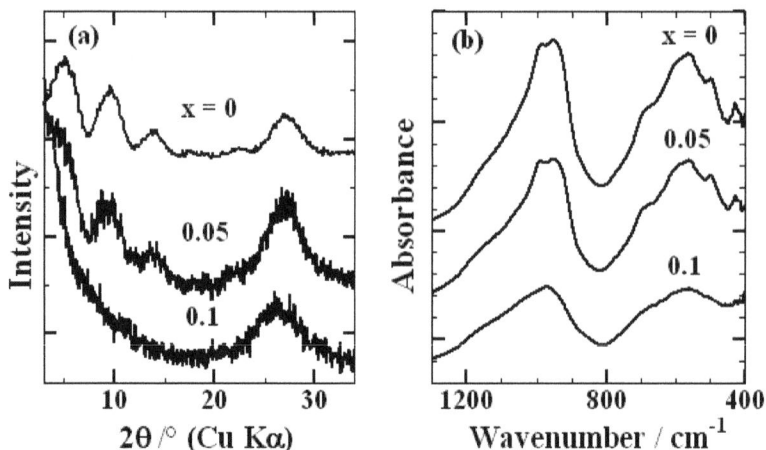

Figure 8. Characterization of Fe containing samples with x (= Fe/Al+Fe, atomic ratio) =0, 0.05, 0.1. XRD profiles (a) and FT-IR spectra (b) [23].

In order to investigate the state of iron ions in Fe-imogolite, it was compared with the Fe^{3+} ion adsorbed on imogolite The adsorbed Fe^{3+} ions sample was prepared by adsorbing $FeCl_3$ onto imogolite from aqueous solution of $FeCl_3$ and is called $FeCl_3$/imogolite. These samples are both reddish brown and the absorption bands of them were observed at the region above 15,000 cm^{-1} in diffuse reflectance ultraviolet visible (UV-VIS) spectra (Figure 9). The absorption bands of Fe^{3+} in many minerals are observed in this region.

It has been known that when the tetrahedral species of Fe^{3+} exist, the pre-edge peak appears strongly in X-ray absorption near edge structure (XANES) spectra. XANES spectra of Fe K-edge (7.111 keV) are shown in Figure 10. The pre-edge peak was not observed in the spectrum of Fe-imogolite nor $FeCl_3$/imogolite or Fe_2O_3. Thus it is clear that the state of iron in Fe-imogolite is octahedral Fe^{3+} ion from the results of UV-VIS and XANES spectra.

Figure 11 shows their Fourier transforms (FT) spectra of Fe-imogolite, $FeCl_3$/imogolite and Fe_2O_3. FT spectrum as radial structure function was obtained by Fourier transformation of k^3-weighted extended X-ray absorption fine structure (EXAFS) function. The FT spectrum of Fe-imogolite is different to that of $FeCl_3$/imogolite or Fe_2O_3. It can be concluded that the state of Fe^{3+} in Fe-imogolite is different to the state of Fe^{3+} ions adsorbed on imogolite. We speculate that Fe^{3+} replaced the Al^{3+} sites in imogolite from these results.

The tapping mode AFM image of Fe-imogolite is shown in figure 12 [39]. The fibrous morphology was observed similar to the synthetic imogolite. It was found that the tube diameter was almost uniform and estimated to be 2.2-2.4 nm from section analysis.

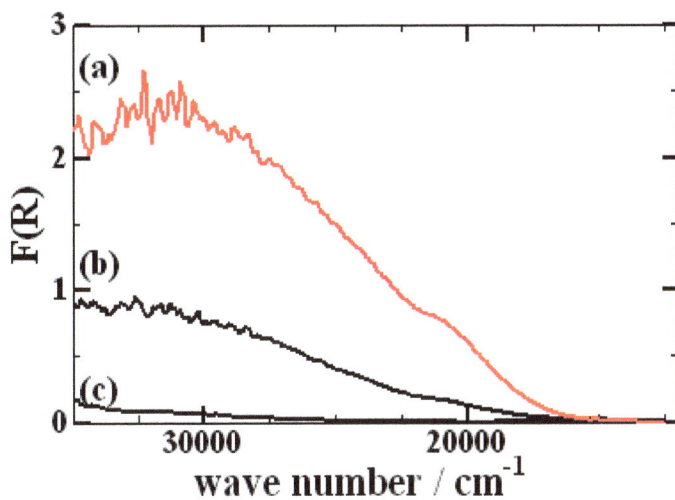

Figure 9. Diffuse reflectance UV-VIS spectra of (a) Fe-imogolite, (b) FeCl₃/imogolite and (c) imogolite [23].

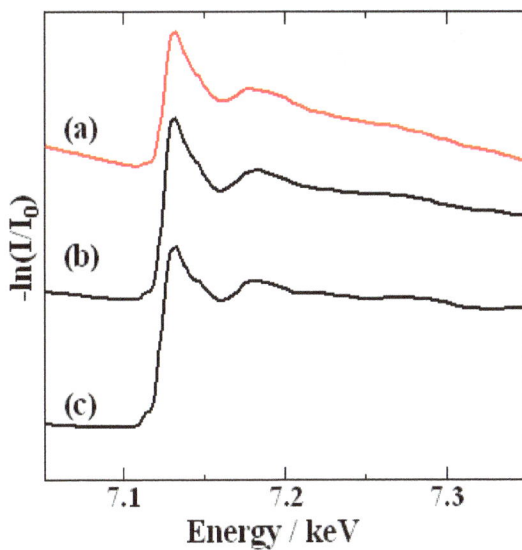

Figure 10. Fe K-edge XANES spectra of (a) Fe-imogolite, (b) FeCl₃/imogolite and (c) Fe₂O₃ (hematite) [23].

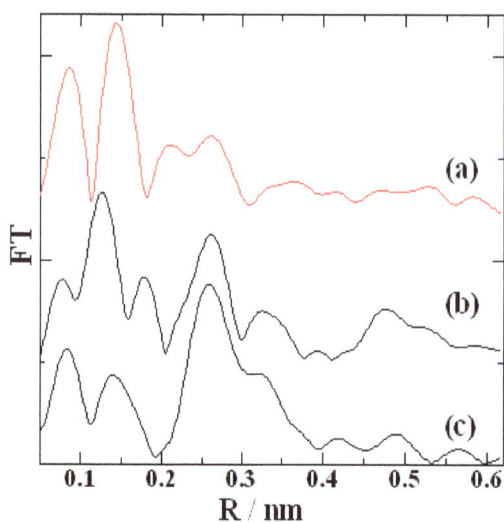

Figure 11. FT spectra of Fe K-edge k^3-weighted EXAFS functions. (a) Fe-imogolite, (b) FeCl$_3$/imogolite and (c) Fe$_2$O$_3$ (hematite) [23].

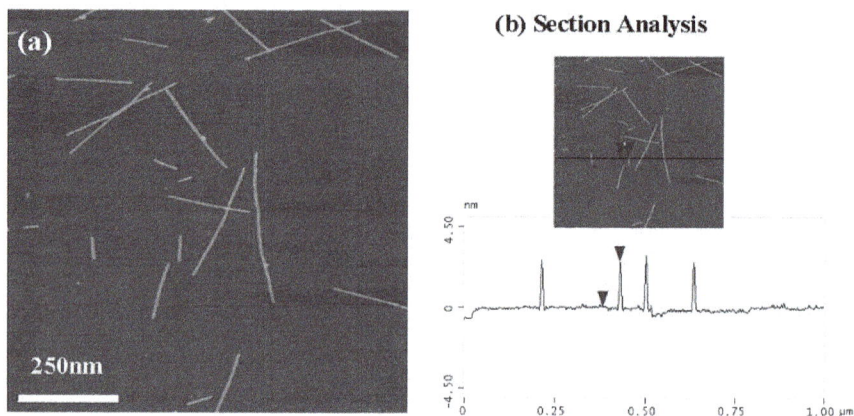

Figure 12. AFM image of Fe-imogolite (a) and section analysis (b). The height profile (right bottom) shows the height on black line in AFM picture (right top) [39].

3.3. Oxidation of hydrocarbons using Fe-imogolite catalyst

3.3.1. Oxidation of cyclohexene [23]

The oxidation of cyclohexene was carried out using cyclohexene, H$_2$O$_2$ and acetonitrile as a solvent with catalysts under being stirred at 50 °C for 24 hours. No product was detected without a catalyst. In the case of imogolite, 2-cyclohexene-1-ol and 1,2-epoxycyclohexane (EP) were produced slightly as stated above. EP was also detected in separate experiments using gibbsite, boehmite and Al_2O_3 as catalysts. Mandelli et al. [40] have reported the epoxidation of cyclohexene using H_2O_2 over Al_2O_3. It was attributed that these oxidation compounds may be produced on the outer surface of the imogolite. The oxidation reaction was promoted by Fe-imogolite and not only alcohol and epoxy compounds but trans-1,2-cyclohexanediol, cis-1,2-cyclohexanediol and 2-cyclohexene-1-one were also obtained as products.

3.3.2. Oxidation of aromatic hydrocarbon [23, 39]

Phenol is one of the most important chemicals in the fields of fiber and medicine manufacturing. More than 90% of phenol is produced by the cumen process, which is a three-step process and produces acetone as a by-product. The development of a one-step process for phenol synthesis by the direct oxidation of benzene is important when concerned with green chemistry as an environment-friendly technique. It was found that benzene or other aromatic hydrocarbons were oxidized by H_2O_2 over Fe-imogolite. These results are described.

The reaction of benzene was carried out using 2 mmol of benzene, 11 mmol of H_2O_2 and 10 mL of acetonitrile as a solvent with 100 mg of catalyst under being stirred at 60 °C. The products were analyzed by GC-MASS. The conversion of H_2O_2 was determined by a volumetric analysis with $KMnO_4$.

Figure 13 shows the results of oxidation reactions of benzene using H_2O_2 and compounds containing Fe. None of the oxidation products were detected without catalysts and with $FeCl_3$/imogolite, Fe_2O_3 (hematite) and α-FeOOH (goethite) as a catalyst. It is an interesting finding that only phenol was obtained as an oxidation product by GC-Mass using Fe-imogolite [23].

The oxidation reactions of benzene using four solvents were examined. The results are summarized in Table 2. None of oxidation products were detected in the case of acetic acid and propionic acid as a solvent. It was found that acetonitrile was the most effective for this reaction among the examined solvents. Although the solution turned palish yellow using acetonitrile as a solvent, after the reaction the absorption due to the Fe^{3+} ion was not detected in UV-VIS absorption spectrum. It was suggested that the origin of the coloring could be phenolic tars that could not detected by GC-Mass. A conversion of H_2O_2 was high with all solvent and H_2O_2 efficiency was 2 % using acetonitrile as a solvent. Thus, the decomposition of H_2O_2 was only caused by using other solvents except for acetonitrile.

The results of the oxidation reaction of aromatic hydrocarbons using Fe-imogolite as a catalyst are summarized in Table 3 [39]. Hydroquinone and catechol were produced as products from phenol by the oxidation with H_2O_2. o-Chlorophenol and p-chlorophenol were produced from chlorobenzene. When the side chain is OH group or Cl group, only the benzene ring was oxidized and the *ortho* and *para* isomers were obtained. Benzaldehyde, o-cresol and p-cresol were produced from toluene. In the case of benzaldehyde, most of the oxidation product was benzoic acid.

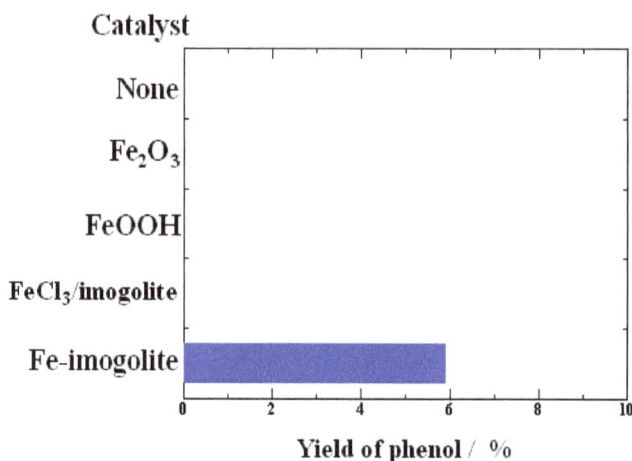

Figure 13. Results of the oxidation reactions of benzene using Fe-imogolite, FeCl₃/imogolite, Fe₂O₃ (hematite) or FeOOH (geothite) as a catalyst [23].

| Solvent | Conversion / % | | Yield of |
	benzene	H_2O_2	Phenol / %
Acetonitrile	10.6	93	10.6
2-Propanol	tr.	99	tr.
Acetic acid	0.0	95	0.0
Propionic acid	0.0	95	0.0

Table 2. The effects of solvents on the oxidation of benzene [39].
Catalyst Fe-imogolite, temperature = 60 °C, time = 6 h, benzene = 2 mmol,
H_2O_2 = 11.5 mmol, solvent = 10 mL

When the side chain is a hydrocarbon group such as methyl, both the benzene ring and the side chain group were oxidized. It was found that the side chain group was more easily oxidized.

Monfared and Amouei [41] have reported direct oxidation reactions of benzene or some aromatic hydrocarbon compounds over Fe^{3+} loaded Al_2O_3 (Fe^{3+}-Al_2O_3) with H_2O_2 in acetonitrile. In their system, however, o-cresol and m-cresol as main products were produced from toluene and no oxidation compounds were produced from phenol. It has been shown that the oxidation property of a Fe-imogolite catalyst is different from that of a Fe^{3+} - Al_2O_3 catalyst.

3.3.3. Oxidation of cyclohexane [42, 43]

Oxidation of cycloxane under mild conditions have been very interesting and have attempted widely [44]. Since the oxidation takes place not only on aromatic rings but on methyl groups by H_2O_2 in acetonitrile over Fe-imogolite catalysts, other organic compounds such as saturated hydrocarbons could be oxidized under this reaction condition. We examined the oxidation of cyclohexane using this catalyst.

Reactant	Products (Yield* / %)	

| phenol | catechol (13.7) | hydroquinone (15.9) |

| chlorobenzene | p- chlorophenol (9.7) | o- chlorophenol (4.8) |

| toluene | benzaldehyde (5.8) | p- cresol (2.2)   o- cresol (1.3) |

| benzaldehyde | benzoic acid (71.7) | 2-hydroxy-benzaldehyde (1.7) |

Table 3. Oxidation reactions of aromatic hydrocarbons with H_2O_2 over Fe-imogolite [39].
*The yield of product is estimated using the ratio of peak area of GC-Mass.
Temperature = 60°C, time = 6 h, reactant = 2 mmol, H_2O_2 = 11.5 mmol, solvent = 10 mL

The reaction of cyclohexane was carried out using 2 mmol of cyclohexane, 10 mmol of H_2O_2 and 10 ml of acetonitrile as a solvent with 50 mg of catalyst under being stirred at 60 °C for 3 hours. The products were analyzed by GC-Mass and GC with FID detector. The oxidation products were hardly detected without catalysts. With Fe-containing imogolite as a catalyst, this reaction was promoted and three oxidation products were created. Two compounds were easily identified by retention time of standard reagents such as cyclohexanone and cyclohexanol among these products. Another is speculated as cyclohexyl hydroperoxide from fragmentation patterns in the mass spectrum. It was reported [45] that cyclohexyl hydroperoxide was prepared efficiently and selectively using cyclohexane and H_2O_2 over Fe^{3+} ion-changed montmorillonite. We identified one of the oxidation products as cyclohexyl hydroperoxide by the compound obtained following this examination. The conversion of cyclohexane was ca. 25%.

The cyclohexyl hydroperoxide was produced immediately as soon as the reaction started. Additionally, it was detected only in oxidation reactions with Fe_2O_3 or α-FeOOH as a catalyst. It was found that cyclohexanone and cyclohexanol were produced via cyclohexyl hydroperoxide as an intermediate which was obtained by reacting cyclohexane and H_2O_2 (Figure 14).

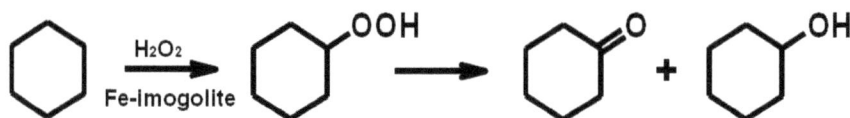

Figure 14. Oxidation reactions of cyclohexane using H_2O_2 over Fe-imogolite [42].

4. Conclusion

Fe-imogolite was synthesized using Na_4SiO_4, $AlCl_3$ and $FeCl_3$ with the atomic ratio Fe / (Al+Fe) = 0.05 and applied as a liquid-phase oxidation catalyst with hydrogen peroxide.

The XRD profile and FT-IR spectrum of this material were similar to the synthetic imogolite. AFM images showed fibrous morphology with ca. 2 nm of diameter. UV-VIS and X-ray absorption spectra revealed the state of Fe^{3+} to be in the octahedral coordination. It was found that Fe-imogolite played as an oxidation catalyst of some hydrocarbons such as cyclohexane, benzene, phenol, toluene and cyclohexane with hydrogen peroxide. The oxidation reaction of cyclohexene was promoted by using Fe-imogolite instead of imogolite as a catalyst. It gave 2-cyclohexene-1-ol, 1,2-epoxycyclohexane, 1,2-cyclohexanediol and 2-cyclohexene-1-one as products. Phenol was produced by the oxidation reaction of benzene. The benzene ring in the aromatic hydrocarbons such as phenol, chlorobenzene, toluene and benzaldehyde was oxidized. Moreover, when the side-chain is a hydrocarbon group, side-chain group was also oxidized. It could be more easily oxidized than the benzene ring. Cyclohexyl hydroperoxide, Cyclohexanone and cyclohexanol were obtained as oxidation products of cyclohexane. It was clarified found that cyclohexanone and cyclohexanol were produced via cyclohexyl hydroperoxide. The possibility of Fe-imogolite as an oxidation catalyst is shown. However, its reaction mechanism has not been clarified yet. It is necessary to investigation the catalytic properties and the structure of Fe-imogolite further.

Author details

Masashi Ookawa

Department of Chemistry and Biochemistry, Numazu National College of Technology, Numazu, Japan

5. References

[1] Yoshinaga N., Aomine S. (1962) Imogolite in some Ando soils, Soil Sci. Plant Nutr. 8: 114–121.

[2] Cradwick P. D. G., Farmer V. C., Rucell J. D., Masson C. R., Wada K., Yoshinaga N., (1972) Imogolite, a Hydrated Aluminum Silicate of Tubular Structure Nature Phys. Sci. 240:187-189.

[3] Bursill L. A., Peng J. L., Burgeois L. N. (2000) Imogolite: an aluminosilicate nanotube materials, Philos. Mag., 80: 105-117.

[4] Koenderink G.H., Kluijtmans S. G J M, Philipse A. P.,(1999) On the Synthesis of Colloidal Imogolite Fibers, J. Colloid. Interface Sci. 216: 429-431.

[5] Hu J., Kannangara G.S. K., Wilson M.A., Reddy N., (2004) The fused silicate rote to protoimogolite and imogolite, J. Non-cryst. Solids, 347:224-230.

[6] Suzuki M., Ohashi F., Inukai K., Maeda M., Tomura S. (2000) Synthesis of Allophane and Imogolite from Inorganic Solution - Influence of Co-Existing Ion Concentration and Titration Rate on Forming Precursor -, Nendo Kagaku (J. Clay Sci. Soc. Japan), 40:1-14.

[7] Suzuki M., Inukai K., (2010) Synthesis and Applications of Imogolite Nanotubes, in: Kijima T. editors, Inorganic and Metallic Nanotubular Materials, Topics in Applied Physics, 117: Springer pp. 159-167, DOI: 10.1007/978-3-642-03622-4_12

[8] Levard C., Masion A., Rose J., Doelsch E., Borschneck D., Dominici C., Ziarelli F., Bottero J. –Y., (2009) Synthesis of Imogolite Fibers from Decimolar Concentration at Low Temperature and Ambient Pressure: A Promising Route for Inexpensive Nanotubes, J. Am. Chem. Soc. 131:17080–17081.

[9] Abidin Z., Matsue N., Henmi T. (2008) A New Method for Nano Tube Imogolite Synthesis, Jpn. J. Appl. Phys. 47:5079-5082.

[10] Mukherjee S., Bartlow V. M., Nair S. (2005) Phenomenology of the Growth of Single-Walled Aluminosilicate and Aluminogermanate Nanotubes of Precise Dimensions, Chem. Mater., 17, 4900-4909.

[11] Donkai N., Miyamoto T., Kokubo T., Tanei H. (1992) Preparation of transparent mullite-silica film by heat-treatment of imogolite, J. Mater. Sci., 27: 6193-6196.

[12] MacKenzie K. J. D., Bowden M. E., Brown I. W. M., Meinhold R. H., (1989) Structure and thermal transformations of imogolite studied by ^{29}Si and ^{27}Al high-resolution solid-state nuclear magnetic resonance, Clays Clay Miner. 37: 317-324.

[13] Tamura, K. And Kawamura K. (2002) Molecular Dynamics Modeling of Tubular Aluminum Silicate: Imogolite, J. Phys. Chem. B, 106: 271-278.

[14] Luciana Guimarães L., Andrey N. Enyashin A. N., Frenzel J., Heine T., Duarte H. A., Seifert G. (2007) Imogolite Nanotubes: Stability, Electronic, and Mechanical Properties, ACS Nano, 1:362-368.

[15] Fernando Alvarez-Ramírez, (2007) Ab initio simulation of the structural and electronic properties of aluminosilicate and aluminogermanate natotubes with imogolite-like structure, Phys. Rev. B 76:125421-125434.

[16] Yamamoto K., Otsuka H., Wada S.-I., Sohn D., Takahara A., (2005) Preparation and properties of [poly(methyl methacrylate)/imogolite] hybrid via surface modification using phosphoric acid ester, Polymer, 46, 12386-12392.

[17] Yah W. O., Yamamoto K., Jiravanichanun N. , Otsuka H., Takahara A. (2010) Imogolite Reinforced Nanocomposites: Multifaceted Green Materials, Materials, 3:1709-1745; doi:10.3390/ma3031709.

[18] Ohashi F., Tomura S., Akaku A., Hayashi S., Wada S. -I., (2004)Characterization of synthetic imogolite nanotubes as gas strage, J. Mater. Sci., 39: 1799-1801.

[19] Ackerman W. C., Smith D. M., Huling J. C., Kim Y. -H., Bailey J. K., Brinker C. J., (1993) Gas/Vapor Adsorption in Imogolite: A Microporous Tubular Aluminosilicate, Langmuir, 9, 1051-1057.

[20] Suzuki M., Ohashi F., Inukai K., Maeda M., Tomura S., Mizota T., (2001) Hydration Enthalpy Measurement and Evaluation as Heat Exchangers of Allophane and Imogolite, J. Ceram. Soc. Japan, 109: 681- 685.

[21] Imamura S., Hayashi Y., Kajiwara K., Hoshino H., Kaito C., (1993) Imogolite: A Possible New Type of Shape-Selective Catalyst, Ind. Eng. Chem. Res., 32: 600-603.

[22] Imamura S., Kokubu K., Yamashita T., Okamoto Y., Kajiwara K., Kanai H., (1996) Shape-Selective Copper Loaded Imogolite Catalyst, J. Catal., 160: 137-139.

[23] Ookawa M., Inoue Y., Watanabe M., Suzuki M., Yamaguchi T. (2006) Synthesis and Characterization of Fe containing Imogolite, Clay Sci., 12, Supplement 2: 280-284.

[24] Farmer V. C., Fraser A. R., Tait J. M. (1977) Synthesis of imogolite: A tubular aluminum silicate polymer, J. Chem. Soc. Chem. Comm., 13: 462-463.

[25] Wada S. -I., Eto A., Wada K. (1979) Synthetic allophane and imogolite, J. Soil Sci., 30: 347-355.

[26] Tani M., Liu C., Huang P. M. (2004) Atomic force microscopy of synthetic imogolite, Geoderma, 118: 209-220.

[27] Ohrai, Y., Gozu, T., Yoshida, S., Takeuchi, O., Iijima, S., Shigekawa, H. (2005) Atomic force microscopy on imogolite, aluminosilicate nanotube, adsorbed on Au(111) surface, Jpn. J. Appl. Phys., 44:5397-5399.

[28] Mccutcheon A., Hu J., Kannangara G. S. K., Wilson M. A., Reddy N., (2005) [29]Si labelled nanoaluminosilicate imogolite, J. Non-Cryst. Solids, 351: 1967-1972.

[29] Wada S. -I., Wada K. (1982) Effects of the substitution of germanium for silicon in imogolite, Clays Clay Miner., 30: 123-128.

[30] Barron P. F., Wilson M. A., Campbell A. S., Frost R. L., (1982) Detection of imogolite in soils using solid state ^{29}Si NMR, Nature 299: 616 – 618.

[31] Ildefonse P., Kirkpatrick R. J., Montez B., Calas G., Flank A. M., Lagarde P., (1994) Investigated structure of imogolite using [27]Al magic-angle-spinning (MAS) NMR and aluminum X-ray absorption near edge structure, Clay. Clay Min., 42:276-287.

[32] Goodman B. A., Russell J. D., Montez B., Oldfield E., Kirkpatrick R. J., Structural studies of imogolite and allophanes by aluminum-27 and silicon-29 nuclear magnetic resonance spectroscopy, Phys. Chem. Minerals, 12:342-346.

[33] Hiradate S., Wada S. -I. (2005) Weathering process of volcanic glass to allophane determined by ^{27}Al and ^{29}Si solid-state NMR, Clay. Clay Min., 53:401-408.

[34] Hatakeyama M., Hara T., Ichikuni N., Shimazu S., (2011) Characterization of Heat-Treated Synthetic Imogolite by [27]Al MAS and ^{27}Al MQMAS Solid-State NMR Bull. Chem. Soc. Jpn., 84: 656–659.

[35] Henmi T., Wada K.,(1974) Surface acidity of imogolite and allopahne, Clay Minerals, 10: 231-245.

[36] Bonelli B., Ilaria Bottero I., Ballarini N., Passeri S., Cavani F., Garrone E., (2009) IR spectroscopic and catalytic characterization of the acidity of imogolite-based systems, J. Catal.,264: 15-30.

[37] Ookawa M., Onishi Y., Fukukawa S., Matsumoto K., Watanabe M., Yamaguchi T., Suzuki M. (2006) Catalytic Property of Synthetic Imogolite, Nendo Kagaku (J. Clay Sci. Soc. Japan), 45: 184-187.

[38] Ohnishi R., Tanabe K., Morikawa S., Nishizaki T. (1974) Isomerization of 2-Pinene Catalyzed by Solid Acids, Bull. Chem. Soc. Jpn. 47: 571 - 574.

[39] Ookawa M., Takata Y., Suzuki M., Inukai K., Maekawa T., Yamaguchi T. (2008) Oxidation of aromatic hydrocarbons with H_2O_2 catalyzed by a nano-scale tubular aluminosilicate, Fe-containing imogolite, Res. Chem. Inter., 34: 679-685.

[40] Mandelli D., van Vliet M. C.A, Sheldon R. A., Schuchardt U., (2001) Alumina-catalyzed alkene epoxidation with hydrogen peroxide, Appl. Cat. A, 219:209-213

[41] Monfared H. H., Amouei Z. (2004) Hydrogen peroxide oxidation of aromatic hydrocarbons by immobilized iron(III), J. Mol. Cat. A, 217: 161-164.

[42] Ookawa M., Nagamitsu Y., Oda M., Takata Y., Yamaguchi T., Maekawa T. (2008) Catalytic properties of Fe-containing imogolite in cyclohexane oxidation, Interfaces Against Pollutions 2008 Programs & Abstracts: 24.

[43] Ookawa M., Nagamitsu, Yamaguchi T., Maekawa T. (2007) Oxidation of cyclohexane Catalyzed by Fe-containing imogolite, Abstracts ISSEM2007 International Symposium on Sustainable Energy & Materials: 47.

[44] Schuchardt U., Cardoso D., Sercheli R., Pereira R., da Cruz R. S., Guerreiro M.C., Mandelli D., Spinacé E. V., Pires E. L. (2000) Cyclohexane oxidation continues to be a challenge, Appl. Cat. A 211:1-17.

[45] Ebitani K., Ide M., Mitsudome T., Mizugaki T., Kaneda K. (2002) Creation of a chain-like cationic iron species in montmorillonite as a highly active heterogeneous catalyst for alkane oxygenations using hydrogen peroxide, Chem. Comm.: 690-691.

Chapter 12

Role of Clay Minerals in Chemical Evolution and the Origins of Life

Hideo Hashizume

Additional information is available at the end of the chapter

1. Introduction

A number of hypotheses have been proposed regarding the origins of life on Earth. In the Russian text of 1924, Oparin (1938) suggested that simple molecules (e.g., CH_4, NH_3) in the early Earth, reacted to form small bio-molecules and complex bio-polymers (e.g., nucleoside, nucleotide, peptide, polynucleotide) which then evolved into multimolecular functional systems, and finally 'life' [1]. A few years later, Haldane (1929) independently proposed a similar hypothesis for the origins of life [2]. It was Bernal (1951), however, who first suggested that clay minerals played a key role in chemical evolution and the origins of life because of their ability to take up, protect (against ultraviolet radiation), concentrate, and catalyze the polymerization of, organic molecules [3]. Indeed, Cains-Smith (1982) has suggested that clay minerals can store and replicate structural defects, dislocations, and ionic substitutions, and act as 'genetic candidates' [4]. Thus, intercalation of organic molecules and monomers into the layer structure of clay minerals, such as montmorillonite and kaolinite, would favor the formation and replication of biopolymers with specified sequences (e.g., enzymes, polynucleotides).

The composition of the primitive atmosphere is an important factor influencing the formation of small biomolecules. Urey (1952) and Miller and Urey (1959) proposed that the early Earth had a reducing atmosphere, and conducted their experiments on chemical evolution accordingly [5, 6]. Computer simulation, however, would indicate that the primitive atmosphere was not reducing. Moreover, it was very difficult to synthesize bio-organic molecules under reducing conditions.

In this review we describe the environment of the primitive Earth, outline the clay-induced formation of small molecules and simple bio-molecules, discrimination of optical isomers, and polymerization of bio-molecules, and then briefly remark on the RNA world and the origin of cells.

2. Environment of the early Earth

In discussing the origins of life, it is important to know the state of the early Earth. Cosmic dust grains, rotating around the primitive Sun, coalesced to form planetesimals, and then larger bodies (e.g., planets) through gravitation, giving rise to the solar system about 4.6 billion years ago [7]. The surface of the primitive Earth was molten to a depth of 1000 km [8]. The light elements had disappeared into space but various gases were retained on the surface by gravitation. As the temperature decreased, the surface of the magma ocean gradually solidified. Water vapor, carbon oxide, nitrogen gas began to cover the Earth surface, forming the primitive atmosphere. Water vapor gave rise to clouds which turned into rain, feeding rivers and oceans. Dissolved metal ions from rocks entered into the primitive ocean. Lightnings and volcanic eruptions often occurred. Small and large meteorites also bombarded the early Earth. These events and light from the sun were conducive to creating simple organic compounds and small bio-molecules.

2.1. Atmosphere

In an early paper, Urey (1952) suggested that Earth's primitive atmosphere was mainly composed of anoxic gases (e.g. NH_3, CH_3) and water vapor [5]. On the other hand, Levine et al. (1982) proposed that non-reductive gases (e.g. CO, CO_2, N_2) made up the paleoatmosphere [9]. More recently, Owen (2008) argued for a composition between anoxic and non-reductive gases [10].

2.2. Meteorite

There are basically two types of meteorites: primitive and fractionated. Table 1 shows a classification scheme for meteorites. Carbonaceous (C-)chondrite is considered to be the earliest type of meteorite, containing a 'memory' of the primitive solar system. C-chondrites are mainly composed of Mg-rich minerals including a hydrous silicate, serpentine. They also contain organic and bio-organic molecules (e.g., amino acids). When such meteorites rained down on the early Earth, the energy of collision would convert simple organic molecules to bio-organic compounds. Table 2 lists the range and variety of organic molecules in space.

2.3. Bombardment by meteorites and asteroids

The craters on Moon were formed about 3.8–4.0 billion years ago through bombardment by meteorites and asteroids [11]. At the same time, huge numbers of meteorite and asteroids would have hit the relatively larger Earth because of its proximity to Moon. As a result, the surface temperature of Earth would markedly increase, and most liquid water would have evaporated. Likewise, many simple organic compounds or large bio-molecules that were present, or formed through meteorite impact, would have volatilized or decomposed.

Stony chondrites
Enstatite chondrites
H(high-Fe) chondrites
L(low-Fe) chondrites
LL(low Fe and low metal) chondrites
Carbonaceous chondrites: Type I No Chondrule, & Type II Chondrule
Achondrites
Ca-poor: Aubrites, Diogenites, Ureilites, Chassignite
Ca-rich: Angrite, Nalhlites, Eucrites, Howardites
Stony- Irons
Pallasites, Mesosiderites
Irons
I AB, II AB, IIAB, IVA, IV B

Table 1. Classification of meteorites [59].

Nitryl and acetylene derivative etc.
HCN, HC_3N, HC_5N, HC_7N, HC_9N, $HC_{11}N$, HC_2CHO, CH_3CN, CH_3C_3N, CH_3C_2H, CH_3CH_2CN, CH_2, CHCN, HNC, HNCO, HNCS, HNCCC, CH_3NC, HCCNC
Aldehyde, Alcohol, Ether, Ketone, Amine, etc.
H_2CO, H_2CS, CH_3CHO, NH_2CHO, H_2CCO, CH_3OH, CH_3CH_2OH, CH_3SH, $(CH_3)_2O$, $(CH_3)_2CO$, HCOOH, $HCOOCH_3$, CH_3COOH, CH_2NH, CH_3NH_2, NH_2CN, H_2C_3, H_2C_4, H_2C_6
Allene
$c-C_3H_2$, $c-SiC_2$, $c-C_3H$, $c-C_2H_4O$
Molecular ions
HCS^+, CO^+, HCO^+, $HOCO^+$, H_2COH^+, $HCNH^+$, HC_3NH^+, HOC^+
Radical
OH, CH, CH_2, NH_2, HNO, C_2H, C_3H, C_4H, C_5H, C_6H, C_7H, C_8H, CN, C_3N, C_5N, CH_2CN, CH_2N, NaCN, C_2O, NO, SO, HCO, MgNC, MgCN, C_2S, NS

Table 2. Organic molecules in space [60]. c- (Allene): circlar.

2.4. Clay minerals

Clay minerals would have formed by weathering of volcanic glass and rocks. Also, when the temperature of land and atmosphere decreased, the highly concentrated cations and anions in the primitive ocean would have precipitated on the primitive ocean floor, and there interacted to yield certain compounds. The oldest rock on Earth is sedimentary in origin, suggesting that land erosion by rivers had already happened. Water would have come into contact with volcanic glass and rocks, opening the way to clay mineral formation. The Mars investigation indicates the occurrence in the planet's surface of clay minerals with an age of > 3.5 Ga, and a chemical composition consistent with Al-Si-O-H and Mg-Si-O-H systems [12]. By analogy, clay minerals would have formed on the early Earth.

3. Simple bio-molecules

The bio-organic compounds of 'life' comprise amino acids, nucleic acid bases, sugars, and lipids (Tables 3 and 4; Figure 1). The role of clay minerals in the synthesis of amino acids, nucleic acid bases and sugars is described below.

Amino acid	Side chaine (R)	Amino acid	Side chaine (R)
Glycine (Gly)	-H	Aspartic acid (Asp)	$-CH_2-\underset{OH}{C}=O$
Alanine (Ala)	$-CH_3$	Glutamic acid (Glu)	$-CH_2-CH_2-\underset{OH}{C}=O$
Valine (Val)	$-\underset{CH_3}{\overset{CH_3}{C}}-H$	Asparagine (Asn)	$-CH_2-\underset{NH_2}{C}=O$
Leucine (Leu)	$-CH_2-\underset{CH_3}{\overset{CH_3}{C}}-H$	Glutamine (Gln)	$-CH_2-CH_2-\underset{NH_2}{C}=O$
Isoleucine (Ile)	$-\underset{CH_3}{\overset{H}{C}}-CH_2-CH_3$	Lysine(Lys)	$-CH_2-CH_2-CH_2-CH_2-NH_2$
Phenylalanine (Phe)	$-CH_2-\bigcirc$	Arginine (Arg)	$-CH_2-CH_2-CH_2-NH-\underset{NH_2}{C}=NH_2$
Tyrosine(Tyr)	$-CH_2-\bigcirc-OH$	Histidine (His)	$-CH_2$ imidazole ring
Serine (Ser)	$-CH_2-OH$	Tryptophan (Try)	$-CH_2$ indole ring
Threonine(Thr)	$-\underset{OH}{CH}-CH_3$	Proline (Pro)	pyrrolidine ring structure
Cysteine (Cys)	$-CH_2-SH$		
Methionine (Met)	$-CH_2-CH_2-S-CH_3$		

Table 3. Twenty bio-amino acids and their side chains. Proline: the side chaine is red. The structure of the amino acid: R-CH(NH2)-COOH

| Phosphatidate |
| Phosphatidylcholine |
| Phosphatidylethnolamine |
| Phosphatidylglycerol |
| Phosphatidylinositol |
| Phosphatidylserine |
| Cardiolipin |
| Sphingomyelin |
| Glycolipid |
| Cholesterol |

Table 4. Main lipids of biologial membrane. [61]

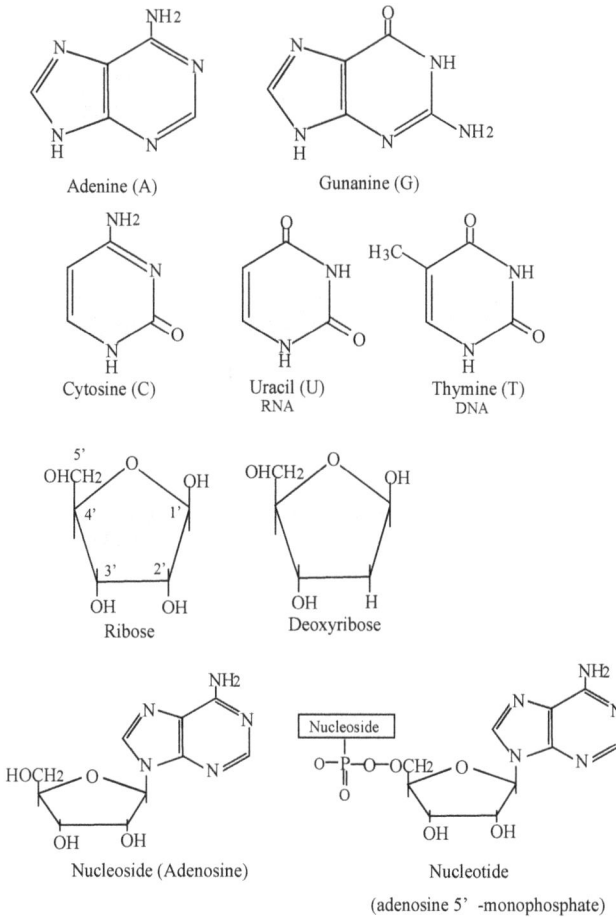

Figure 1. Components of RNA (or DNA) and nucleoside and nucleotide.

3.1. Amino acids

Miller and Urey (1959) were able to synthesize bio-molecules from simple precursors (e.g., NH_3, CH_3, water) by circulating the mixture past an electric discharge ("spark"), simulating a lightning strike. Table 5 lists the compounds obtained abiotically under reducing atmospheric conditions. When montmorillonite was added to Miller's (1953) system, Shimoyama et al. (1978) found that the yield of amino acids with an alkylated side chain increased [13, 14] (Table 5). Subsequently, Yuasa (1989) conducted the sparking experiment using HCN and $NH_4(OH)$ in the presence of montmorillonite [15], obtaining glycine, alanine, and aspartic acid as the main products.

Glycine	Succinic acid
Glycolic acid	**Aspartic acid**
Sarcosine	**Glutamic acid**
Alanine	Iminodiacetic acid
Lactic acid	Iminoacetic-propionic acid
N-Methylalanine	Formic acid
α-Amino-n-butyric acid	Acetic acid
α-Aminoisobutyric acid	Propionic acid
α-Hydroxybutyric acid	Urea
β-Alanine	N-Methy urea

Table 5. Organic compounds dected in Miller's experiment [5,13]. Bold: bio-amino acid. Red; more of these compounds were obtained in the presence of monmtrillonite [13].

Some components of the primitive atmosphere are soluble in water. Under hydrothermal conditions, as would pertain in a thermal vent, the dissolved components would react to form various amino acids (e.g., glycine, alanine, lysine, isoleucine) as Marshall (1994) has reported [16].

The primitive atmosphere might not have been reducing, however. Further, bio-organic molecules are difficult to obtain under the conditions used by Miller (1953) [14]. In an attempt to make for favorable conditions, Kobayashi et al. (1990) used proton irradiation to produce a reducing atmosphere from an oxidizing one [17]. Nevertheless, the role of clay minerals in the formation of bio-molecules remains uncertain. Infrared spectroscopy suggests that the dust in the diffuse interstellar medium contains aliphatic hydrocarbons [18]. Again, it is uncertain whether clay minerals are involved in their formation.

3.2. Adsorption of amino acids by clay minerals

Clay minerals would be capable of adsorbing bio-organic molecules from the early ocean. The resultant clay-organic complexes would partly be deposited on the ocean floor.

Greenland et al. (1962, 1965) investigated the interactions of various amino acids with H-, Na-, and Ca-montmorillonites [19, 20]. Arginine, histidine, and lysine adsorbed to Na- and Ca- montmorillonites by cation exchange. Other amino acids (alanine, serine, leucine, aspartic acid, glutamic acid, phenylalanine) adsorbed to H-montmorillonite by proton transfer. The adsorption of glycine and its oligo-peptides by Ca-montmorillonite and Ca-illite increased with the degree of oligomerization (molecular weight). Hedges and Hare (1987) suggested that the amino and carboxyl groups of the amino acids were involved in their adsorption to kaolinite [21], while Dashman and Stotzky (1982, 1984) reported that kaolinite adsorbed less amino acids and peptides than did montmorillonite [22, 23].

Serpentine is a clay mineral, formed by the weathering of olivine and pyroxene. As such, serpentine would be expected to occur on the surface of the early Earth. Serpentine, however, has a limited capacity for taking up amino acids (Hashizume, 2007) although it can adsorb measurable amounts of aspartic and glutamic acids [24]. On the other hand, allophane can take up appreciable amounts of alanine [25]. The adsorption isotherms showed three distinct regions as the (equilibrium) concentration (C_e) of alanine increased: a nearly linear rise at low C_e, a leveling off to a plateau at intermediate C_e, and a steep linear increase at high C_e. The oligomers of alanine were also adsorbed by allophane but the extent of adsorption did not vary greatly with solute molecular weight [26].

3.3. Optical discrimination

Amino acids can exist in two enantiomeric (chiral) types, namely, D (dextrorotatory) and L (levorotatory) (Figure 2). Both enantiomers would have formed, in equal amounts under abiotic conditions, giving a racemic mixture with a D/L molar ratio of 1/1. The amino acids in living organisms, however, are generally of the L-type. This finding is one of the problems associated with the origins of life.

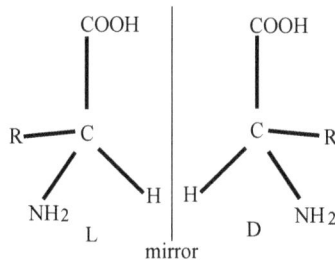

Figure 2. D- and L-enantiomers ("mirror" images) of amino acids. R represents the side chain (Table 3).

The question arises whether clay minerals can discriminate between D- and L-amino acids when placed in contact with a racemic mixture. Using Na-montmorillonite and a racemic mixture of several amino acids, Friebele et al. (1981) did not observe any difference in adsorption between the D- and L-enantiomers [27]. This finding is not altogether surprising since clay minerals have no chirality in their bulk structures although the layer structure of kaolinite may be chiral due to the presence and positioning of vacancies (Figure 3). The edge surface of a montmorillonite layer may also be structurally chiral due to the presence of defects. These chiral structures, however, are not individually separable.

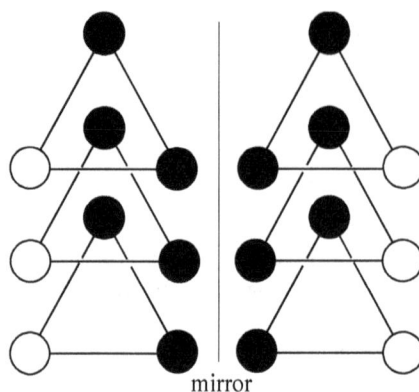

Figure 3. Overlap of Al atoms in stacked layers of kaolinite. Black circles indicate occupied Al site, and white circles indicate vacant Al sites [62].

On the other hand, quartz crystals are intrinsically chiral and, as such, can show stereo-specific effects. Bonner et al. (1974), for example, found that L-quartz preferred L-alanine to its D-enantiomer, while D-quartz adsorbed more D- than L-alanine. The difference in preference was about 1%. Interestingly, Siffert and Naidja (1992) reported that montmorillonite showed stereo-selectivity in the adsorption and deamination of aspartic and glutamic acids. Likewise, Hashizume et al. (2002) reported that an allophane from New Zealand, extracted from a volcanic ash soil, showed a clear preference for L-alanyl-L-alanine over its D-enantiomer [30]. They suggested that the size, intramolecular charge separation, and surface orientation of L-alanyl-L-alanine zwitterions combined to confer 'structural chirality' to the allophane-amino acid complex. Although the allophane sample was purified before use, the presence of trace amounts of organic matter might have left a chiral 'imprint'.

3.4. Nucleic acid bases

Nucleic acids contain two purine bases (adenine, guanine), and three pyrimidine bases (cytosine, uracil, thymine) (Figure 1). Uracil and thymine are found in RNA and DNA, respectively.

Adenine could be prebiotically synthesized from hydrogen cyanide, and cytosine from cyanoacetylene, while uracil could arise from cyanoacetylene via malic acid. Chittenden and Schwartz (1976) reported that the addition of montmorillonite increased the rate of adenine formation [32].

No nucleic acid bases were found in the Miller experiment [6]. Adenine was formed, however, when a mixture of HCN and montmorillonite was added to the reaction vessel, and exposed to lightning [15]. Similarly, uracil could be synthesized from CO, N_2 and H_2O by proton irradiation [17].

3.5. Adsorption of nucleic acid bases by clay minerals

The adsorption of nucleic acid bases by montmorillonite has been widely investigated. Lawless et al. (1984) and Banin et al. (1984) reported that the adsorption of adenosine monophosphate (AMP) by montmorillonite, containing different exchangeable cations (Zn, Cu, Mn, Fe, Ca, Co, Ni), generally increased as solution pH decreased [33, 34]. In the case of Zn-montmorillonite, adsorption of 5′-AMP reached a maximum at pH ~7. The extent of adsorption was primarily influenced by the acid dissociation constant of the nucleic acid base. Winter and Zubay (1995) investigated the relative ability of montmorillonite and hydroxylapatite in adsorbing adenine and adenine-related compounds [35]. They found that montmorillonite adsorbed more adenine than the other compounds (adenosine, 5′-AMP, 5′-ADP, 5′-ATP), while hydroxylapatite preferred adenosine phosphate to adenine and adenosine. The extent of adsorption depended on solution pH, and might also be affected by the buffer used.

More recently, Hashizume et al. (2010) investigated the adsorption of adenine, cytosine, uracil, ribose, and phosphate by Mg-montmorillonite [36]. At comparable concentrations in the equilibrium solution, adsorption decreased in the order adenine > cytosine > uracil, while ribose was hardly adsorbed. Hashizume and Theng (2007) found that allophane had a greater affinity for 5′-AMP than for adenine, adenosine, or ribose [37]. Again, very little ribose was adsorbed. The strong adsorption of 5′-AMP accords with the high phosphate-retention capacity of allophane [38].

The adsorption of nucleic acid bases to clay mineral surfaces has also been assessed by computer simulation. An *ab initio* study by Michalkova et al. (2011) suggests that uracil was adsorbed perpendicularly to the kaolinite surface [39]. With montmorillonite, on the other hand, nucleic acid bases tend to adsorb in a face-to-face orientation with respect to the basal siloxane plane [40].

3.6. Sugar

Sugars may be synthesized from formaldehyde through the Formose reaction. Clay and layered minerals (e.g., montmorillonite, brucite) can catalyze the self-condensation of formaldehyde. Further, the sugar oligomers formed are stabilized by adsorption to montmorillonite [41, 42].

4. Polymerization of bio-polymers

4.1. Peptides

Peptides are polymers of amino acids (Table 3; Figure 4). On the early earth, peptides may have formed at places where energy is produced, such as thermal vents on the sea floor. The primitive ocean may have contained small bio-molecules, including amino acids. As already mentioned, clay minerals would have played an important role in concentrating and polymerizing such molecules on their surfaces.

$$
\begin{array}{cccc}
O & R1 & & O & R2 \\
\| & | & & \| & | \\
HO - C - C - NH2 & + & HO - C - C - NH2 \\
| & & | \\
H & & H
\end{array}
$$

$$
\longrightarrow \quad HO - C - C - N - C - C - NH2 \; + H2O
$$

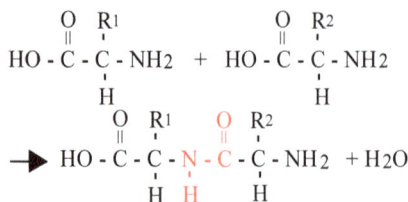

Figure 4. Polymerization of amino acids. Red indicates peptide bonding. R1 and R2 represent amino acid side chains (Table 3).

According to the thermal vent model, organic molecules sink to the sea floor around a thermal vent, and polymerize under conditions of high pressure and temperature [43]. The polymers formed would then move away from the thermal vent. Imai et al. (1997) have attempted to synthesize oligopeptides in a flow reactor, simulating a submarine hydrothermal system [44]. Details of the instrumentation have been given by Matsuno (1997) [45]. With glycine as the monomer, both di- and tri-glycine were formed. The effect of metal cations on amino acid oligomerization was also investigated but that of clay minerals has not been assessed.

The temperature of seawater on the early Earth is expected to be appreciably higher than that at present, a condition that would favor organic molecule polymerization. The thermal copolymerization of various amino acids (aspartic acid, glutamic acid, glycine, alanine, leucine) has been reported by Fox and Harada (1958). Indeed, they were able to synthesize a protenoid microsphere [46].

Plate tectonics would have been operative in the early Earth. When organic-rich sediments moved into a trench where the temperature and pressure are higher than at the surface, the water in the sediments would be depleted. As a result, the concentration of organic molecules would increase, promoting their polymerization [47]. The synthesis of glycine peptides with montmorillonite under trench-like hydrothermal conditions (5–100 MPa pressure; 150 °C temperature) has been reported by Ohara et al. (2007) who obtained up to 10-mers of glycine [48].

Clay mineral particles on the beach undergo repeated drying and wetting, being dried at low tide, and wetted at high tide. This condition would favor polymerization of the clay-associated organic molecules. Using kaolinite and bentonite as the clay minerals, and glycine as the organic species, Lahav et al. (1978) obtained measurable amounts of glycine oligomers up to the 5-mer [49], as shown in Table 6. Ferris et al. (1996) obtained about 50-mers of glutamic acid [50] by incubating (activated) glutamic acid with illite.

In a "shock" experiment, simulating collision of meteorite and asteroids with Earth, by Blank et al. (2001) amino acids were polymerized into oligo-peptides (mostly dimers and trimmers) [51].

Cycles No.	Net heating period (days)	Yields (nmol/mg clay)			
		Dimer	Trimer	Tetramer	Pentamer
Kaolinite					
11	33.7	2.27	0.45		
21	55.0	1.99	0.79	0.29	trace
27	67.4	2.25	1.01	0.33	trace
33	77.3	2.21	0.83	0.32	trace
27	67.4	0.97	0.38	0.10	trace
27	67.4	1.31	0.55	0.15	trace
27	67.4	3.50	1.58	0.60	trace
Bentonite					
11	32.8	6.37	0.20	n.d.	n.d.
21	55.0	7.99	0.60	n.d.	n.d.
27	67.4	4.92	0.61	trace	n.d.
27	67.4	2.92	0.20	n.d.	n.d.
27	67.4	12.7	1.90 trace	n.d.	
11	57.0	36.7	8.2	2.5	
1	10.6	11.9 trace	n.d.		
5	25.4	26.9	1.9	n.d.	
11	57.0	40.1	7.9	1.2	0.8

Table 6. Yields of glycine oligomers in the presence of kaolinite and bentonite, subjected to wetting and drying cycles [49]. n.d. = not detected

4.2. Nucleotide polymers (RNA world)

One hypothesis concerning the origins of life involves the 'RNA world' in which RNA molecules acted as both enzyme-like catalysts and genetic materials [52]. The four nucleic acid bases in RNA have a complementary function. Thus. RNA would be able behave like DNA, although uracil (U) and ribose were used in RNA instead of thymine (T) and deoxyribose in DNA [53]. The molecule of RNA is composed of a nucleic acid base, ribose and phosphate. Combination of a nucleic acid base with 1'-ribose gives rise to a nucleoside, and the addition of phosphate at the 3'- and 5'- positions of ribose yields a nucleotide (Figure 1). RNA is therefore a polynucleotide.

The clay-catalyzed synthesis of polynucleotides has been investigated by Ferris and coworkers. Using the 5-phosphorimidazolide of adenine (ImpA) as the activated RNA monomer, Ferris and Ertem (1993) were able to obtain oligomers containing 6–14 monomer units in the presence of montmorillonite [54]. The formation of RNA oligomers, however, is but the first step towards preparing RNA with more than 40 monomers that are theoretically required for the initiation of the RNA world. Long-chain (elongated) RNA can be obtained using the "feeding" procedure; that is, by daily addition of ImpA to the decanucleotide (10-mer primer) adsorbed to Na-montmorillonite. Polynucleotides containing more than 50-mers are formed after 14 feedings although the principal products contain 20–40 monomer units [55]. Using activated adenosine-, uridine-, guanosine- or cytosine-5′-phospho-1-methyladenine, Joshi et al. (2009) obtained the corresponding 40 to 50-mers [56].

5. Cell origin

Lipids make up part of the living cell (Table 4). In water lipids form a micelle structure where that the outer hydrophilic part is in contact with water, and the hydrophobic part is turned inside (Figure 5). The cell wall has a trans-membrane protein through which nutrients enter the cell.

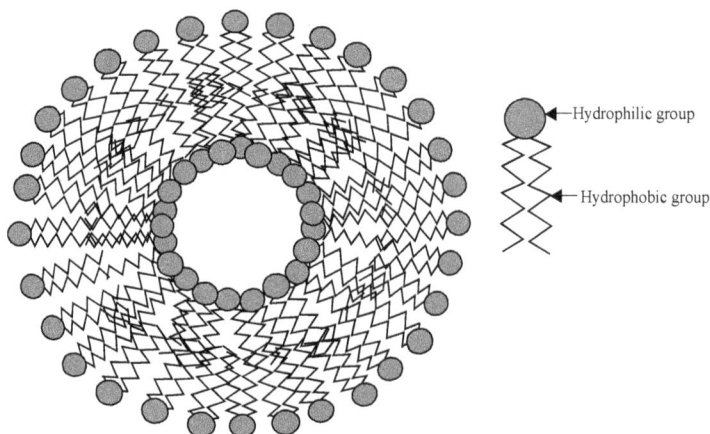

Figure 5. Micelle structure. Hydrophilic group: phospate group, choline group, phosphocholine group etc. Hydrophobic group: stearate, oleate, linoleiate etc.

Clay minerals might function as a primordial cell [4]. When clay minerals are deposited on the ocean floor (or dried), the particles form a pile, enclosing small spaces (Figure 6). It is conceivable that the small spaces behave like cells. Further, when clay minerals are dispersed in water, bubbles form in water or the surface of water, while the clay particles gather at the boundary between water and air, as shown in Figure 7 [57]. In such a case, clay minerals make a cell-like spherule.

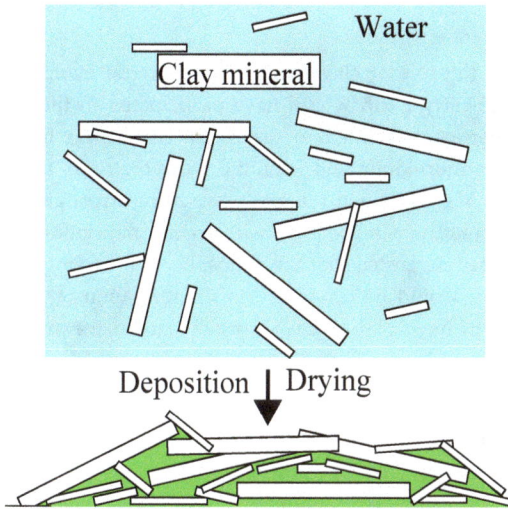

Figure 6. Schematic figure of the room (green) to function like the cell. Clay mineral layers were dispersed in water and then dry water but small rooms play the role like the cell [4].

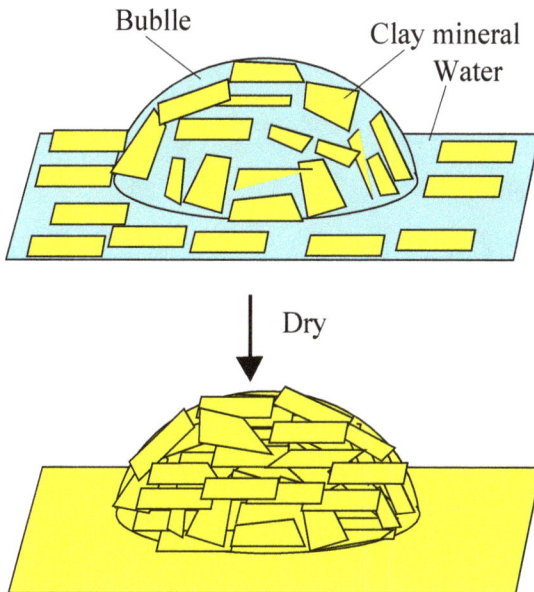

Figure 7. Schematic figure of a bubble shape sheet clay mineral micelle [57].

Unlike surfactants, lipids are difficult to synthesize. Surfactants may transform into lipids. Apatite has been reported to be capable of catalyzing the formation of a proto-lipid [58].

6. Conclusions

This article describes the role of clay minerals in chemical evolution although various other materials in the early Earth would have participated in the formation of life-like structures. Most experiments related to the origins of life on Earth use specific clay minerals, such as montmorillonite and kaolinite. Volcanic rocks from the magma ocean would be enriched in Mg^{2+} ions. On this basis, we have investigated the interactions of Mg-rich clay minerals (e.g., talc, serpentine) with organic molecules, including bio-organic compounds. It is further suggested that the atmosphere of the early Earth contained little oxygen. This condition would be conducive to the formation of Fe^{2+}-rich clay minerals which, therefore, might have played an important part in the synthesis of simple bio-organic molecules.

Besides being able to concentrate organic molecules, clay minerals can also control the surface arrangement of adsorbed nucleic acid bases or amino acids. By using a mixture of different clay mineral species, it may be possible to select a given bio-molecule over another for adsorption and polymerization. Although there is an element of trial and error in investigating the role of clay minerals in chemical evolution and the origins of life, we may yet be surprised by the outcome.

Author details

Hideo Hashizume
National Institute for Materials Science, Tsukuba, Japan

Acknowledgement

We thank the editor Dr. M. Valaskova and the co-editor Dr. G. S. Martynkova for a chance to write this review. We also thank Dr. B. K. G. Theng (Landcare Research, New Zealand) for useful discussions and polishing English.

7. References

[1] Oparin A. I (1938) The Origin of Life. New York: MacMillan. (Orginal: (1924) Proiskhozhdenie zhizny. Moscow: Izd. Moskovski Rabochii.

[2] Haldane J. B. S (1929) The Origin of Life. The Rationalist Annual 148: 3–10.

[3] Bernal J. D (1951) The Physical Basis of Life. London: Routledge and Kegan Paul. 364p.

[4] Cains-Smith A. G (1982) Genetic Takeover. Cambridge: Cambridge University Press. 133p.

[5] Urey H. C (1952) On the Early Chemical History of the Earth and the Origin of Life. Proc. Nat. Acad. Sci. USA 38: 351–363.

[6] Miller S. L, Urey H.C (1959) Organic Compound Synthesis on the Primitive Earth. Science 130: 245–251.

[7] Taylor S. R, Norman M. D (1990) Accretion of Differentiated Planetesimals to the Earth. in Newsom H. E, Jones J. R, editors, Origin of the Earth. New York: Oxford University Press, pp. 29–43.

[8] Davies G. F (1990) Heat and Mass Transport in the Early Earth. in Newsom H. E, Jones J. R, editors, Origin of the Earth. New York: Oxford University Press, pp. 175–194.

[9] Levine J. S, Augustsson T.R, Natarajan M (1982) The Prebiological Paleoatmosphere: Stability and Composition. Origin Life 12: 245–259.

[10] Owen T (2008) The contributions of Comets to Planets, Atmospheres, and Life: Insights from Cassini-Huygens, Galileo, Gitto, and Inner Planet Missins. Space Sci. Rev. 138: 301-316.

[11] Chyba C, Sagan C (1992) Endogenous Production, Exogenous Delivery and Impact-Shock Synthesis of Organic Molecules: an Inventory for the Origins of Life. Nature 355: 125–132.

[12] Bristow T. F, Milliken R.E (2011) Terrestrial Perspective on Authigenic Clay Mineral Production in Ancient Martian Lakes. Clays Clay Miner. 59: 339–358.

[13] Shimoyama A, Blair N, Ponnamperuma C (1978) Synthesis of Amino Acids under Primitive Earth Conditions in the Presence of Clay. In Noda N. H editor, Origin of Life. Tokyo: Center for Academic Publisher, pp. 95–99.

[14] Miller S. L (1953) A Production of Amino Acids under Possible Primitive Earth Conditions. Science 117: 528–529.

[15] Yuasa S (1989) Polymerization of Hydrogen Cyanide and Production of Amino Acids and Nucleic Acid Bases in the Presence of Clay Minerals –In Relation to Clay and the Origin of Life. Nendo Kagaku 29: 89–96 (in Japanese).

[16] Marshall W. L (1994) Hydrothermal synthesis of amino acids. Geochim. Cosmochim. Acta 58: 2099–2106.

[17] Kobayashi K, Tsuchiya M, Oshima T, Yanagawa H (1990) Abiotic Synthesis of Amino Acids and Imidazole by Proton Irradiation of Similated Primitive Earth Atmospheres. Origin Life 20: 99–109.

[18] Sandford S. A, Pendleton Y. J, Allamandola L. J (1995) The Galactic Distribution of Aliphatic Hydrocarbons in the Diffuse Interstellar Medium. Astrophys. J. 440: 697–705.

[19] Greenland D. J, Laby R. H, Quirk J. P (1962) Adsorption of Glycine and Its Di-, Tri-, and Tetra-Peptides by Montmorillonite. Trans. Farad. Soc. 58: 829–841.

[20] Greenland D. J, Laby R. H, Quirk J. P (1965) Adsorption of Amino-Acids and Peptides by Montmorillinite and Illite. Trans. Farad. Soc. 61: 2024–2035.

[21] Hedges J. I, Hare P. E (1987) Amino Acid Adsorption by Clay Minerals in Distilled Water. Geochim. Cosmochim. Acta 51: 255–259.

[22] Dashman T, Stotzky G, (1984) Adsorption and Binding of Peptides on Homoionic Montmorillonite and Kaolinite. Soil Biol. Biochem. 16: 51–55.

[23] Dashman T, Stotzky G (1982) Adsorption and Binding of Amino Acids on Homoionic Montmorillonite and Kaolinite. Soil Biol. Biochem. 14: 447–456.

[24] Hashizume H (2007) Adsorption of Some Amino Acids by Chrysotile. Viva Origino 35: 60–65.

[25] Hashizume H, Theng B. K. G (1999) Adsorption of DL-Alanine by Allophane: Effect of pH and Unit Particle Aggregation. Clay Minerals 34: 233–238.

[26] Hashizume H, Theng B. K. G (1998) Adsorption of L-Alanine Monomer, Dimer, Tetramer, Pentamer by Some Allophanes In: Arehart G.B. Hulston J.R editors, Proc. 9th Intern. Symposium Water-Rock Interaction Rotterdam: A.A. Balkema. pp.105-107

[27] Friebele E, Shimoyama A, Hare P. E, Ponnamperuma C (1981) Adsorption of Amino Acid Entantiomers by Na-Montmorillonite. Origin Life 11: 173–184.

[28] Bonner W. A, Kavasmaneck P. R, Martin F. S (1974) Asymmetric Adsorption of Alanine by Quartz. Science 186: 143–144.

[29] Siffert B, Naidja A (1992) Stereoselectivity of Montmorillonite in the Adsorption and Deamination of Some Amino Acids. Clay Minerals 27: 109–118.

[30] Hashizume H, Theng B. K. G, Yamagishi A (2002) Adsorption and Discrimination of Alanine and Alanyl-Alanine Enantiomers by Allophane. Clay Minerals 37: 551–557.

[31] Fraser D. G, Greenwell H. C, Skipper N. T, Smalley M. V, Wilkinson M. A, Demè B, Heenan R. K (2011) Chiral Interactions of Histidine in a Hydrated Vermiculite Clay. Phys. Chem. Chem. Phys. 13: 825–830.

[32] Chittenden G. J. F, Schwartz A.W (1976) Possible Pathway of Prebiotic Uracil Synthesis by Photodehydrogenation. Nature 263: 350–351.

[33] Lawless J. G, Banin A, Church F. M, Mazzurco J, Huff R, Kao J, Cook A, Lowe T, Orenberg J. B (1985) pH Profile of the Adsorption of Nucleotides onto Montmorillonite I. Selected Homoionic Clays. Origin Life 15: 77–87.

[34] Banin A, Lawless J. G, Mazzurco J, Church F. M, Margulies L, Orenberg J. B (1985) pH Profile of the Adsorption of Nucleotides onto Montmorillonite II Adsorption and Desorption of 5′-AMP in Iron-Calcium Montmorillonite System. Origin Life 15: 89–101.

[35] Winter D, Zubay G (1995) Binding of Adenine and Adenine-Related Compounds to the Clay Montmorillonite and the Mineral Hydroxylapatite. Origin Life Evol. Biosphere 25: 61–81.

[36] Hashizume H, van der Gaast S, Theng B. K. G (2010) Adsorption of Adenine, Cytosine, Uracil, Ribose, and Phosphate by Mg-exchanged Montmorillonite. Clay Minerals 45: 413–419.

[37] Hashizume H, Theng B. K. G (2007) Adenine, Adenosine, Ribose and 5′-AMP Adsorption to Allophane. Clays Clay Miner. 55: 599–605.

[38] Theng B. K. G, Russell M, Churchman G. J, Parfitt R. L (1982) Surface Properties of Allophane, Halloysite, and Imogolite. Clays Clay Miner. 30: 143–149.

[39] Michalkova A, Robinson T. L, Leszezynski J (2011) Adsorption of Thymine and Uracil on 1:1 Clay Mineral Surfaces: Comprehensive Ab Initio Study on Influence of Sodium Cation and Water. Phys.Chem. Chem. Phys. 13: 7862–7881.

[40] Mignon P, Ugliengo P, Sodupe M (2009) Theoretical Study of the Adsorption of RNA/DNA Bases on the External Surfaces of Na^+-Montmorillonite. J. Phys. Chem. C 113: 13741–13749.

[41] Gabel N. W, Ponnamperuma C (1967) Model for Origin of Monosaccharides. Nature 216: 453–455.

[42] Saladino R, Neri V, Crestini C (2010) Role of Clays in the Prebiotic Synthesis of Sugar Derivatives from Formamide. Philos. Mag. 90: 2329–2337.

[43] Mulkidjanian A. Y, Bychkov A. Y, Dibrova D. V, Galperin M. Y, Koonin E. V (2012) Origin of First Cells at Terrestrial, Anoxic Geothermal Fields. PNAS: http://www.pnas.org/cgi/doi/10.1073/pnas.1117774109. Accessed 2012 Jan 20.

[44] Imai E. - I, Honda H, Hatori K, Matsuno K (1997) Autocatalytic Oligopeptide Synthesis in a Flow Reactor Simulating Submarine Hydrothermal Vents. Viva Origino 25: 291–295.

[45] Matsuno K (1997) A Design Principle of a Flow Reactor Simulating Prebiotic Evolution. Viva Origino 25: 191-204.

[46] Fox S. W, Harada K (1958) Thermal Copolymerization of Amino Acids to a Product Resembling Protein. Science 128: 1214.

[47] Nakazawa H, Yamada H, Hashizume H (1993) Origin of Life in the Earth's Crust, a Hypothesis: Probable Chemical Evolution Synchronized with the Plate Tectonics of the Early Earth. Viva Origino 21: 213–222 (In Japanese with English abstract).

[48] Ohara S, Kakegawa T, Nakazawa H (2007) Pressure Effects on the Abiotic Polymerization of Glycine. Origin Life Evol. Biosphere 37: 215–223.

[49] Lahav N, White D, Chang S (1978) Peptide Formation in the Prebiotic Era: Thermal Condensation of Glycine in Fluctuating Clay Environments. Science 201: 67–69.

[50] Ferris J. P, Hill Jr. A. R, Liu R, Orgel L. E (1996) Synthesis of Long Prebiotic Oligomers on Mineral Surfaces. Nature 381: 59–61.

[51] Blank J. G, Miller G. H, Ahrens M. J, Winans R. E (2001) Experimental Shock Chemistry of Aqueous Amino Acid Solution and the Cometary Delivery of Prebiotic Compounds. Origin Life Evol. Biosphere 31: 15–51.

[52] Cech T. R (1986) A Model for the RNA-Catalyzed Replication of RNA. Proc. Nat. Acad. Sci. USA 83: 4360–4363.

[53] Gilbert W (1986) The RNA World. Nature 319: 618.

[54] Ferris J. P, Ertem G (1993) Montmorillonite Catalysis of RNA Oligomer Formation in Aqueous Solution. A Model for the Prebiotic Formation of RNA. J. Am. Chem. Soc. 115: 12270–12275.

[55] Ferris, J. P, Hill Jr. A. R, Liu R, Orgel L. E (1996). Synthesis of Long Prebiotic Oligomers on Mineral Surfaces. Nature 381: 59–61.

[56] Joshi P. C, Aldersley M. F, Delano J. W, Ferris J. P (2009) Mechanism of Montmorillonite Catalysis in the Formation of RNA Oligomers. J. Am. Chem. Soc. 131: 13369–13374.

[57] Subramanian A. B, Wan J, Gopinath A, Stone H. A (2011) Semi-Permeable Vesicles Composed of Natural Clay. Soft Matter 7: 2600–2612.

[58] Ourisson G, Nakatani Y (1998) Molecular Evolution and Origin of Biomembranes. Protein Nucleic Acid Enzyme 43: 1953–1962 (in Japanese).

[59] Henderson P (1982) Inorganic Geochemistry. Oxford: Pergamon Press, 353p.

[60] National Atronomical Observatory (2000) Rika Nenpyo (Chronological Scientific Table 2000) Tokyo:Maruzen Co. Ltd. p. 167.

[61] Vote D, Vote J. G (1990) Biochemistry New York: John Wiley & Sons. (Translated into Japanese by Tamiya N, Muramatsu M, Yagi T, Yoshida H (1992) Vote Seikagaku (Jo), Tokyo: Tokyo Kagaku Dojin, 546p.

[62] The Clay Science Society of Japan (2009) Handbook of Clays and Clay Minerals, 3rd edition. Tokyo:Gihodo. pp. 28–35 (in Japanese).

Chapter 13

Application of Electrochemistry for Studying Sorption Properties of Montmorillonite

Zuzana Navrátilová and Roman Maršálek

Additional information is available at the end of the chapter

1. Introduction

Electrochemical methods have been used for studying clay minerals to a limited extent in comparison with X-ray diffraction, infrared spectroscopy, thermal analysis etc. However, so called "clay electrodes" have become individual and indisputable part of electrochemistry [1 - 8] since the first electrode modified with clay mineral was described in 1983 [1]. Voltammetry on clay electrodes has found its role in the research of clay minerals and their properties, especially ion-exchange and sorption. A thin layer of inorganic material – clay mineral – on the electrode surface does not possess the significant isolating properties thus charge transport proceeds on the clay electrode. The electrode covered with the clay mineral film enables to study the electrode processes and surfaces. By means of the standard electrochemical methods, transport of charge through the clay layer, the sorption and ion-exchange processes in the clay minerals structure can be studied, too. Accumulation of the electroactive compounds into the clay mineral can be successfully used in electroanalysis [7, 9].

Possibilities of electrochemistry in the study of clay minerals by the clay modified electrodes have been in detail stated in the reviews of A. Fitch and her colleagues [3 - 5]. Very interesting opinion concerning the clay mineral structure is presented in the work dealing with study of flow and transport of compounds through the clay film [5]. The clay structure due to its layers charge forms an electrically charged interphase clay – liquid, thus electric double-layer exists on the surface of the clay minerals particles. The processes taking place in the double-layer are considered to be analogous to those in the interphase electrode – solution. The transport mechanisms in the charged media can be studied by the similar way – for example electroosmosis, electromigration or conductivity. These phenomena study has a practical significance in the electrochemical renewal of the soils contaminated with metals [5], for example the technology for elimination of As, Cu, Cd, Cr, Pb, and Zn from the localities impacted with the dangerous wastes. Of course, the technology is suitable in the case of the metals ions able to participate in the ions reactions and to migrate. An advantage of the electrochemical removing of metals consists in the low financial costs and the insignificant environmental impact in comparison with the technologies based on extraction vaporization or exhaustion.

Cyclic voltammetry on the clay modified electrodes has been used to study the sorption properties of clay minerals. Repetitive (multisweep) cyclic voltammetry on the clay modified electrode exhibited dependences similar to the sorption isotherms [8]. A consecutive occupation of the ion-exchange sites in the structure of clay mineral by an appropriate compound results in a potential shift in comparison with the unmodified electrode. With increasing concentration of sorbate the potential shift exhibits curves in the shape of the sorption isotherms which can be used to evaluation of an extent of the ion-exchange or sorption process. Sorption of metals cations on montmorillonite, vermiculite and kaolinite was studied by means of multisweep cyclic voltammetry on the carbon paste electrodes modified with these clays [10]. Similarly to [8] the current response dependences on the cycling time exhibited the same course as the sorption isotherms. The dependences enable to distinguish an extent of sorption of the individual cation and an ability of clay mineral to adsorb the given cation, of course only in the first approximation. For example, the highest sorption of copper was found in the case of montmorillonite, which was used for determination of Cu [11]. The current vs. time dependences obtained by multisweep cyclic voltammetry on the montmorillonite modified carbon paste electrodes were used for determination of the Cu(II) adsorption kinetics [12]. Adsorption of Cu(II) on the various types of montmorillonite was found to be in accordance with the second order model, the experimental values of the maximum current correlate to those calculated from the supposed equation of the kinetics. Cation exchange of Ag(I) and Ca(II) studied on the carbon paste electrode modified with vermiculite showed to be a dominant process of the cations sorption; the simplified model was worked out and equilibrium constant of the Ag(I) ion exchange was determined [13]. The equilibrium constant value was in a good agreement with the constant determined by other method.

In spite of the lower anion exchange capacity of clay minerals in comparison with the cation exchange capacity the exchange of the complex anions $[Hg(ac)_4]^{2-}$, $[HgCl_4]^{2-}$, and $[HgCl_3]^-$ (ac – acetate) was proved on the carbon paste electrodes modified with montmorillonite and vermiculite [14] and it was used for determination of Hg [15]. The same mechanism was found in the case of $[Au(Cl)_4]^-$ on the montmorillonite modified carbon paste electrode [16], which was also used in the electroanalysis [16, 17]. The lower anion exchange ability of clay minerals is caused by presence of the negative charge of layer. It is supposed, that the anion forms of compounds are "repelled" and they are not gripped in the interlayer [4]. A suitable chemical modification of clay minerals can enhance their affinity to anions. This so called "tunning charge selectivity" has been applied in the field of clay electrodes [18]. For example, smectite with bound propylamine groups exhibited the higher ability to accumulate anion $[Fe(CN)_6]^{3-}$ due to protonization of amine groups. The originally cation-exchange smectite was "tunned" to anion-exchange.

Clay minerals represent the significant natural matrix in the soil medium, which participate in many geochemical processes both natural and those connected with transport and behaviour of the anthropogenic compounds in the soils. Interactions of clay minerals with metal and organic compounds influence their activity, transport, and biological availability. The electrodes modified with clay minerals can serve as a model suitable to study some soil processes connected with clay minerals. The carbon paste electrode modified with vermiculite was used as a model of soil fraction to study the binding interactions of Cu(II) with vermiculite [19]. The selected pesticides and their influence on sorption of Cu(II) on vermiculite were studied. The noncomplexing ligands such as fenamiphos, fenmedipham, and atrazine did not exhibited any influence on the Cu(II) ions sorption. The compounds such as desethylatrazine, desizopropylatrazine, and desethyldesizopropylatrazine do not bind on vermiculite, but they decrease the Cu(II) sorption due to formation the coordination compounds with Cu(II). Apart from the determination of these influences on the metal sorption on clay mineral kinetic and thermodynamic aspects of the sorption processes can be characterized by this way. A soil organo-clay complex – clay humate – is formed by interaction of natural organic matter based on humic and fulvic acids with the clay particles surfaces. The humic adsorbates significantly change properties of clay minerals which influenced their reactions with both natural and anthropogenic substances. The carbon paste electrode modified with the prepared clay humates was used for characterization of the clay minerals reactions with Cu(II) in comparison with the origin clay minerals [20]. Cyclic voltammetry on these electrodes distinguished various types of the clay humates, the obtained results were proved by X-ray diffraction study of the clay humate structures.

Organo-clay modified electrodes represent a new type of clay modified electrodes similar to those with clays grafted with the suitable organic function groups [18, 21, 22]. Similarly as the above mentioned "tunning charge selectivity", cation-exchange ability of clay can be changed to anion-exchange ability due to the cationic surfactants adsorbed onto the clay structure [23]. Clay minerals intercalated with alkylammonium cations (cationic surfactants) exhibit the higher affinity to organic compounds. For example, montmorillonite intercalated with hexadecyltrimethylammonium as a modifier in the carbon paste electrode was able to adsorb pesticides isoproturon, carbendiazim, and methyl parathion [24], which showed to be suitable for stripping voltammetric determination of these pesticides in soil and water. Preconcentration of phenol on glassy carbon electrode modified with film of hydrotalcite-like clay containing surfactant sodium octyl sulfate, sodium dodecyl sulfate, or sodium dodecylbenzenesulfonate [25] as well as octylphenoxypolyethoxyethanol or cetylpyridinium bromide [26] was studied. The electrodes exhibited good sensitivity and reproducibility of phenol determination [25]. Carbon paste electrode modified with montmorillonite exchanged with hexadecyltrimethylammonium bromide was successfully used to determine 4-chlorphenol in water samples [27]. Sorption of Hg, Cd, Pb, Cu, and Zn on montmorillonite intercalated with hexadecyltrimethylammonium cations resulted in use of this organo-montmorillonite as a carbon paste modifier [28]. This organo-montmorillonite loaded with 1,3,4-thiadiazole-2,5-dithiol exhibited an excellent selectivity for Hg(II) ions in presence of other ions. The carbon paste electrode modified with these 1,3,4-thiadiazole-2,5-dithiol-organo-montmorillonite provides a selective sensor for the mercury determination.

The examples mentioned above have at least one common denominator: processes are accompanied by changes of charge on or inside materials. The measurement of zeta potential is one of the methods which provide to obtain imagination about character of the particle surface itself and then also about the processes running on this surface (e.g. adsorption, ion exchange, modification). The experiments connected to the zeta potential measuring represent a factor helping to explain the principles of interactions between surface and its surroundings. As an example heavy metals adsorption on clay minerals (heavy metals removing) or surfactant adsorption on carbonaceous materials (flotation of coals) can be mentioned. The zeta potential knowledge can be also applied in the field of oxidative catalysts, pigments, waste slurries, etc. [29 – 31].

Clay minerals are very often characterized by measurement of the zeta potential. One of the most common measurements is monitoring the zeta potential changes with changing the pH value. This monitoring is performed by titration of montmorillonite, illite, and chlorite by hydrochloric acid and sodium hydroxide. The dependence of the zeta potential on pH exhibited a typical form. The acid addition led to increase of the zeta potential, on the contrary increase in pH (addition of base) caused decrease of the zeta potential. The individual clay minerals exhibited differences arising from the structure and chemical composition of the studied samples. The most significant change of pH came up in the case of chlorite and it was also the only one clay mineral where the isoelectric point was determined (pH=5) [32]. Knowledge of the zeta potential value at the given pH is necessary for understanding the processes running on the surfaces.

The adsorption of heavy metals and metal oxides on the surface of clay minerals plays an important role as well. Sorption of iron and aluminium on the surface of illite, montmorillonite and kaolinite led to reduction of the negative charges on the particle surface so that the isoelectric point of these minerals was shifted to the higher values of pH [33].

The zeta potential of zeolites was examined in connection to the sorption of heavy metals on these adsorption materials. Dependence of the zeta potential on pH was influenced by concentration of a bulk electrolyte ($NaNO_3$). As the concentration of $NaNO_3$ was increasing the value of the zeta potential was increasing as well. That fact is explained by change of thickness of double-layer caused by ionic strength of solution. These changes consequently influenced the adsorption of heavy metals (Pb, Cu, Cd, and Zn). The highest adsorption capacity was found in water [34].

The surfactant molecules generally adsorb in the interfaces between two bulk phases such as solid-liquid or electrode-solution [35]. When adsorbing on solid an ionic surfactant exhibits the surface charge. Zeta potential is one of few effective techniques for characterization of the surface charge as well as the surface chemical properties of solids in solution and for understanding the changes on the solid surfaces. The zeta potential values correspond to the quantity and quality of functional groups on the surface [36].

The work deals with use of the montmorillonite modified carbon paste electrodes for studying of the Cu(II) sorption on two types of montmorillonite - montmorillonite SAz-1 and montmorillonite SWy-2 - and their organo-derivatives containing alkylammonium cations - hexadecyltrimethylammonium, benzyldimethylhexadecylammonium, and hexadecylpyridinium. The zeta potential measurement was used to characterize the Cu(II) and hexadecyltrimethylammonium cation sorption on montmorillonite.

2. Experimental part

2.1. Materials and chemicals

Montmorillonites of two types – montmorillonite SAz-1 (MMT,SAz-1) (Apache County, USA) and montmorillonite SWy-2 – (MMT,SWy-2) (Crook County, USA) were provided from The Clay Minerals Society, Source Clays Repository (USA). The fraction used for all experiments consisted of 80 % of particles below 5 μm (Fritsch Particle Sizer Analysette 22, Fritsch GmbH, Idar-Oberstein, Germany). Another sample of montmorillonite MMT,Wy (deposit Wyoming) was obtained from an older collection of colleagues from Institute of Geonics, CAS Ostrava.

Cation exchange capacity (CEC) was calculated as the sum of cations exchanged with NH_4^+ ions during leaching per gram of montmorillonite [14]. The CEC values were 56 cmol(+)/kg and 76 cmol(+)/kg for MMT,SAz-1, resp. MMT,SWy-2. Mineralogical characterisation was performed by infrared spectrometry and X-ray diffraction [37]. The montmorillonite samples were classified as pure montmorillonites without any admixture of other minerals including quartz.

Organo-montmorillonites were prepared by intercalation of three alkylammonium cations. Hexadecyltrimethylammonium bromide (HDTMABr) (Sigma-Aldrich), benzyldimethyl hexadecylammonium chloride (BDHDACl) (Fluka), and hexadecylpyridinium bromide (HDPBr) (Sigma-Aldrich) of analytical reagent grade were used to prepare the modified montmorillonites MMT,SAz-1–HDTMA, MMT,SAz-1-BDHDA, MMT,SAz-1-HDP), MMT,SWy-2–HDTMA, MMT,SWy-2-BDHDA, MMT,SWy-2-HDP.

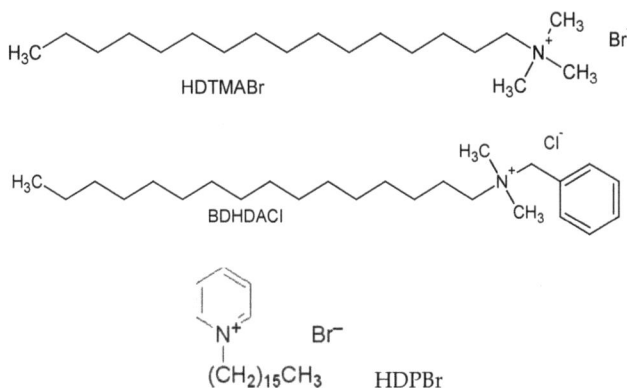

Figure 1. Structures of the used alkylammonium cations

All the chemicals used (sodium acetate and acetic acid for preparation of the background electrolyte as well as sodium hydroxide and hydrochloric acid for measurement of zeta potential) were of analytical grade (Merck, Darmstadt, Germany). The sorption solutions of copper were prepared from $Cu(NO_3)_2 \cdot 3H_2O$ (Lachema Neratovice). The Cu standard for AAS (Cu) (Fluka) was used for AAS analysis. Stock standard solutions of Cu for voltammetry were prepared from Titrisol standards (Merck, Darmstadt, Germany).

3. Procedures

3.1. Preparation of carbon paste electrodes

Carbon paste electrodes (CPEs) modified with either montmorillonites or their organo-derivatives were prepared by the standard procedure [7]. Flake graphite and paraffin oil (Nujol) were thoroughly mixed; in the case of the modified CPEs, an appropriate amount of modifier was added to the graphite before mixing it with oil. A ratio of graphite or the admixture of graphite–modifier to oil was 2.5. The modifier content in the prepared carbon pastes was 10 %, (w/w). The electrodes modifier was before mixing to the carbon paste previously saturated with water vapour, which ensured that the modifier was sufficiently wet but without excess water. No activation and regeneration of the electrode surface prepared in such a way was necessary. The surface was easily renewed by extruding a very small amount of paste and by polishing it on a plastic sheet or scratchboard.

The following carbon paste electrodes were prepared:

- CPE(0) – unmodified carbon paste electrode
- CPE(MMT,SAz-1) – carbon paste electrode modified with 10 wt.% (mass %) of montmorillonite SAz-1
- CPE(MMT,SAz-1-HDTMA) – carbon paste electrode modified with 10 wt.% (mass %) of HDTMA-montmorillonite SAz-1
- CPE(MMT,SAz-1-BDHDA) – carbon paste electrode modified with 10 wt.% (mass %) of BDHDA-montmorillonite SAz-1
- CPE(MMT,SAz-1-HDP) – carbon paste electrode modified with 10 wt.% (mass %) of HDP-montmorillonite SAz-1
- CPE(MMT,SWy-2) – carbon paste electrode modified with 10 wt.% (mass %) of montmorillonite MMT,SWy-2
- CPE(MMT,SWy-2-HDTMA) – carbon paste electrode modified with 10 wt.% (mass %) of HDTMA-montmorillonite SWy-2
- CPE(MMT,SWy-2-BDHDA) – carbon paste electrode modified with 10 wt.% (mass %) of BDHDA-montmorillonite SWY-2
- CPE(MMT,SWy-2-HDP) – carbon paste electrode modified with 10 wt.% (mass %) of HDP-montmorillonite SWy-2

3.2. Cyclic voltammetry

Multisweep cyclic voltammetry (MCV) on the modified CPEs was performed on EKO-TRIBO-Polarograph (EKOTREND, Prague, Czech Republic). A three-electrode cell was equipped with a carbon paste electrode (CPE) (working), an Ag/AgCl (saturated KCl) reference electrode, and a Pt wire auxiliary electrode. MCV at a scan rate of 20 mV s^{-1} was applied with a potential range from –0.6 V to +0.2 V.

3.3. Sorption of copper

An appropriate amount of montmorillonite or its organo-derivative and a volume of the Cu(II) solution (concentrations in the range 0,5 – 10 mmol . l^{-1}) was inserted into an Erlenmeyer flask (ratio solid : liquid = 1 : 100). The suspension was shaken at the laboratory temperature for 24 h. The amount of the adsorbed Cu(II) was determined as a difference between its concentration before and after the sorption (equation 2 below).

3.4. Zeta potential measurement

The Coulter Delsa 440 SX (Coulter Electronic, USA) instrument was used to measure the zeta potential. Delsa 440 SX uses the scattering effect of Doppler light to determine the electrophoretic mobility. The zeta potential was obtained from the electrophoretic mobility by the Smoluchowski equation:

$$\zeta = \frac{\mu.\eta}{\varepsilon} \tag{1}$$

ζ is the zeta potential (V), η represents dynamic viscosity (Pa.s), and ε stands for the dielectric constant. The fixed conditions of measuring were the following ones: temperature (298 K), electric field (15 V), frequency (500 Hz), and the properties of the samples – viscosity (0.0089 kg.m^{-1}.s^{-1}), refraction index (1.333), and dielectric constant (78.36). The samples were sonicated for 1 minute before zeta potential analysis. All zeta potential measurements were at least duplicated; the mean relative standard deviation of the values reported usually did not exceed 5 %. All the solutions were made in distilled water. Analytical grade chemicals were used. Zeta potential measurements consisted from three steps.

At first a dependence of zeta potential on pH was measured. An amount of 0.1g montmorillonite MMT,Wy was added to the flask with 50 ml of distilled water. The pH value of each suspension was adjusted by adding either NaOH or HCl; pH of the solution was measured using the combination single-junction pH electrode with Ag/AgCl reference cell (LP Prague, model MS 22 pH meter).

The second and the third step of the zeta potential measurements were in principle the same. The zeta potential changes were monitored after adsorption of Cu(II) and HDTMA on MMT,Wy. The clay fraction with the particle size below 5 µm was used for the adsorption experiments. The clay amount of 0.1 g of was weighed in the flask and 100 ml of the Cu(II) or HDTMA solution of a known concentration was added. The suspensions were inserted into a thermostatic bath (25°C) and flasks were permanently shaken. As it was found in the previous experiments, a 24 hours´ period is needed to reach equilibrium. The zeta potential of the adsorption suspensions was measured. Then, the clay sample was separated by filtration with paper filter.

The amount of Cu(II) and HDTMA adsorbed (a) was determined from the change in the solution concentration before and after equilibrium, according to:

$$a = \frac{(c_0 - c_e)V}{m} \tag{2}$$

where c_0 is the initial concentration of the HDTMA solution, c_e the concentration of the HDTMA solution at the adsorption equilibrium, V the volume of the HDTMA solution and m the mass of the clay.

The HDTMA concentration of the filtered solutions was determined by UV/VIS spectrophotometry (ALS laboratory group, CZ_SOP_D06_07_N03).

3.5. Preparation of organo-montmorillonites

An amount of 1 g of montmorillonite and 100 ml 7.5 mmol . l^{-1} solution of alkylammonium cation was shaken at the laboratory temperature for 2.5 h (the time was found in the previous experiments). After the sorption the suspension was centrifuged (9000 rev min^{-1}) for 10 min. The supernatant was removed, the solid was washed out with 5 ml solution of ethanol : water = 2 : 1 and the suspension was again centrifuged at the same rate. The washing out was repeated with 5 ml of ethanol and the preparative of organo-montmorillonite was air dried after centrifugation.

3.6. X-ray diffraction

X-Ray diffraction (XRD) was carried out at Nanotechnology Centre, VSB-Technical University Ostrava. XRD patterns of the tested samples were measured by diffractometer INEL equipped with Cu anode, generator (2000 sec, 35 kV, 20 mA) and detector CPSD 120, samples were measured in a flat rotation holder.

3.7. Infrared spectroscopy

Infrared spectra were recorded on Nicolet Avatar 320 FTIR spectrometer (ThermoNicolet, USA) equipped with the DTGS/KBr detector for the middle IR range. The KBr pressed-disc (13 mm diameter) technique (1 mg of sample and 200 mg of KBr) was used. The spectra were measured in the spectral range from 4000 to 400 cm^{-1} (64 scans, 4 cm^{-1} resolutions).

3.8. Thermal analysis

Thermal analysis was carried out using multimodular thermal analyser SETSYS 12-SETARAM equipped with a measurement head TG/DTA rod (Institute of Geonics, CAS, Ostrava). The TG/DTA curves were recorded under an air environment from 25 to 1200 °C, the heating rate was 10 K min^{-1}.

3.9. Analysis of metals

The metal amount in the supernatant after sorption was found by means of atomic absorption spectrometry (AA240FS Varian, USA) by flame atomization air-acetylene (flow rate 13.5 l min⁻¹, λ_{Cu} 249.2 nm, and slit width 0.5 nm for Cu.

3.10. Analysis of alkylammonium cations

The HDTMA concentrations in the supernatant after sorption were determined by UV-VIS spectrophotometry (ALS laboratory group, CZ_SOP_D06_07_N03, Ostrava, Czech Republic). The BDHDA and HDP concentrations in the supernatant after sorption were determined by UV VIS spectrophotometry (Varian Cary 50) at 264 nm (BDHDA) and 260 nm (HDP). The amount of the adsorbed metal was determined as a difference between its concentration before and after the sorption. The amount of the adsorbed alkylammonium cation was determined as a difference between its concentration before and after the sorption (equation 2).

4. Results

4.1. Characterization of prepared organo-montmorillonites

The obtained XRD patterns of the organo-montmorillonites were analysed for d-values of the basal spacing (001) and compared with those of the original montmorillonites (Table 1).

montmorillonite	d(001) [nm]	montmorillonite	d(001) [nm]
MMT,SAz-1	1.47	MMT,SWy-2	1.36
MMT,SAz-1–HDTMA	1.65	MMT,SWy-2–HDTMA	1.68 – 2.40
MMT,SAz-1–BDHDA	1.78 – 2.40	MMT,SWy-2–BDHDA	1.79
MMT,SAz-1–HDP	1.69 – 2.20	MMT,SWy-2–HDP	1.72

Table 1. Basal spacing values d(001)

The organo-montmorillonites exhibited an evident increase of the basal spacing in comparison with those of the unmodified which indicates an intercalation of alkylammonium cations in their interlayer [38 - 40]. The obtained values about 1.7 nm can be judged to bilayer arrangement of cations, the higher values about 2.2 – 2.4 nm probably correspond to a paraffin-type of arrangement – the ammonium groups are attached to the silicate layer, the nonopolar chains are oriented under the tilt angle [41, 42].

The infrared spectra of the original montmorillonites have been already studied [37] and interpreted according to [43]. The infrared spectra of the organo-montmorillonites were interpreted with help of the spectra of the pure alkylammonium salts to distinguish new absorption bands in the organo-montmorillonites. All infrared spectra exhibited characteristic absorption bands of the presented alkylammonium cations. The absorption bands at 2920 cm⁻¹ correspond to antisymmetric stretching vibrations and the bands at 2850 cm^{-1} to symmetric stretching vibrations of C – H bounds reflecting alkyl chains of alkylammonium cations. The presence of benzene ring in BDHDA and HDP is confirmed by symmetric stretching vibration of C – H bounds in aromates at 3050 cm^{-1} and by symmetric stretching vibration of C – C bounds of conjugated system at 1620 and 1471 cm^{-1}. The absorption band at 1487 cm^{-1} corresponds to bending vibration of the N – H bounds of ammonium groups. The measured infrared spectra of MMT,SWy-2 and its derivatives with HDTMA and BDHDA are shown at Figure 2. The other organo-montmorillonites exhibited the similar infrared spectra. The presence of all cations in the organo-montmorillonites is evident due to either intercalation or adsorption process.

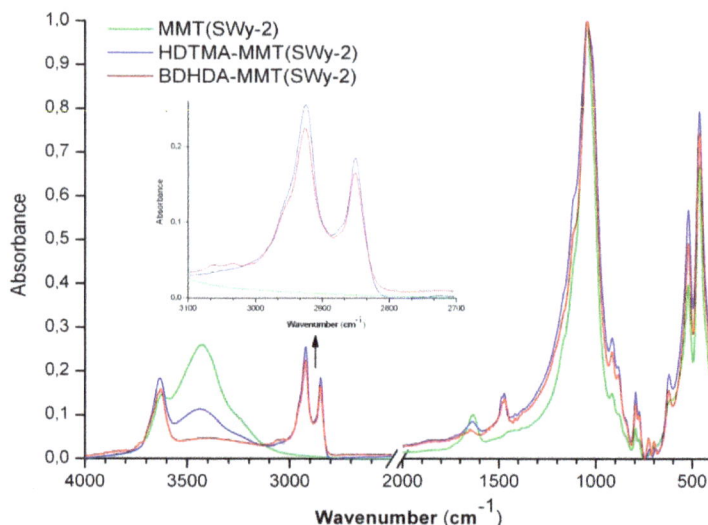

Figure 2. Infrared spectra of montmorillonite SWy-2 and its organo-derivatives

Thermogravimetric (TG) and differential thermal (DTA) curves of the original montmorillonites were found in the curve library of Clay Minerals (Institute of Geonics, CAS, Ostrava). The peak temperatures for montmorillonites obtained from DTA curve are: 166 °C and 234 °C for dehydration, 673 °C and 884 °C for dehydroxilation/melting and 1027 °C for recrystallization/transformation. In the case of the organo-montmorillonites, the shape of TG/DTA curves corresponded to the original montmorillonites, but the thermal effects exhibited the slightly different temperatures and intensities. The organo-montmorillonites exhibited the higher values of temperatures related to the total melting. The temperatures of exothermic effects connected to recrystallization and transformation increased with the increasing amount of the alkylammonium cations. The temperature and intensity of the first two peaks related to the dehydration process decreased with the increasing amount of the added surfactant. In addition, the organo-montmorillonites are subjected to a thermal effect at the temperature interval 270–450 °C corresponding to the alkylammonium decomposition [44]. As expected, the mass loss in the whole temperature interval 25-1200 °C increased with increasing amount of the alkylammonium cations.

The results obtained by X-ray diffraction, infrared spectroscopy and thermal analysis proved the presence of the alkylammonium cations in the organo-montmorillonites. As it was already stated [38, 45] intercalation of alkylammonium cations takes place due to both ion-exchange and induced and π-π interactions forming a double-layer in the interlayer space of montmorillonites. Although the previous work [38] supposed the higher intercalation due to a content of benzene ring in the case of BDHDA, the values of the parameter d(001) (Table 1) do not prove this suggestion.

4.2. Characterization of montmorillonite by zeta potential measurement

4.2.1. The pH influence on zeta potential of montmorillonite suspension

The most important factor that affects the zeta potential is pH. The zeta potential value on its own without a stated pH is only a virtually meaningless number. Generally, the zeta potential versus pH curve will be positive at low pH and lower or negative at high pH. The point where the plot passes through the zero value of the zeta potential is called the isoelectric point and it is very important from a practical consideration. It is normally the point where the colloidal system is stable to a lesser extent.

The following figure shows a typical curve for the zeta potential value on the pH value in the case of the montmorillonite particles.

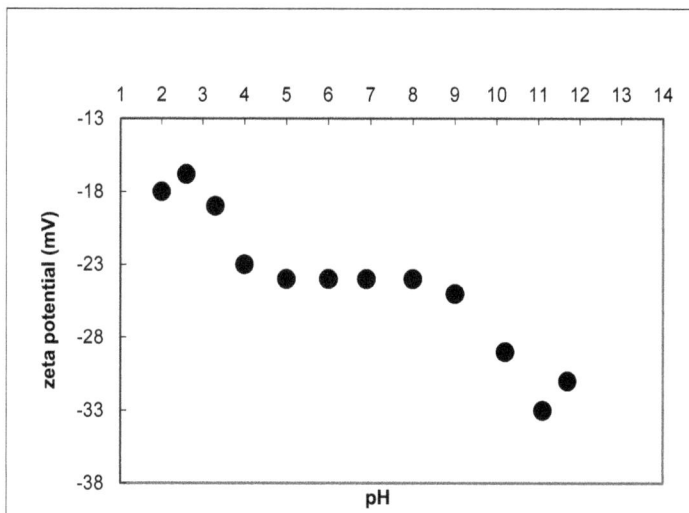

Figure 3. The influence of pH on zeta potential of the MMT,Wy particles

The zeta potential of the montmorillonite particles in distilled water (pH 6) reaches approximately -24 mV, the zeta potential is negative. With increasing addition of alkali to the suspension of pH 6 the particles tends to acquire a more negative charge and with increasing addition of acid a charge is negative to a lesser extent.

At least two main results can be mentioned from the previous picture:

-   the zeta potential of the montmorillonite particles did not change significantly in the pH range 4 - 9. In this range particles have tendency to coagulate.
-   in the strongly acid solution the zeta potential became positive but the isoelectric point was not reached.

Next, the zeta potential of the montmorillonite particles in the copper solutions was determined during the Cu(II) sorption. The values of the zeta potential before and after the Cu(II) adsorption on the montmorillonite were compared (Figure 4).

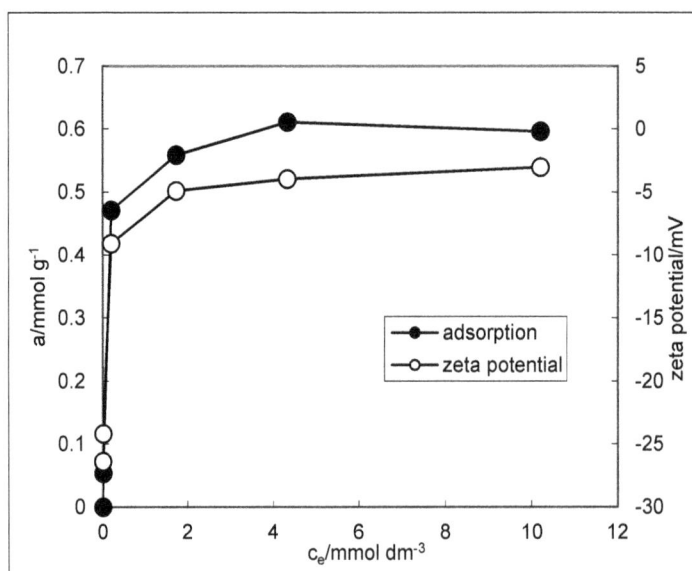

Figure 4. Figure 4. Dependence of zeta potential and adsorption of Cu(II) on MMT,Wy on the Cu(II) equilibrium concentration (c_e)

The adsorption of the copper ions caused the change of the zeta potential of the clay particles. The zeta potential became more positive. On the contrary to Figure 3 where the isoelectric point was not reached with increasing H^+, in the case of addition of the cooper ions the zeta potential was very close to 0 mV. It just confirmed well known fact that the valencies of the ions have great impact on the electrokinetic behaviour of the suspensions.

The zeta potential changes during the Cu(II) sorption can be used as an additional parameter for characterization of the sorption on montmorillonite. The excellent correlation of the Cu(II) adsorption isotherm with the zeta potential dependence on the Cu(II) concentration in the sorption solutions demonstrated on Figure 4 proved this conclusion.

4.2.2. Sorption of alkylammonium cations – coherence of classical batch experiments and zeta potential measurement

The following research has been performed in order to show the above mentioned possibility of zeta potential for evaluation of the sorption processes. The adsorption of HDTMA on the montmorillonite SAz-1 was studied by conventionally measured adsorption isotherms and the zeta potential was measured simultaneously in the sorption suspensions. Figure 5 demonstrates the typical adsorption isotherm which shape indicates the adsorption isotherm of Langmuir model of monolayer coverage of an adsorbent.

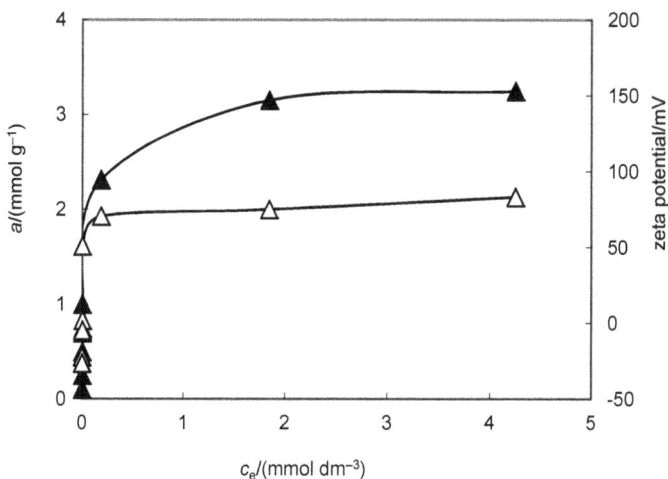

Figure 5. Dependence of zeta potential (blank symbols) and adsorption of HDTM (full symbols) on MMT,SAz-1 on the equilibrium concentration of the HDTMA solution (c_e) [45]

Figure 5 shows the remarkable same course of the adsorption isotherm and the changes of zeta potential of the adsorption system indicating a change of the surface charge due to the HDTMA adsorption. In comparison to other studied sorbent (e.g. coal), the zeta potential is influenced by adsorption to a lesser extent on montmorillonite (about 100 mV). However, an amount of the adsorbed HDTMA is much more higher on montmorillonite and it even exceeds its cation exchange capacity, which proves the concept of the double- or triple-layer arrangement of the adsorbed alkylammonium cations. Thus, adsorption of HDTMA on montmorillonite probably takes place by cation exchange into its interlayer space as well on the external surface. Subsequently, the HDTMA adsorption proceeds via van der Waals interactions [45].

4.3. Comparison of Cu(II) sorption on montmorillonites and their organo-derivatives

4.3.1. Batch technique study

The adsorption isotherms of Cu(II) measured by the batch technique on montmorillonites and their alkylammonium-derivatives are demonstrated on Figures 6 and 7. It is evident that the presence of all three alkylammonium cations in the montmorillonites caused a decrease of the Cu(II) sorption. No significant differences were found in the case of individual alkylammonium cation, the sorption was decreased on about 50 % in comparison with the original montmorillonites. The same decrease of sorption was also found in the case of HDTMA [46] and tetrabutylammonium cations [47]. The sorption sites of clay mineral are occupied by a relatively great alkylammonium cation, which inhibits sorption of the metal cations.

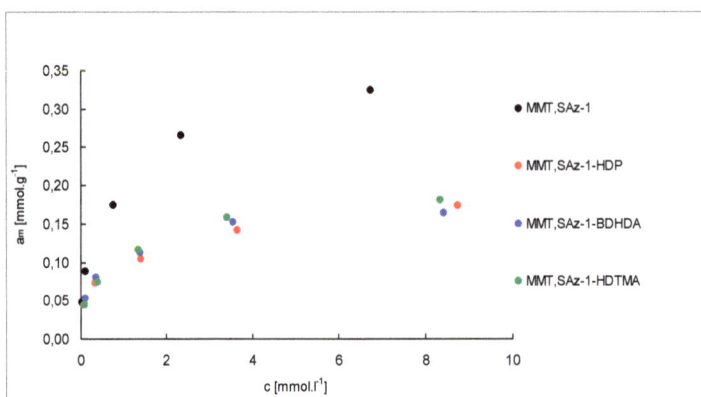

Figure 6. Adsorption isotherms of Cu(II) on MMTA,SAz-1 and its organo-derivatives

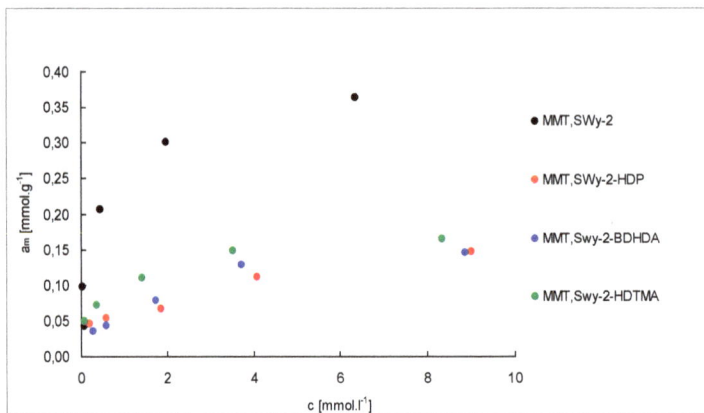

Figure 7. Adsorption isotherms of Cu(II) on MMTA,SWy-2 and its organo-derivatives

The linearized forms of the adsorption isotherms proved that all obtained adsorption isotherms exhibited the Langmuir model of sorption. The linearized forms of the adsorption isotherms were used to calculate a maximum adsorbed amount of Cu(II) named a_{Cu} (Table 2). The values of parameter a_{Cu} indicate no significant differences between the studied organo-montmorillonites. This fact corresponds to the found changes of the parameter d(001) that indicated very similar increase of the basal spacing in all prepared organo-montmorillonites (Table 1).

Montmorillonite	a_{Cu} [mmol . g^{-1}]	montmorillonite	a_{Cu} [mmol . g^{-1}]
MMT,SAz-1	0.34	MMT,Swy-2	0.38
MMT,SAz-1–HDTMA	0.17	MMT,Swy-2–HDTMA	0.17
MMT,SAz-1–BDHDA	0.17	MMT,Swy-2–BDHDA	0.17
MMT,SAz-1–HDP	0.18	MMT,Swy-2–HDP	0.16

Table 2. Maximum absorbed amount of Cu(II) on montmorillonites and their organo-derivatives

4.3.2. Cyclic voltammetry study

Multisweep cyclic voltammetry (MCV) represents a suitable technique to study adsorption of metals onto a modifier in the carbon paste. In the case of the montmorillonite modifier the obtained current increases with successive occupation of the ion-exchange sites of its structure until a constant, maximum value of current (steady state current) is achieved. The obtained dependences of the current response on a number of cycling (on time) can be used as a characteristic feature for the metals sorption on montmorillonites. The typical multisweep cyclic voltammograms are shown on Figure 8 that depicts MCV of Cu(II) performed on the carbon paste electrode modified with MMT,SAz-1. The successive occupation of the ion-exchange sites of montmorillonite with increasing number of cycling (time) caused the current increase corresponded to the adsorbed amount of Cu(II) [8,10] (the underneath voltammetric peak on Figure 8).

The Cu(II) sorption on two types of montmorillonite – SWy-2 and SAz-1 - was studied by means of multisweep cyclic voltammetry. The obtained current responses on the carbon paste electrodes CPE(MMT,SWy-2) and CPE(MMT,SAz-1) exhibited the time dependences that correspond to the Cu(II) sorption on the montmorillonites (Figure 9). These dependences enable to distinguish the Cu(II) sorption on the various types of montmorillonite. It is seen, that the higher sorption capacity was found in the case of MMT,SWy-2. These finding closely corresponds to the results obtained by the batch technique (Table 2) that proved the slightly higher sorption on the MMT,SWy-2, too.

The multisweep voltammetric study of the Cu(II) sorption on the montmorillonite and its organo-derivative has already demonstrated that the organo-derivative MMT,SAz-1-HDTM exhibited the lower steady state current due to a lower sorption of Cu(II) (Figure 10). The cation exchange sites of the MMT,SAz-1-HDTMA are occupied with the HDTMA cations, which inhibits sorption of the cationic forms Cu^{2+} and $[Cu(ac)]^+$ (ac – acetate) in comparison with the unmodified montmorillonite. HDTM incorporated into the interlayer of MMT,SAz-1 decreased the Cu(II) sorption approximately to   65 % [48].

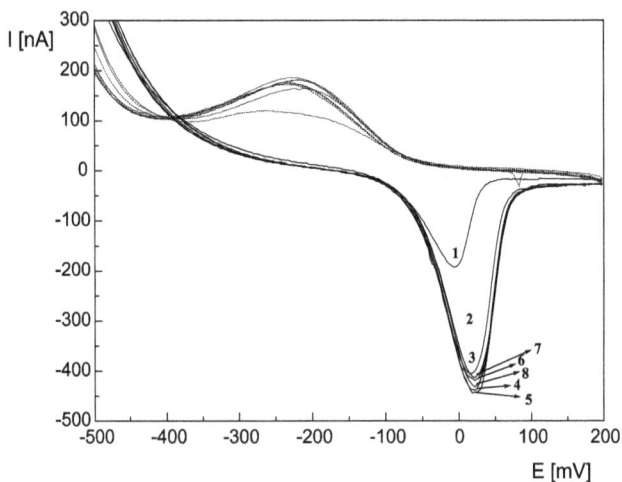

Figure 8. Multisweep cyclic voltammetry of Cu(II) (2.5 x 10^{-5} mol . l^{-1}) in acetate buffer pH 3.6 on CPE(MMT,SAz-1)

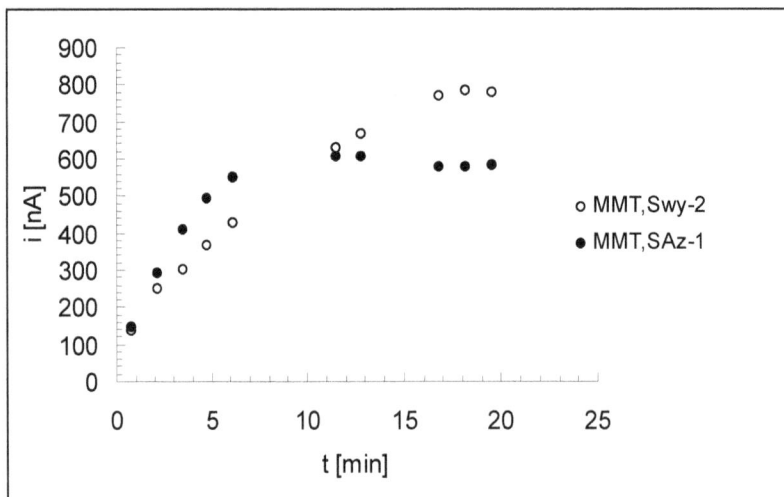

Figure 9. Multisweep cyclic voltammetry of Cu(II) (2.5 x 10^{-5} mol .l^{-1}) in acetate buffer pH 3,6 on CPE(MMT,SAz-1) and CPE(MMT,SWy-2)

In this study MCV of the Cu(II) was performed by the same procedure [48] on the carbon paste electrodes modified with the prepared organo-montmorillonites MMT,SAz-1-BDHDA, MMT,SAz-1-HDP, MMT,SWy-2-BDHDA, and MMT,SWy-2-HDP. The MCV voltammograms of Cu(II) performed on the CPEs modified with the organo-montmorillonites in the medium of acetate buffer pH 3.6 – 5.2 were used to construct the dependence of the maximum current response (steady state current) on the cycling time. The typical dependences in comparison to the above mentioned results [48] are shown on Figure 10. The dependences also indicate the decrease of the steady state current on all organo-montmorillonites in comparison to the original montmorillonite.

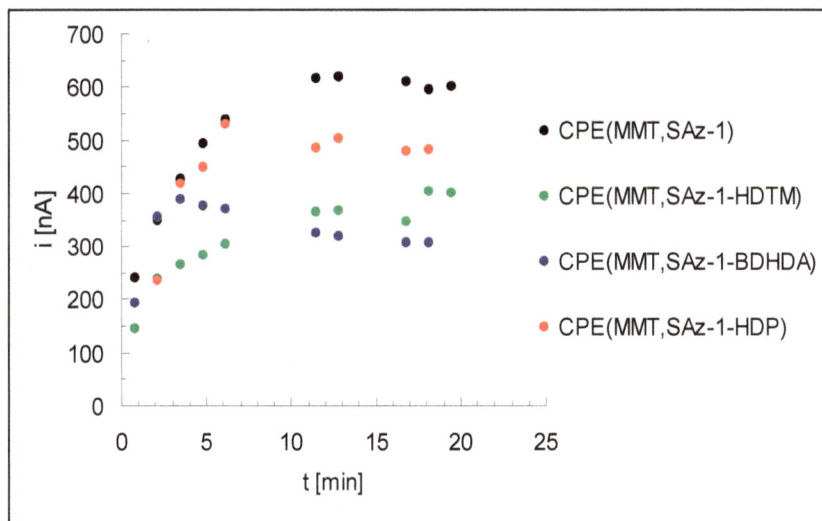

Figure 10. Current vs. t dependences for MCV of Cu(II) (2.5×10^{-5} mol . l^{-1}) in acetate buffer pH 4.0 on CPEs modifed with organo-montmorillonites

The steady state current decrease corresponds to the decrease of the Cu(II) sorption onto the modifier in the carbon paste which is caused by the presence of the alkylammonium cations in the interlayer structure or on the surface of the montmorillonite modifier in the carbon paste electrode. It is seen form Figure 10 that the Cu(II) sorption is decreased on about 80 % and 63 % in the case of MMT,SAz-1-HDP, resp. MMT,SAz-BDHDA.

Although the demonstrated dependences (Figure 9 and 10) cannot be considered as the classical sorption isotherms, they indicates the same characteristics of the Cu(II) sorption:

- the higher sorption capacity to Cu(II) was found for MMT,SWy-2 by means of the classical sorption isotherms as well as by the multisweep cyclic voltammetry on the carbon paste electrodes modified with montmorillonites
- the highest sorption was found on the original montmorillonites by the batch technique as well as by the multisweep cyclic voltammetry on the carbon paste electrode modified with the original montmorillonite
- all prepared organo-montmorillonites exhibited significantly lower sorption capacity for Cu(II) calculated from the adsorption isotherms as well as measured by means of the multisweep cyclic voltammetry on the carbon paste electrode modified with the organo-montmorillonites.

5. Conclusions

The applied electrochemical techniques – measurement of zeta potential and multisweep cyclic voltammetry – offer possibility to study and characterize properties of clay minerals connected with the sorption processes on their surfaces.

The zeta potential measurement is a method suitable for characterization of the clay minerals particles from the point of view of their surface charge that is one of the significant parameters influencing sorption on clay minerals. As it was shown the measurement of zeta potential enable to determine the pH range where the montmorillonite particles did not change significantly the surface charge and where they tend to coagulate. The zeta potential measurement during sorption of Cu(II) exhibited the excellent agreement of the zeta potential dependence on the Cu(II) equilibrium concentration with the adsorption isotherm measured by the classical sorption experiments (batch techniques). Analogously, the same course of the adsorption isotherm and the zeta potential changes in the sorption system of montmorillonite – hexadecyltrimethylammonium cation was found. The zeta potential changes can be compared with its changes on other sorbents (coals). The alkylammonium adsorption way can be evaluated from the zeta potential changes [45].

Multisweep cyclic voltammetry represents another method to characterize the metals sorption using the electrodes modified with a studied adsorbent – clay mineral. The current response dependences on time give the typical curves suitable to describe the sorption. The following general conclusions can be obtained by this method:

- comparison of the individual clay minerals in terms of their sorption capacity
- influence of the alkylammonium presence in the clay mineral on the metals sorption
- influence of other important parameters on the sorption – for example pH, presence of other cations, temperature

The advantage of multisweep cyclic voltammetry consists in relatively fast performance providing the first idea about sorption. For example, measurement on carbon paste electrode takes about 20 – 30 min in comparison with the   time-consuming batch sorption experiments – minimum 24 hours. On the other hand, multisweep cyclic voltammetry enable only a semi-quantitative evaluation of sorption.

The described electrochemical methods can be successfully used as the additional methods of study and characterization of clay minerals.

Author details

Zuzana Navrátilová and Roman Maršálek

Department of Chemistry, Faculty of Science, University of Ostrava, Ostrava, Czech Republic

Acknowledgement

The contribution has been done in connection with the project Institute of Environmental Technologies, reg. no. CZ.1.05/2.1.00/03.0100 supported by Research and Development for

Innovations Operational Programme financed by Structural Founds of Europe Union and from the means of state budget of the Czech Republic.

The authors are also grateful to The Specific university research Ostrava University, project No. sgs18/PrF/2011.

6. References

[1] Ghosh P K, Bard A J (1983) Clay-modified Electrodes. J. am. chem. soc. 105: 5691 - 5693.

[2] Bard A J, Mallouk T (1992) Electrodes Modified with Clays, Zeolites, and Related Microporous Solids. In: Murray R W, editor. Molecular Design of Electrode Surfaces. New York: J. Wiley, New York. pp. 271 - 311.

[3] Fitch A (1990) Clay-modified Electrodes: A Review. Clay. clay miner. 38: 391 - 400.

[4] Macha S M, Fitch A (1998) Clays as aAchitectural Units at Modified Electrodes. Microchim. acta. 128: 1 - 18.

[5] Macha S, Baker S, Fitch A (2002) Clays and Electrochemistry: An Introduction. CMS workshop lectures. 10: 1 - 62.

[6] Villemure G (2002) Electron Transport in Electrodes Modified with Synthetic Clays Containing Electrochemically Active Transition Metal Sites. CMS workshop lectures 10: 149 - 184.

[7] Navratilova Z, Kula P (2003) Clay Modified Electrodes: Present Applications and Prospects. Electroanalysis. 15: 837 - 846.

[8] Fitch A (1990) Apparent Formal Potential Shifts in Ion Exchange Voltammetry. J. electroanal. chem. 284: 237 - 244.

[9] Mousty Ch (2004) Sensors and Biosensors Based on Clay-modified Electrodes – New Trends. Appl. clay sci. 27:159 - 177.

[10] Kula P, Navratilova Z (1994) Voltammetric Study of Clay Minerals Properties. Acta u. carol. geol. 38: 295 - 301.

[11] Kula P, Navratilova Z (1996) Voltammetric Copper(II) Determination with a Montmorillonite-modified Carbon Paste Electrode. Fresenius j. anal. chem. 354: 692 - 695.

[12] Navrátilová Z, Hranicka Z (2008) Montmorillonite Modified Electrodes for Study of Cu Adsorption Kinetics. Sensing in Electroanalysis. 3: 55 – 64.

[13] Kalcher K, Grabec I, Raber G, Cai X, Tavcar G, Ogorevc B (1995) The Vermiculite-modified Carbon Paste Electrode as a Model System for Preconcentrating Mono- and Divalent Cations. J. electroanal. chem. 386: 149 - 156.

[14] Navratilova Z, Kula P (2000) Cation and Anion Exchange on Clay Modified Electrodes. J. solid state electrochem. 4: 342 - 347.

[15] Kula P, Navratilova Z, Kulova P, Kotoucek M (1999) Sorption and Determination of Hg(II) on Clay Modified Carbon Paste Electrodes. Anal. chim. acta. 385: 91 - 101.

[16] Kula P, Navratilova Z (2001) Anion Exchange of Gold Chloro Complexes on Carbon Paste Electrode Modified with Montmorillonite for Determination of Gold in Pharmaceuticals. Electroanalysis. 13: 795 - 798.

[17] Navratilova Z, Kula P (2000) Determination of Gold Using Clay Modified Carbon Paste Electrode. Fresenius j. anal. chem. 367: 369 - 372.

[18] Tonle I K, Ngameni E, Walcarius A (2004) From Clay- to Organoclay-film Modified Electrodes: tuning chargé selectivity in Ion Exchange Voltammetry. Electrochim. Acta. 49: 3435 - 3443.

[19] Grabec-Svegl I, Ogorevc B, Hudnik V (1996) A Methodological Approach to the Application of a Vermiculite Modified Carbon Paste Electrode in Interaction Studies: Influence of Some Pesticides on the Uptake of Cu(II) from a Solutions to the Solid Phase. Fresenius z. anal. chem. 354: 770 - 773.

[20] Kula P, Navratilova Z, Chmielova M, Martinec P, Weiss Z, Klika Z (1996) Modified Electrodes: Sorption of Cu(II) on Montmorillonite - Humic Acid System. Geologica carpathica – series clays 5: 49 - 53.

[21] Tonle I K, Ngameni E, Njopwouo D, Carteret C, Walcarius A (2003) Functionalization of Natural Smectite-type Clays by Grafting with Organosilanes: Physico-chemical Characterization and Application to Mercury(II) Upteke. Phys. chem. chem. phys. 5: 4951 - 4961.

[22] Tonle I K, Ngameni E, Walcarius A (2005) Preconcentration and Voltammetric Analysis of Mercury(II) at a Carbon Paste Electrode Modified with Natural Smectite-type Clays grafted with Organic Chelating Groups. Sensors and actuators B – chemical. 110: 195 - 203.

[23] Ngameni E, Tonle I K, et al. (2006) Permselective and Preconcentration Properties of a Surfactant-intercalated Clay Modified Electrode. Electroanalysis. 18: 2243 - 2250.

[24] Manisankar P, Selvanathan G, Vedhi C (2006) Determination of Pesticides Using Heteropolyacid Montmorillonite Clay-Modified Electrode with Surfactant. Talanta. 68: 686 - 692.

[25] Fernandez L, Borrás C, Carrero H (2006) Electrochemical Behavior of Phenol in Alkaline Media at hydrotalcite-like Clay/Anionic surfactants/Glassy Carbon Modified Electrode. Electrochim. Acta. 52: 872 - 884.

[26] Fernandez M, Fernández L, Borras C, Mostany J, Carrero H (2007) Characterization of Surfactant/Hydrotalcite-like Clay/Glassy Carbon Modified Electrodes: Oxidation of Phenol. Anal. chim. Acta. 597: 245 - 256.

[27] Yang H, Zheng X., Huang W, Wu K (2008) Modification of Montmorillonite with Cationic Surfactant and Application in Electrochemical Determination of 4-chlorophenol. Colloid and surfaces B: biointerfaces. 65: 281 -284.

[28] Newton Dias Filho L, Ribeiro do Carmo D (2006) Study of an Organically Modified Clay: Selectvie Adsorption of Heavy Metal Ions and Voltammetric Determination of Mercury(II). Talanta. 68: 919 - 927.

[29] Marsalek R. (2009) The influence of surfactants on the zeta potential of coals. Energ Source Part. A. 31: 66-75.

[30] Kaluza L, Gulkova D, Vit Z, Zdrazil M. (2007) Tailored Distribution of MoO3 in the TiO2 and ZrO2 Supported Catalysts by Water-Assisted Spreading. Proceedings of European Congress of Chemical Engineering (ECCE-6). Copenhagen, 16-20 September 2007.

[31] Marsalek R. (2011) The Reduction of Zinc using Goethite Process and Adsorption of Pb+II, Cu+II and Cr+III on Selected Precipitate. International journal of environmental science and development (IJESD). Available http://www.ijesd.org/abstract/133-C013.htm. Accessed 2011 APR 8.

[32] Sondi I, Biscan J, Pravdic V (1996) Electrokinetics of Pure Clay Minerals Revisited. Journal of colloids and interface science. 178: 514-522.

[33] Zhuang J, Yu G.R. (2002) Effects of Surface Coating on Electrochemical Properties and Contaminant Sorption of Clay Minerals Chemosphere. 49: 619-628.

[34] Lv L., Tsoi G., Zhao X.S. (2004). Uptake Equilibria and Mechanisms of Heavy Metal Ions on Microporoous Titanosilicate ETS-10.Ind. Eng. chem. res. 43: 7900-7906.

[35] Vittal R, Gomathi H, Kim K J (2006) Beneficial Role of Surfactants in Electrochemistry and in the Modification of Electrodes. Adv. colloid. interface. sci. 119: 55-68.

[36] Wu S F, Yanagisawa K., Nishizawa T (2001) Zeta Potential on Carbons and Carbides. Carbon. 39: 1537-1541.

[37] Navratilova Z, Vaculikova L (2006) Electrodeposition of Mercury Film on Electrodes Modified with Clay Minerals. Chemical Papers. 60: 348–352.

[38] Navratilova Z, Wojtowicz P, Vaculikova L., Sugarkova V (2007). Sorption of Alkylammonium Cations on Montmorillonite. Acta Geodynamica et Geomaterialia. 4: 59–65.

[39] He H, Ma Y, Zhu J, Yuan P, Qing Y (2010) Organoclays Prepared from Montmorillonites with Different Cation Exchange Capacity and Surfactant Configuration. Applied clay science 48: 67 - 72.

[40] Volzone G, Rinaldi J O, Ortiga J (2006) Retention of Gases by Hexadecyltrimethylammonium-montmorillonite Clays. Journal of environmental management 79: 247 - 252.

[41] Betega de Paiva L, Morales A R, Valenzuela Díaz F R (2008) Organoclays: Properties, Preparation and Applications. Applied clay science, 42: 8–24.

[42] Kooli F, Liu Y, Alshahateet S F, Messali M, Bergaya F (2009) Reaction of Acid Activated Montmorillonites with Hexadecyltrimethylammonium Bromide Solution. Applied clay science. 43: 357–363.

[43] Madejova J (2003) FTIR Techniques in Clay Mineral Studies. Vibrational Spectroscopy. 31: 1-10.

[44] Delbem M F et al. (2010) Modification of a Brazilian Smectite Clay with Different Quaternary Ammonium Salts. *Química Nova*. 33: 309-315.

[45] Marsalek R, Navratilova Y (2011) Comparative Study of CTAB Adsorption on Bituminous Coal and Clay Mineral. Chemical Papers. 65: 77 – 84.

[46] Zeng Z, Jiang J (2005) Effects of the Type and Structure of Modified Clays on Adsorption Performance. International journal of environmental studies. 62: 403 – 414.

[47] Gupta S S, Bhattacharyya K G (2006) Adsorption of Ni (II) on Clays. Journal of colloid and interface science. 295: 21 – 32.

[48] Navratilova Z, Hranicka Z (2009) Carbon Paste Electrode Modified with Alkylammonium-clay Composite. Sensing in elecroanalysis 4: 39 – 46.

Chapter 14

Application of Clay Mineral-Iridium(III) Complexes Hybrid Langmuir-Blodgett Films for Photosensing

Hisako Sato, Kenji Tamura and Akihiko Yamagishi

Additional information is available at the end of the chapter

1. Introduction

There has been an extensive interest in developing photo-responsive devices based on luminescent transition metal complexes (Sato & Yamagishi, 2007). As a promising applicant for emitting composites, cyclometalated iridium(III) complexes are attracting a wide attention due to their highly emitting properties in a visible region (Lo et al., 2011, Ulbricht et al., 2009). The lifetime of the excited triplet states is very long (ca. 1 μs) and the quantum yield attains a value as high as 10 ~ 100 %. These iridium(III) complexes are used for photo-responsive molecular devices such as photo-diodes and oxygen sensors (Lowry & Benhard, 2006, Sajoto et al., 2009). The attempts are based on the fact that energy transfer takes place efficiently from the triplet excited state of an iridium(III) complex to semiconductors or an oxygen molecule in the triplet ground state.

Clay is an environmentally-friendly ubiquitous material. They are characterized by layered structures with cation-exchange properties (Ogawa & Kuroda, 1995). Cationic molecules are intercalated in the narrow galleries between aluminosilicate layers. The materials are used as a host for various types of photochemical reactions. We recently studied the interactions of cationic iridium(III) complexes with a colloidally dispersed clay (Sato et al., 2009, 2011a). The adsorption of iridium(III) complexes by a clay was found to result in the drastic enhancement of emission intensity in an aqueous solution. The attempts demonstrate that emission behavior often provide a key to monitoring the delicate change of adsorption structures.

Recently the application of clay minerals for photochemical reactions was further extended to thin-film systems. For such purposes, luminescent Langmuir-Blodgett (LB) films were prepared by depositing the monolayers of amphiphilic iridium(III) complexes onto a glass substrate (Sato et al., 2010). The emission properties of a single layered film were studied under vacuum or under the atmosphere of various gases. As far as our literature survey is concerned, it was the first report on the Langmuir-Blodgett films consisting of iridium(III) complexes with no additives. This pioneering work, however, revealed several problems concerning the low stability and poor reproducibility in sensing functions due to their fragile properties.

In order to overcome the above disadvantages, we attempted to construct a hybrid film of an amphiphilic iridium(III) complex with a clay (Sato et al., 2011b). In these years, the inclusion of layered materials such as layered niobates, titania and clays has been attempted to enhance the mechanical strength of a molecular film and stabilizing its sensor function (Acharya et al. 2009). When clay minerals were used, it was expected that the diversity of elemental compositions of clay sheets might enable us to tune the sensitivity and selectivity of sensing towards a wide range of target molecules. Motivated by these backgrounds, a LB film was constructed by hybridizing an amphiphilic cationic iridium(III) complex with various clays such as synthetic saponite, synthetic hectorite, and natural montmorillonite. As a result, a single layered hybrid LB film was shown to exhibit emission intense enough to study the interaction of the film with gaseous molecules (Sato et al., 2011b). This work would be a benchmark to explore a gas sensor based on cyclometalated iridium(III) complexes.

2. Interaction of clays with luminescent iridium(III) complexes

2.1. Metal ion sensing by luminescence

Cationic cyclometalated iridium(III) complexes were used as an emitting adsorbate by a clay. We synthesized an iridium(III) complex, [Ir(ppy)$_2$dmbpy]PF$_6$ (ppyH = 2-phenylpyridine and dmbpy = 4,4'-dimethylbipyridine: Chart 1) (denoted by [Ir(III)L$_1$] complex), according to the reported method (Lowry & Benhard, 2006). Synthetic saponite (Kunimine Ind. Co.; (Si$_{7.20}$Al$_{0.80}$)(Al$_{0.03}$Mg$_{5.97}$)O$_{20}$(OH)$_4$Na$_{0.77}$ (CEC: 75 meq/100g) or sodium montmorillonite (Kunipia-P, Kunimine Ind. Co.; (Si$_{7.70}$Al$_{0.30}$)(Al$_{3.12}$Mg$_{0.68}$Fe$_{0.19}$)O$_{20}$(OH)$_4$) (Na$_{0.49}$Mg$_{0.14}$) (CEC: 115 meq/100 g) was used as a host material. Adsorption was carried out by mixing a solution of the [Ir(III)L$_1$] complex with a clay suspension within 10 milliseconds by means of a stopped-flow apparatus. This procedure guaranteed the uniform adsorption of the metal complexes over clay particles particularly at low loading.

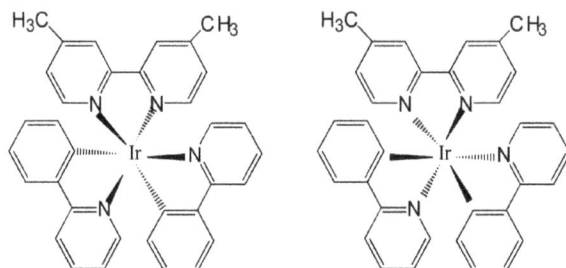

Δ-[Ir(ppy)$_2$(dmbpy)]$^+$ (left) and Λ-[Ir(ppy)$_2$(dmbpy)]$^+$ (right) (ppyH = 2-phenylpyridine and dmbpy = 4,4'-dimethyl-2,2'-bipyridine).

Chart 1. Chiral structures of [Ir(ppy)$_2$dmbpy]$^+$

In case of synthetic saponite, the luminescence spectra were measured under air on an aqueous dispersion containing [Ir(III)L1] complex and various amounts of a clay. Notably quantum yield (Φ) increased from 0.04 to ca.1.0 with the increase of an added clay even in an aqueous dispersion as shown in Fig. 1. The main cause for the increase of Φ might lie in the elimination of water molecules in the vicinity of the [Ir(III)L1] complexes located on a clay surface. The structural fixation of a flexible ligand (dmbpy) in the [Ir(III)L1] complex could be another factor. The introduction of air had little effects on Φ. The fact was in marked contrast with the homogeneous media, in which oxygen molecules quench the excited complexes efficiently. Thus a clay provided such a site as protected from quenching by oxygen molecules. The emission intensity continued to increase even after the equivalent amount of a clay and attained the maximum value around [clay]/[Ir(III)L1] = 10. This might reflect that the adsorbed complexes were in an isolated state, being free from the self-quenching among them.

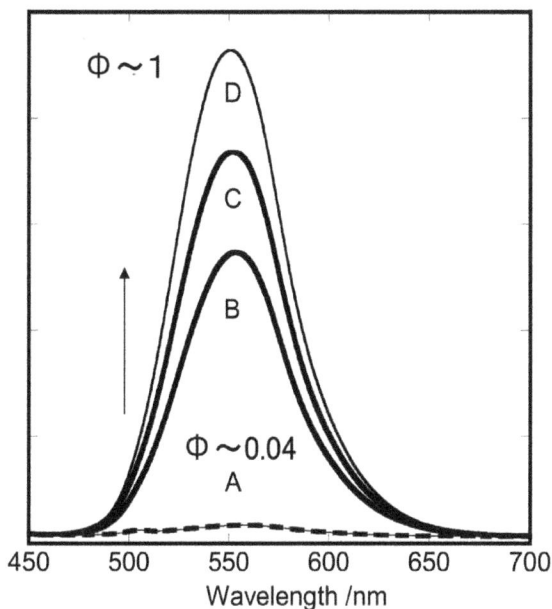

Figure 1. The effect of synthetic saponite on the luminescence spectra of [Ir(bpy)2dmbpy]+ under degassed condition. The excitation wavelength was 430 nm. The concentration of [Ir(III)L1] complex was 1×10^{-5} M and clay (A) 0.0 M, (B) 1.0×10^{-5} M, (C) 4.0×10^{-5} M and (D) 1.0×10^{-4} M. The lowest dotted curve was an emission spectrum for the absence of a clay. A solvent was 3:1(v/v) H_2O/CH_3OH.

The effect of a clay on the transient behavior of excited [Ir(III)L1] complexes was studied by the lifetime measurements under various conditions. Under air, the decay profile was composed of at least two components. This suggested that there were more than two kinds of adsorption states. For example, a part of the complexes were in the interlayer space and the other on the external surface of a clay. If it was the case, the latter state was more easily quenched by oxygen molecules in correspondence to the shorter component of life time. Under argon atmosphere, the decay profile for a clay dispersion changed to a single exponential curve whose lifetime was nearly equal to the longer component under air. This was reasonable since the Ir(III) complexes on an external surface were no more quenched by oxygen molecules.

In case of sodium montmorillonite, the emission quantum yield (Φ) of the complex decreased by adsorption on a clay particle. The behavior was ascribed to the quenching by Fe(III) ions located in a layer and partly by water molecules as shown in Fig. 2. Interestingly Φ recovered by adding alkali or alkaline-earth metal ions to a clay suspension. The results were rationalized in terms of the model that the quenching by Fe(III) ions. The effect of metal ions on the recovery of luminescence indicated that bound metal ions diminished the quenching ability of water molecules. It was suggested that the adsorption of metal ions hydrated water molecules on the clay surface. Such hydration might deprive water molecules of quenching ability towards the [Ir(III)L$_1$] complexes. If that is the case, the effect is thought to be critically dependent on the charge of the metal ion, because the hydration is stronger for metal ions of higher valence. The fact that alkaline earth metal ions were more effective than alkali metal ions was in accord with this view. It should be emphasized that the influence of metal ions as observed here appeared at concentrations as low as 10^{-5} M. No work has ever revealed such hydration effects by metal ions at such a low concentration. Highly emitting properties of the present [Ir(III)L$_1$] complex enabled us to detect the effect under those extreme conditions. From a practical point of view, the present finding may open the possibility of developing the sensing of metal ions by use of the emission from a clay-metal complex adduct.

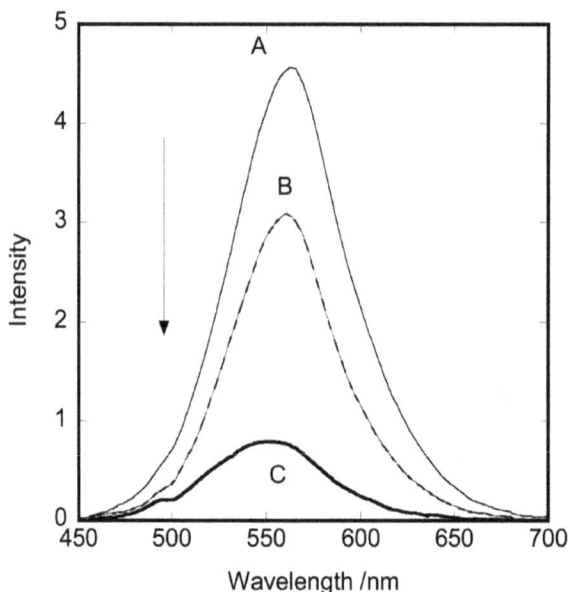

Figure 2. Luminescence spectra of an aqueous dispersion containing [Ir(ppy)$_2$dmbpy]$^+$ ($6.5{\times}10^{-6}$ M) and various amounts of clay ((A) 0.0 M, (B)$1.5\text{x}10^{-5}$ M (C) $1.8\text{x}10^{-4}$ M). The maximum loading of [Ir(III)L$_1$] complex was 5.4 % with respect to the CEC of clay.

2.2. Enantioselective sensing by luminescence

Clay minerals have been also applied as a host in the photochemical reactions involving optically active molecules (Fujita et al., 2006). It should be noted that enantioselective luminescence quenching is a dynamical recognition phenomenon (Inoue, 1992, Tsuchiya et al., 2009). The discrimination of chirality is accomplished within the short lifetime of an excited molecule. Clay may assist the emitter to orient preferably for the stereoselective attack by a quencher. If the emitting properties of the [Ir(III)L1] complexes are connected with their chiral structures, it may open a possibility for luminescent chiral sensing. Under these backgrounds, a clay mineral is used as a host to fix the orientation of an iridium(III) complex towards a quencher (Sato et al., 2011a).

The highly emitting properties of the iridium complex bound by a clay prompted us to investigate the possibility of stereoselective energy transfer. The optical resolution of a cationic iridium(III) complex, [Ir(ppy)2dmbpy]+ (Chart 1), was attempted by several ways such as anionic resolving reagent and chiral adsorbents (Chen et al., 2007). The only successful method was to use an ion-exchange adduct of a clay and chiral [Ru(phen)3]2+ (phen = 1,10-phenanthroline) as a resolving agent. As a chiral quencher, a tris(β-diketonato)ruthenium(III), [Ru(acac)3] (Chart 2), was chosen.

The emission intensity at 650 nm was compared between two systems, clay/Δ-[Ir(III)L1] /Δ-[Ru(acac)3] (pseudo-enantiomeric combination) and clay/Δ-[Ir(III)L1]/Λ-[Ru(acac)3] (pseudo-racemic combination), in 3:1 (v/v) water-methanol. In both cases, the intensity of emission decreased on adding [Ru(acac)3], indicating that Ru(III) complex acted as an efficient quencher in these systems. The quenching effect was analyzed in terms of the Stern-Volmer plots (Eq. (1)). It was apparent that luminescence quenching was more efficient for the clay/Δ-[Ir(III)L1] /Δ-[Ru(acac)3] system than for the clay/Δ-[Ir(III)L1] /Λ-[Ru(acac)3] over the whole concentration range. The plots showed the tendency of leveling off at the higher concentration of the quenchers. The curves were fitted by the two-site model as given by Eq . (2):

$$\frac{I_0}{I} = 1 + \frac{K_q}{K_F}[P_{O_2}]$$

(1)

Here, k_q and k_F are the bimolecular rate constant of quenching and the unimolecular rate constant of spontaneous luminescence, respectively.

$$\frac{I_0}{I} = [\frac{f_1}{1 + K_{sv1}P_o} + \frac{f_2}{1 + K_{sv2}P_o}]^{-1}$$
$$f_1 + f_2 = 1$$
$$K_{sv_0} = f_1 \times K_{sv1} + f_2 \times K_{sv2}$$

(2)

in which I_0, I, f_1, f_2, P_0, K_{SV1} and K_{SV2} denote the emission intensities in the absence of and in the presence of a quencher, the fractions of processes 1 and 2, the concentration of a quencher and the Stern-Volmer constants for the processes 1 and 2, respectively. K_{sv0} is the overall Stern-Volmer constant.

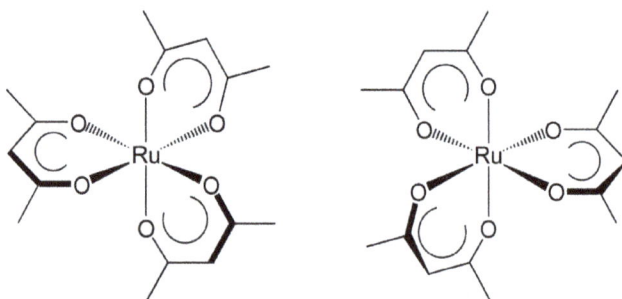

Chart 2. Chiral structures of [Ru(acac)₃] as a quencher: Δ–[Ru(acac)₃] (left) and Λ–[Ru(acac)₃] (right)

In order to confirm the existence of stereoselectivity, we performed the same experiments for the opposite emitter/quencher combinations or the clay/Λ-[Ir(III)L₁] /Λ-[Ru(acac)₃] (pseudo-enantiomeric combination) and the clay/Λ-[Ir(III)L₁] /Δ-[Ru(acac)₃] (pseudo-racemic combination). From the K_{sv0} obtained from Eq. 2, the overall selectivity factor, which is defined to be the ratio of $K_{sv0}(\Delta-\Delta$ or $\Lambda-\Lambda)/ K_{sv0}(\Delta-\Lambda)$, was obtained to be 1.84 in favor of the pseudo-enantiomeric combination. The quenching reaction was not a simple collisional process, but it might involve the process of molecular association on a clay surface. It was added that no stereoselectivity was detected in methanol for the same emitter/quencher pairs. Thus the fixation of the iridium(III) complex on a clay surface was concluded to be a crucial step for chiral recognition as shown in Scheme 1.

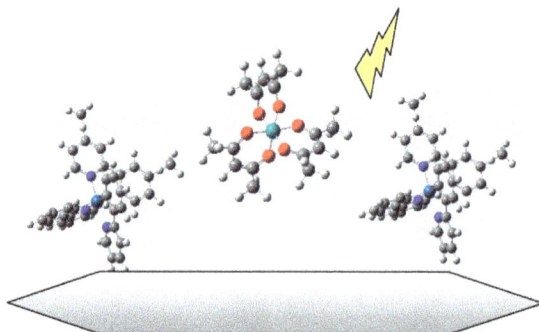

Scheme 1. A model of chiral sensing by [Ir(III)L₁] complexes adsorbed on a clay surface

3. Preparation of thin films of clays by the Langmuir-Blodgett (LB) method

The photochemical reactions involving clay minerals were further extended to thin film systems. In such attempts, the preparation of thin films with uniform properties is essentially important to achieve well-defined reaction systems. Yamagishi *et al.* first reported the nanometer-thick films of an ion-exchange adduct of a clay (synthetic saponite) and an alkylammonium cation (trimethylstearylammonium) as prepared by the Langmuir-Blodgett method. (Inukai et al., 1994). For preparing such a film, the ion-exchanged adduct of a clay-alkylammonium is dispersed in chloroform and spread over the surface of pure water. According to the method, a layer-by-layer film was prepared in such a way as donor and acceptor molecules were intercalated in an alternative order. It was revealed that a single clay layer acted as an efficient barrier in the transfer of photon energy. For example, the photoinduced electron-transfer was studied from an amphiphilic polypyridyl-Ru(II) complex (electron donor) to an amphiphilic acetylacetonato-Ru(III) complex (electron acceptor). Recently the method was called as " Clay LB Method"(Tamura et al. , 1999, Ras et al., 2009). We have been attempting to improve the " Clay LB Method" in order to develop nano-structured photodevices based on clay minerals (Sato et al., 2005).

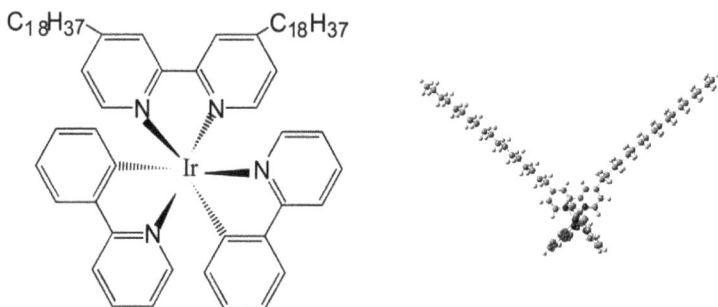

Chart 3. The structure of [Ir(ppy)₂(dc18bpy)]⁺

An amphiphilic cyclometalated iridium(III) complex, [Ir(ppy)₂(dc18bpy)]ClO₄ (ppy = 2-phenylpyridine; dc18bpy = 4,4'-dioctadecyl-2,2'-bipyridine) (denoted by [Ir(III)L₂] (Chart 3)), was prepared by refluxing [Ir(ppy)₂Cl]₂ with an equal amount of dc18bpy in glycerol at 170 °C for 8 hours. The compound was purified chromatographically by being eluted on an HPLC column (MG (Shiseido Inc. Ltd.)) with chloroform. The Langmuir-Blodgett (LB) method has been applied for preparing a thin clay film as shown in Scheme 2. The details of preparation is described below. A LB trough with an area of 10.0 cm × 13.0 cm is maintained at 20°C by circulating water. The clays used can be synthetic saponite or sodium montmorillonite or synthetic hectorite $(Si_{8.00})(Mg_{3.50}Li_{0.30})O_{20}(OH)_4)$ $(Na_{0.70})$. A chloroform solution of an amphiphilic cationic iridium(III) complex, ([Ir(ppy)₂(dc18bpy)]ClO₄ $(3.2 \times 10^{-5}$ molL^{-1}), is spread onto an aqueous suspension of a clay at various concentrations. As a reference, the same solution is spread over pure water. A floating monolayer is formed on the surface of a subphase. The surface pressure versus molecular area (π-A) curves is obtained by compressing the monolayer. Figure 3 shows the example for π-A curves with 0 mgL^{-1}, 10 mgL^{-1} and 20 mgL^{-1} of synthetic saponite. In all cases, surface pressure levels off from zero in the region of the molecular area below 0.5 -1.5 nm² per molecule. A critical molecular area (Sc) is obtained by extrapolating the linear portion of each π-A curve to zero surface pressure. Both Sc changes significantly, when a subphase of pure water is replaced with a clay suspension. Moreover the slope of π-A curve becomes steeper by this replacement, indicating that the floating films are more rigid on hybridization with clay particles. These facts support the occurrence of hybridization of a molecular film of [Ir(III)L2] complex with clay particles at an air-water interface. In these cases, 10 mgL^{-1} of clay is concluded to be the best condition for the rigid films. Since the sectional area of the head groups of the present complex is estimated to be ca. 1 nm^2 on a molecular model, a floating film is concluded to be composed of the monolayer of the metal complex.

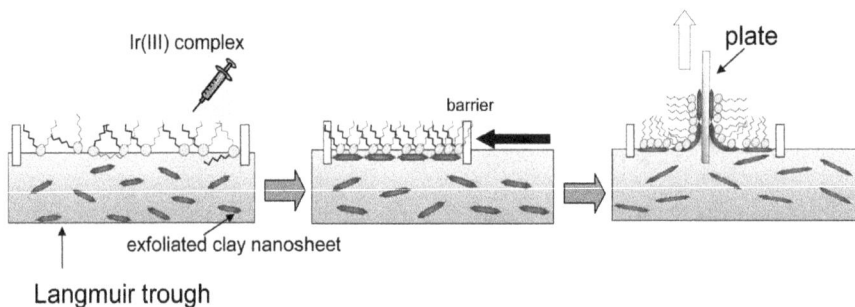

Scheme 2. *Clay LB* method (vertical deposition)

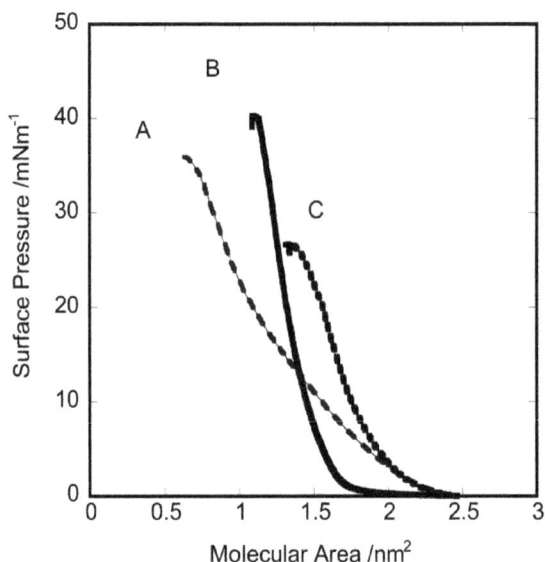

Figure 3. The surface pressure versus molecular area (π-A) curves when a chloroform solution of [Ir(ppy)2(dc18bpy)]ClO4 was spread over a subphase of (A) pure water, or (B) synthetic saponite (10 mg L^{-1}) or (C) synthetic saponite (20 mgL^{-1}).

4. Structures and properties of clay LB films

This section describes how the structure of a LB clay film is studied. After 30 min, the surface was compressed at a rate of 10 cm^2 min^{-1} until the surface pressure reached 10 mNm^{-1}. A floating film was transferred at 10 mNm^{-1} onto a hydrophilic glass plate or silicon by the vertical method at a dipping rate of 10 mm min^{-1}. The transfer ratio was estimated to be 0.9 ± 0.1 for all cases. The AFM images of the deposited film showed the presence of particles with the characteristic shape depending on the kind of clay. They definitely demonstrated the inclusion of clay particles in the deposited films. In case of synthetic saponite, for example, the film was composed of spherical domains with the diameter of ca. 50 nm, which indicated the presence of saponite particles. [Ir(III)L$_2$] complexes were thought to be attached uniformly by the particles. In case of synthetic hectorite, the flat regions with the height of ca. 2 nm were observed, indicating the inclusion of hectorite particles. Small domains were observed on such a flat region, which corresponded to the aggregated states of [Ir(III)L$_2$] complexes. In case of montmorillonite, the films were covered with flat particles in various shapes. The thickness of the flat particle was estimated to be ca. 2 nm. Subtracting the thickness of one clay layer (1 nm) from this value, the height of an iridium(III) complex was estimated to be 1 nm. This was less than one-half of the molecular length of the iridium complex along the long alkyl chains. Thus these complexes were thought to be adsorbed with their alkyl chains declined from a clay surface.

5. Application of clay-iridium(III) complex LB films for photo-sensing

5.1. Emission properties of the deposited hybrid LB films

The emission behavior was investigated on a hybrid film of an [Ir(III)L$_2$] complex and a clay as prepared by the LB method. For measurement of emission spectra from a LB film, a glass substrate was placed in a quartz cell at 45 degrees with respect to the incident light (Scheme 3(a)). A gas was introduced into the cell after it was evacuated below 0.1 m torr. The emission spectra was measured under vacuum at room temperature when these substrates were irradiated by a light at 430 nm (Scheme 3(b)).

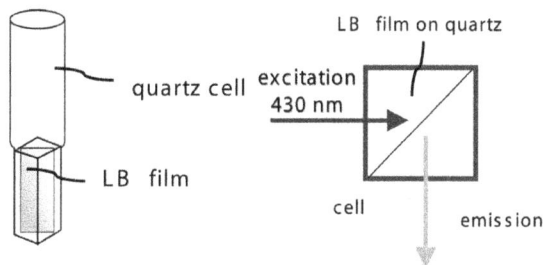

Scheme 3. (a) A quartz cell containing a substrate modified with a LB film; (b) Experimental set-up of measuring an emission spectrum from a film in a quartz cell

The emission peak was slightly dependent on the kind of hybridized clay: 550 nm for synthetic saponite, synthetic hectorite and montmorillonite, and 535 nm for without clay, respectively. The emission intensity was nearly doubled for hybridization with saponite in comparison to that of without clay. Since these films contained nearly the same amount of [Ir(III)L2] complexes as in the film without clay, the increase was thought to be caused by the interaction with a clay surface. Figure 4 shows the emission spectra of [Ir(III)L2]/synthetic saponite and [Ir(III)L2]/pure water. The complexes formed a mono-molecular layer on a clay surface, while they were coagulated to form a multi-molecular layer in the pure LB film. Thus the self-quenching by neighboring molecules would be reduced on a clay surface in comparison to the pure LB film. Among the hybrid films, no enhancement of emission intensity was observed for montmorillonite, probably because Fe(III) ions in a clay layer had an effect of quenching excited iridium(III) complexes.

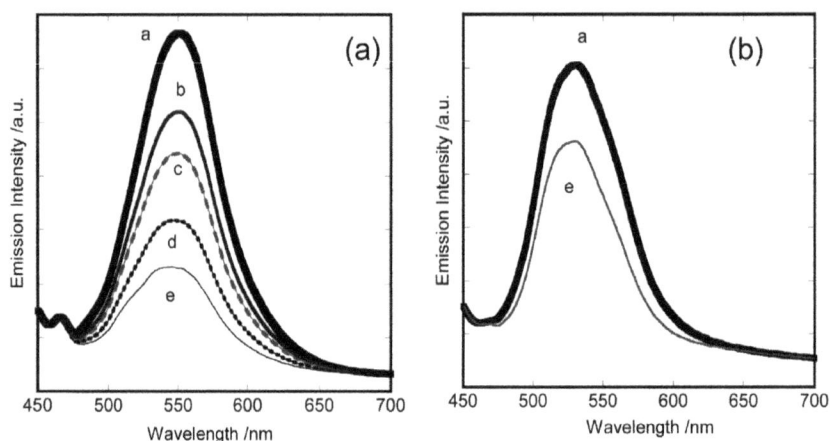

Figure 4. Emission spectra from (a) [Ir(III)L2]/synthetic saponite and (b) [Ir(III)L2]/no clay: oxygen pressure was (a) vacuum, (b) 3.7, (c) 9.0, (d) 31.4 and (e) 101.3 kPa, respectively. The films were prepared for an aqueous dispersion containing 10 mg L^{-1} of synthetic saponite. The excitation wavelength was 430 nm.

5.2. Quenching effects by oxygen and other gaseous molecules

In order to pursue the possibility of applying a clay hybrid films for oxygen sensing, the effect of oxygen gas was studied on the emission behavior. The emission of [Ir(III)L2]/montmorillonite in air was shown in Fig.5. An oxygen gas was introduced into a quartz cell containing a glass substrate modified with a single-layered hybrid LB film. The emission decreased rapidly until it attained the stationary value within a few seconds. It recovered to the initial value by evacuating the oxygen gas. The results implied that the electronic energy of the [Ir(III)L2] complex in the excited triplet state transferred efficiently to an oxygen molecule in the triplet ground state, leading to the formation of a singlet oxygen molecule. As shown in the Figures 4 and 6, the quenching phenomena were observed at the various oxygen pressures. The decrease of emission intensity was evaluated as a function of oxygen pressure. I_0/I is plotted as a function of oxygen pressure ($[P_{O2}]$). Here I and Io denote the luminescence intensity at 550 nm with and without a quencher, respectively. The plots were analyzed according to the equation (Stern-Volmer plots Eq. (1)) The experimental plot did not obey a linear relation but curved downwards at the higher pressure region. The effect was interpreted in terms of the presence of different types of oxygen quenching sites. Assuming that there were two sites for quenching, the curves are fitted by Eq. (2). Comparing the weighted quenching constant among four films, hybrid saponite LB film showed the highest sensitivity towards O_2.

Figure 5. Emission from an [Ir(III)L$_2$]/montmorillonite LB film deposited on a quartz substrate in air.

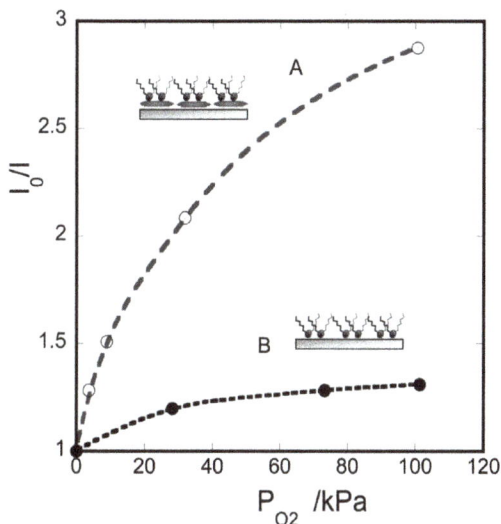

Figure 6. Dependence of the change of emission intensity on the vapor pressure of oxygen gas for (A) [Ir(III)L$_2$]/synthetic saponite and (B) [Ir(III)L$_2$]/no clay. The excitation and emission wavelengths were 430 nm and 550 nm, respectively. The luminescence intensity was recorded at 535 nm for [Ir(III)L$_2$]/no clay. Curves were fit by two-site model proposed by Eq. (2).

For other gases such as water, ethanol, acetone and chloroform, similar experiments were performed on the hybrid LB films of various clays. All these gases acted as a quencher in deactivating the excited iridium(III) complexes. Since these molecules were in the singlet states in contrast to an oxygen molecule, they were assumed to relax the electronic energy of an excited iridium(III) complex non-radiatively through their vibration energy levels. The time course of the emission intensity was dependent on the kinds of clays remarkably (Fig.7). For the case of [Ir(III)L2]/montmorillonite, for example, the signal was reversible for introducing and evaporating gases. The results were consistent with the uniform adsorption of iridium(III) complexes on the clay particle as observed in the AFM image. For the case of [Ir(III)L2]/synthetic hectorite, however, methanol and acetonitrile increased emission intensity instead of acting as a quencher. The results suggested that the self-quenching among [Ir(III)L2] complexes decreased by the inclusion of the gas molecules. The disordering of the alkoxy chains might result in the decrease of quenching among the neighboring [Ir(III)L2] complexes in the film. It was noted that a small molecule with functional group such as -OH, >C=O, CN and -Cl quenched the excited [Ir(III)L2] complexes efficiently, while molecules with no-functional group such as cyclohexane affected little the emission from the hybrid films. Thus the energy relaxation into vibration energy levels occurred exclusively through the specific interaction of the [Ir(III)L2] complexes with these functional groups.

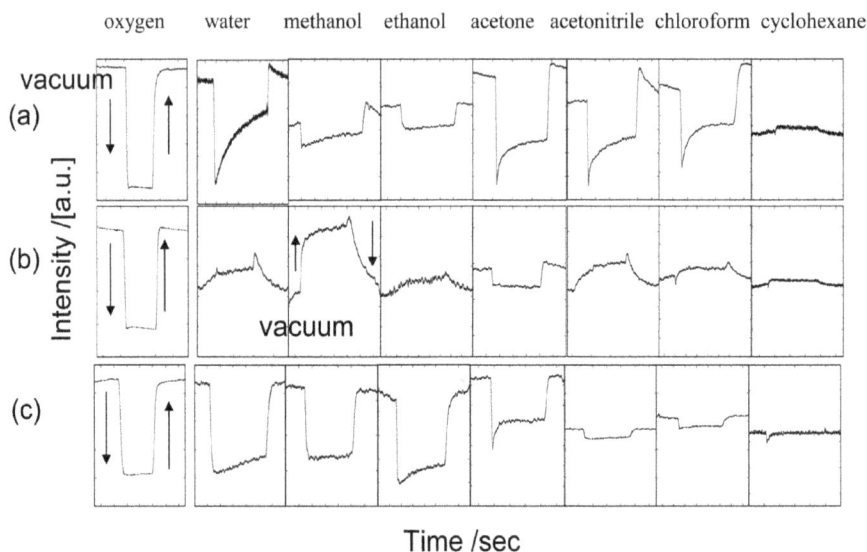

Figure 7. Effects of gases on the time course of the emission intensity for the singly deposited hybrid LB films: (a) [Ir(III)L2]/synthetic saponite, (b) [Ir(III)L2]/synthetic hectorite and (c) [Ir(III)L2]/montmorillonite. The luminescence intensity was measured at 550 nm, respectively. The following gases were used: oxygen, water, methanol, ethanol, acetone, acetonitrile, chloroform and cyclohexane. 101.3 kPa of oxygen was introduced. In other gases, about one-third of their saturation vapor pressure was introduced. The films were prepared for an aqueous dispersion containing 10 mg L^{-1} of clay. The excitation wavelength was 430 nm.

For comparison, the effect of an oxygen gas on the luminescence was studied for the cast film of saponite ion-exchanged with the [Ir(III)L2] complex. The quenching effect was much less efficient than that for the LB films. The results were reasonable, considering the situations that only a small portion of [Ir(III)L2] complexes were located on the external surface of the film. In this sense, the sensitivity for sensing gas molecules was remarkably enhanced by constructing a LB film with nanometer thickness.

6. Conclusion

The hybrid Langmuir-Blodgett (LB) films of an amphiphilic iridium(III) complex, $[Ir(ppy)_2(dc18bpy)]^+$, and clays (synthetic saponite, synthetic hectorite, and sodium montmorillonite) were prepared. A glass substrate was modified with a single layered LB film and placed into a quartz cell. Luminescence was monitored under the atmosphere of various gases. An oxygen gas, for example, quenched the emission from excited iridium(III) complexes linearly in the pressure range of 0 - 30 kPa, while the quenching effect was saturated above 30 kPa, The results indicated the occurrence of adsorption saturation of oxygen molecules into the film. Other gases with functional groups also quenched the luminescence efficiently. These results demonstrated the potentiality of the present hybrid LB films as a gas sensing device.

Author details

Hisako Sato
Department of Chemistry, Graduate School of Science and Engineering,
Ehime University, Matsuyama, Japan

Kenji Tamura
National Institute of Materials Science, Tsukuba, Japan

Akihiko Yamagishi
School of Medicine, Toho University, Ota-ku, Tokyo, Japan

Acknowledgement

This work has been financially supported by MEXT KAKENHI Grant-Aid-for Scientific Research (B) Number 23350069 of Japan. The part of work was financially supported by Nippon Sheet Glass Foundation of Materials and Science and Engineering of Japan.

7. References

Acharya, S., Hill, J. P., & Ariga, K. (2009). Soft Langmuir-Boldgett Technique for Hard Nannomaterials. *Adv. Mater.* , Vol. 21, pp.2959-2981.

Chen, X., Okamoto, Y., Yano, T., & Otsuki, J. (2007). Direct Enantiomeric Separations of Tris (2-phenylpyridine) Iridium (III) Complexes on Polysaccharide Derivative-based Chiral Stationary Phases. *J. Sep. Sci.,* Vol. *30,* pp. 713-716.

Fujita, S., Sato, H., Kakegawa, N., & Yamagishi, A. (2006). Enantioselective Photooxidation of a Sulfide by a Chiral Ruthenium (II) Complex Immobilized on a Montmorillonite Clay Surface: The Role of Weak Interactions in Asymmetric Induction. *J. Phys. Chem. B,* 110, pp. 2533-2540.

Inoue, Y. (1992). Asymmetric Photochemical Reactions in Solution. *Chem. Rev.*, Vol. 92, pp. 741-770.

Inukai, K. , Hotta, Y., Taniguchi, M., Tomura, S., & Yamagishi, A. (1994). Formation of a Clay Monolayer at an Air-Water Interface. *J. Chem. SOC. Chem. Commun.*, pp.959-960.

Lo, K. K.-W., Li, S.P.-Y., & Zhnag, K. Y. (2011). Development of Luminescent Iridium(III) Polypyridine Complexes as Chemical and Biological Probes. *New J. Chem.*, Vol. 35, pp.265-287.

Lowry, M. S., & Bernhard, S. (2006). Synthetically Tailored Excited States: Phosphorescent, Cyclometalated Iridium(III) Complexes and Their Applications. *Chem. Eur. J.* , Vol. 12, pp.7970-7977.

Ogawa, M., & Kuroda, K. (1995). Photofunctions of Intercalation Compunds. *Chem. Rev.*, Vol. 95, pp. 399-438.

Ras, R. H. A. , Umemura, Y. , Johnston, C. T. , Yamagishi, A. , & Schoonheydt, R. A. (2007).Ultrathin Hybrid Films of Clay Minerals. *Phys. Chem. Chem. Phys.*, Vol. 9, pp. 918-932.

Sajoto, T., Djurovich, P. I., Tamayo, A. B., Oxgaard, J., Goddard III., W. A. & Thompson, M. E. (2009). Temperature Dependence of Blue Phosphorescent Cyclometalated Ir(III) Complexes. *J. Am. Chem. Soc.*, Vol. 131, pp.9813-9822.

Sato, H., Hiroe, Y., Tamura, K., & Yamagishi, A. (2005). Orientation Tuning of a Polypyridyl Ru(II) Complex Immobilized on a Clay Surface toward Chiral Discrimination. *J. Phys. Chem. B*, Vol. 109, pp.18935-18941.

Sato, H., & Yamagishi, A. (2007). Application of the ΔΛ Isomerism of Octahedral Metal Complexes as a Chiral Source in Photochemistry. *J. Photochem.Photobiol. C; Photochem. Rev.*, Vol. 8, pp.67-84.

Sato, H., Tamura,K., Taniguchi, M., & Yamagishi, A. (2009). Metal Ion Sensing by Luminescence from an Ion -exchange Adduct of Clay and Cationic Cyclometalated Iridium (III) Complex. *Chem Lett.*, Vol. 38, pp.14-15.

Sato, H., Tamura, K., Taniguchi , M., & Yamagishi, A. (2010). Highly Luminescent Langmuir-Blodgett Films of Amphiphilic Ir(III) Complexes for Application in Gas Sensing. *New J. Chem.*, Vol. 34, pp.617-622.

Sato, H., Tamura, K., Aoki, R., Kato, M., & Yamagishi, A. (2011a). Enantioselective Sensing by Luminescence from Cyclometalated Iridium (III) Complexes Adsorbed on a Colloidal Saponite Clay. *Chem. Lett.*, Vol. 40, pp.63-65.

Sato, H., Tamura, K., Ohara, K., Nagaoka, S. , & Yamagishi, A. (2011b). Hybridization of Clay Minerals with the Floating Film of a Cationic Ir(III) Complex at an Air-water Interface. *New J. Chem.*, Vol. 35, pp.394-399.

Tamura, K., Setsuda, H., Taniguchi, M., Yamagishi, A. (1999). A Clay-Metal Complex Ultrathin Film as Prepared by the Langmuir-Blodgett Technique. *Chem. Lett.* , pp.121-122.

Tsuchiya, K., Ito, E., Yagai, S., Kitamura, A., & Karatsu, T. (2009). Chirality in the Photochemical mer->fac Geometrical Isomerization of Tris(1-Phenyloytrazolato, N, C$^{2'}$) Iridium(III). *Eur. J. Inorg. Chem.*, pp. 2104-2109.

Ulbricht, C., Beyer, B., Friebe, C., Winter, A., & Schubert, U. S. (2009). Recent Developments in the Application of Phosphorescent Iridium (III) Complex Systems. *Adv. Mater.*, Vol. 21, pp.4418-4441.

Chapter 15

Methods of Determination for Effective Diffusion Coefficient During Convective Drying of Clay Products

Miloš Vasić, Željko Grbavčić and Zagorka Radojević

Additional information is available at the end of the chapter

1. Introduction

Drying research is an outstanding example of a very complex field where it is necessary to look comprehensively on simultaneous energy and mass transfer process that takes place within and on the surface of the material. In order to get the full view of drying process, beside previously mentioned, researchers have to incorporate and deal with highly non linear physical phenomena inside drying clay products, non-homogenous distribution of temperature and humidity inside dryers, equipment selection, design, control and final product quality [1]. That is the reason why a unique theoretical setting of drying has to be determined through the balance of the heat flow, temperature changes and moisture flow. Simultaneous heat and mass process are related, regarding to the fact that all phases have to remain in thermodynamic balance established on a local temperature value [2]. In the economy that is becoming increasingly global, laboratory drying process analyses should ensure enough data which are necessary for optimal drying regime establishment. In order to find optimal drying regime it is necessary to understand transport mechanisms which takes place within and on the surface of the clay product. The drying process is characterized by the existence of several internal transport mechanisms such as pure diffusion, surface diffusion, Knudsen diffusion, capillary flow, evaporation and condensation, thermo-diffusion, *etc*. Moisture diffusivity, viewed as a transport of matter due to the random motion of molecules, is the most important mass transport mechanisms, essential for the calculation and modelling of various clay processing operations. Moisture transfer within the solid clay body at a certain temperature is realized due to the different moisture content in the interior and on the surface of a solid body. The mass transfer rate by pure diffusion is therefore proportional to the concentration gradient of the moisture content, with the diffusion coefficient being the proportionality factor. Determination of the diffusion coefficient is essential for a credible description of the mass transfer process, described by the Fick's equation [3]. It is a common practice to describe complete mass transfer with same equations as pure diffusion and to take the correction, for all secondary types of mass transfer into account simply by replacing the pure diffusion coefficient with an effective diffusion coefficient.

Relatively small number of research papers that describe the drying process of ceramic and especially clay materials are available. Some data can be found in the papers of Efremov [4] (bricks), Vasić [5] (heavy clay tiles) Chemkhi [6], Zagrouba [7, 8] (clays), Skansi [9, 10] (brick, hollow brick, heavy clay tiles, tiles) and others. In his paper Efrem [4] gave an analytical solution of diffusion differential equation with boundary conditions in the form of flux. Relying on these studies M. Vasić and colleagues [11] have developed a drying model based on the modification of Efremov's equation and the computer program for determining the effective diffusion coefficient.

Chemki and Zagrouba [6] have estimated the coefficient of moisture diffusivity from drying curve. F. Zagrouba and colleagues [7, 8] have developed a mathematical model of transfer phenomena which has involved at the same time heat, mass and momentum transfer during the convective drying of clay tiles. In their study a method for determination of the heat transfer coefficient and effective diffusion coefficient is presented. Zanden and Kerkhof [12] have performed extensive research on isothermal mass transport mechanisms during the convective drying of clay products. They presented a model which describes moisture transport inside a porous clay material during drying.

M. Vasić and colleagues [5] have developed two computer programs for determination of effective diffusion coefficient, based on mathematical calculation of the second Fick's law and Cranck diffusion equation. Skansi and colleagues [9, 10] were investigating the kinetics of conventional drying of flat tiles in experimental and industrial tunnel dryer. They presented several thin layer models such as exponential one which correlates the kinetics of the whole tile-drying process well and has physical significance. They also presented a method for determining heat transfer coefficient, effective diffusion coefficient and drying constant.

2. Materials and methods

2.1. Theoretical development

In drying studies performed on clay materials, diffusion is generally accepted as the main mechanism of moisture transport from the material interior to its surface. The restriction to one-dimensional diffusion gives a good approximation in many practical systems. Analytical solution of Fick's equation is given for various geometrical shapes, assuming that the transport of moisture occurs by diffusion, that sample shrinkage is neglected and that diffusion coefficient and temperature have constant values. For the case of "thin plate" geometry, a solution is given by Cranck [3] which is represented by the expression:

$$MR = \frac{8}{\pi^2} \sum_{n=0}^{\infty} \frac{1}{(2n+1)^2} \left(-\frac{(2n+1)^2}{4} \pi^2 \frac{D_{eff}t}{l^2} \right) \tag{1}$$

In equation (1) X_0, X and X_{eq}, represent respectively, the initial, current and equilibrium moisture content, kg moisture/kg of dry material, D_{eff} is the effective diffusion coefficient, m²/s, l is the half plate thickness, m and t is time, s. MR represents moisture ratio and has no unit. There is a large body of literature comparing predicted results of drying models that considered as well as neglect shrinkage [16]. Most published drying models do not take shrinkage into account in the balance equations. The drying model equations are typically borrowed from corresponding non-shrinking models, frequently without appropriate physical and mathematical consideration, and are applied to a shrinkage medium. A few studies, describing the sample dimensional correction, can be found in literature. Some data can be found in the papers [17-20]. Silva [21] presented, a way of solving the diffusion equation for the case of spherical samples. Since clay products show dimensional change during drying it was necessary to develop a model that would take this phenomenon into account. By introducing into equation (1) the expression $l_{(t)}$, which represents the experimental dependence of the thickness of the tiles in time, equation (1) is corrected. It should be kept in mind that this type of correction is not mathematically one hundred percent accurate because the resulting equation (1) was obtained using the assumption of unchangeable sample thickness. Formally speaking, a mathematically accurate correction can be obtained by entering the expression $l_{(t)}$ into the equation for the case of constant sample thickness, after an integration step.

2.2. Description of Model A

2.2.1. Model A1 - The case when shrinkage is not included

In order to solve the equations (1) it is necessary to dispose with the experimental results and to have the experimentally determined dependence MR_{eks} - t. MR_{eks} represents the experimentally determined value of MR obtained by calculation from the experimentally measured data X_0, X and X_{eq}. Equation (1) can be converted into the form:

$$MR = \frac{8}{\pi^2} \sum_{n=N+1}^{\infty} \frac{1}{(2n+1)^2} \exp\left(-\frac{(2n+1)^2}{4} \pi^2 \frac{D_{eff}t}{x^2} \right) + \frac{8}{\pi^2} \sum_{n=0}^{N} \frac{1}{(2n+1)^2} \left(-\frac{(2n+1)^2}{4} \pi^2 \frac{D_{eff}t}{x^2} \right) \tag{2}$$

If the value of ε is defined as the relative error of neglecting terms higher then N in equation (2), the value of N can be determined and equation (2) is transformed form an infinite sum into a finite sum of N terms given by equation (3):

$$MR = \frac{8}{\pi^2} \sum_{n=0}^{N} \frac{1}{(2n+1)^2} \left(-\frac{(2n+1)^2}{4} \pi^2 \frac{D_{eff}t}{l^2} \right) \tag{3}$$

The value of $\varepsilon = 0.05$ is accepted for the further calculations in this paper. When $t=0$, $MR=1$, and equation (2) is transformed into equation (4). The value of N used in equation (3) can be determined from equation (4):

$$1 = \frac{8}{\pi^2} \sum_{n=0}^{N} \frac{1}{(2n+1)^2} + 0.05$$

(4)

MR_{an} represents the analytically determined value calculated from equation (3). It is necessary to introduce the concept of a numerical counter i, which can have only an integer value. The numerical counter i is defined for each value of the experimental pairs (MR_{eks}, t). It starts form the value zero and increases by one until it reaches a final value which is related to the last experimental pairs (MR_{eks}, t). This concept enables the number of experimental pairs (MR_{eks}, t) from its first to its last value to be countered. In order to work properly, the program requires the initial value of the effective diffusion coefficient D_{eff}, and the ε value to be entered. Let the initial value of the effective diffusion coefficient D_{eff} be given the value of $1 \cdot 10^{-20}$ /m²/s. Then, for each numerical counter value i, the program calculates the value χ^2 from equation (5);

$$\chi^2 = \sum_{1}^{i} \left(MR_{eksi} - MR_{an_i} \right)^2$$

(5)

In the first cycle, $MR_{an\ i}$ is calculated according to equation (3) using the previously determined value of N and the initial value of D_{eff}. In the next cycle the value of D_{eff} is doubled giving a new value for $MR_{an\ i}$ which is now used to calculate a new χ^2 according to equation (5). The program then compares the value χ^2 obtained in the first cycle and the newly obtained χ^2 value. If the statement $\chi^2_{first} < \chi^2_{second}$ is satisfied, the program will continue previously described cycle, otherwise the program will temporarily stop.

Note: χ^2_{first} and χ^2_{second} refer to the last and the penultimate value of the cycle in which χ^2 is determined.

The last three values for D_{eff} and χ^2 are then recorded. Then, the recorded D_{eff} interval is divided into 100 parts. A hundredth part of this interval is defined as a step s. The program commences a cycle again using the initial value for D_{eff} as D_{eff} third from end + s. The cycle is repeated until the statement $\chi^2_{first} < \chi^2_{second} < 1 \cdot 10^{-10}$ is satisfied. In other words, the cycle is interrupted when the difference $\chi^2_{second} - \chi^2_{first}$ reaches $1 \cdot 10^{-10}$. The final D_{eff} value is then recorded. This value represents the finally calculated effective diffusion coefficient in m²/s.

2.2.2. Model A2 - The case when there shrinkage is included

For materials which shows shrinkage during drying equation (3) needs to be changed by the introduction of the expression $l_{(t)}$ into it. This expression represents the experimentally determined time dependence of the sample thickness. When this correction is entered, the previously described optimizing concept for the determination of the effective diffusion coefficient is applied.

2.3. Description of model B

2.3.1. Model B1 - The case when shrinkage is not included

If parameters of drying medium are kept constant during convective drying of solid bodies, moisture transfer could be treated on macro level as quasi diffusion with appropriate effective diffusion coefficient D_{eff}. The general expression for mass conductivity (Fick's second law) can be presented as a partial differential equation of diffusivity.

$$\frac{\partial X}{\partial t} = div(D_{eff} \cdot gradX) ; \quad \rightarrow \quad \frac{\partial X}{\partial t} = D_{eff} \frac{\partial^2 X}{\partial x^2} \tag{6}$$

The exact solution for drying kinetics can be obtained by applying Laplace transform method in time t for equation of isotropic diffusion with boundary conditions in a form of mass flux J. This flux is proportional to the difference between an equilibrium concentration in the pores of the material X_{eq} and the current concentration X on the material surface.

$$J = -D_{eff} \cdot \frac{\partial X}{\partial x}\Big|_{x=0} = k \cdot \left(X_{eq} - X \right) \tag{7}$$

Kinetic desorption coefficient k (m/s) in equation (7) can be calculated as a ratio l (characteristic thickness value) and time $\left(k = \frac{l}{t} \right)$.

By applying Laplace transform method to equation (7) Efremov in his PhD thesis [13] presented the solution given by equation (8).

$$\frac{X - X_{eq}}{X_0 - X_{eq}} = erf\left(\frac{l}{2\sqrt{D_{eff}t}} \right) + exp\left(\frac{k}{D_{eff}} l + \frac{k^2}{D_{eff}} \cdot t \right) \cdot erfc\left(k\sqrt{\frac{t}{D_{eff}}} + \frac{l}{2\sqrt{D_{eff}t}} \right) \tag{8}$$

X_0, X and X_{eq}, represent respectively, the initial, current and equilibrium moisture content, kg moisture/kg of dry material, D_{eff} is the effective diffusion coefficient, m²/s, and t is time, s. The mass flux on the material surface (x=0) can be calculated through the use of the concentration ratio which is given in equation (9)

$$MR = \frac{X - X_{eq}}{X_0 - X_{eq}} = exp\left(\frac{k^2}{D_{eff}} \cdot t \right) \cdot erfc\left(k\sqrt{\frac{t}{D_{eff}}} \right) \tag{9}$$

Equation (8) was obtained for the process of molecular diffusion. If we analyze equation (9) at the beginning of the drying process (t=0; X=X_0) MR=1 and for long times (t→∞; X=X_{eq}) MR=0 will see that it has a real physical meaning. In order to get the drying equation which is valid for convective mass transport processes it is necessary to introduce the power function of the argument in equation (9), thus the drying equation becomes (10).

$$\frac{X - X_{eq}}{X_0 - X_{eq}} = exp\left(\frac{1}{\pi}\left(\frac{\pi \cdot l^2}{t \cdot D_{eff}} \right)^n \right) \cdot erfc\left(\sqrt{\frac{1}{\pi}\left(\frac{\pi \cdot l^2}{t \cdot D_{eff}} \right)^n} \right) \tag{10}$$

Simple approximation formula for function erf (A) is defined by equation (11) and can be found in Sergei Winitzki [14, 15] papers. The relative precision of this approximation is higher than $4 \cdot 10^{-3}$, uniformly for all real A.

$$erf(A) = \left[1 - \exp\left(-A^2 \frac{1.27 + 0.14 A^2}{1 + 0.14 A^2} \right) \right]^{1/2} \tag{11}$$

After some mathematical manipulation, knowing that erfc (A) = 1 – erf (A), the final drying kinetic equation (12) is obtained.

$$MR = \exp\left(\frac{1}{\pi}\left(\frac{\pi \cdot l^2}{tD_{eff}}\right)^n\right) \cdot \left(1 - \left[1 - \exp\left(-\frac{1}{\pi}\left(\frac{\pi \cdot l^2}{tD_{eff}}\right)^n \cdot \frac{1.27 + 0.14\frac{1}{\pi}\left(\frac{\pi \cdot l^2}{tD_{eff}}\right)^n}{1 + 0.14\frac{1}{\pi}\left(\frac{\pi \cdot l^2}{tD_{eff}}\right)^n} \right) \right]^{1/2} \right) \tag{12}$$

Efremov [4] has calculated the power function of the argument n for clay materials as 1.95. In order to calculate D_{eff} optimization concept is applied. Drying equation (3) is replaced by equation (12). When this correction is entered, the previously described optimizing concept for the determination of the effective diffusion coefficient can be applied.

2.3.2. Model B2 - The case when shrinkage is included

For materials which shows shrinkage during drying equation (12) needs to be changed by the introduction of the expression $l_{(t)}$ into it. This expression represents the experimentally determined time dependence of the sample thickness. When this correction is entered, the previously described optimizing concept for the determination of the effective diffusion coefficient is applied.

2.4. Program algorithm

The algorithm presented below is the same for any software program. Program named "Drying calculator" was written in the Borland C program language on a standard Pentium IV computer (AMD 1200 MHz, 80GB HDD, 256 MB ram memory). Program has the ability to calculate effective diffusion coefficient using the calculation method A1, A2, B1 and B2.

Program algorithm for models A1 and B1, which neglects shrinkage, contains the following steps:

1. Read the values from database: the time (s), MR_{eks}.
2. Enter number ε (Usually ε =0.05). *(Exists only in case of A1)*
3. Enter the initial value of D_{eff} (D_{eff} =1·10⁻²⁰)

4. Enter the characteristic dimension l (samples half thickness, m)
5. Enter the value n form equation (12) (n=1.95) (*Exists only in case of B1*)
6. For each value from the database using equation *(3) in case of A1* or *(12) in case of B1* MR_{an} will be determined.
7. For each value from database χ^2 will be determined using equation (5)
8. In next cycle step starting value D_{eff} is doubled and a new value MR_{an} will be determined and initially used for determination of new χ^2.
9. If the statement $\chi^2_{first} < \chi^2_{second}$ is satisfied, the program will continue previously described cycle, otherwise the program will temporarily stop.
10. The last three values for D_{eff} and χ^2 are then recorded. Then, the recorded D_{eff} interval is divided into 100 parts. A hundredth part of this interval is defined as a step s. The program commences a cycle again using the initial value for D_{eff} as D_{eff} third from end + s. The cycle is repeated until the statement $\chi^2_{first} < \chi^2_{second} < 1\cdot10^{-10}$ is satisfied. In other words, the cycle is interrupted when the difference $\chi^2_{second} - \chi^2_{first}$ reaches $1\cdot10^{-10}$
11. The final D_{eff} value is then recorded.
12. Result will be saved as database: time (s), MR_{eks}, MR_{an}, and value of average D_{eff}.
13. On the base of this database a graphical view can be displayed

Program algorithm for models A2 and B2, which includes shrinkage, is obtained from previously presented algorithm after a few modifications: in steps 1, and 6 are made.

1. Read the values from database: the time (s), MR_{eks}, and characteristic l (m).
6. In equation *(3) for the case A2* or *(12) for the case B2* l is a function of time; l is provided from database where values of l were determined by experimental measuring of thin plate sample shrinkage vs. time.

2.5. Description of the slope model

For long drying times, equation (1) is transformed into equation (13).

$$MR = \frac{8}{\pi^2}\exp(-\pi^2\frac{D_{eff}t}{l^2}) \quad \text{or} \quad \ln(\frac{\pi^2 MR}{8}) = -\pi^2\frac{D_{eff}t}{l^2} \tag{13}$$

From the equation (13) slope D_{eff} coefficient can be calculated.

3. Results and discussion

3.1. Clay characterization and sample preparation

Three raw masonry clays from the locality Banatski Karlovac (I), Ćirilkovac (II) and Orlovat (III) were analyzed. Characterization of raw masonry clays has included chemical, mineralogical, granulometrical, XRD, DTA and TGA examination. Results of chemical analysis are presented in table 1, while granulometric analyze is presented at fig. 1 and 2.

Composition	Clay (I) %	Clay (II) %	Clay (III) %
Loss ignition on 1000°C	11.71	7.15	6.71
SiO_2	53.23	53.08	54.49
Al_2O_3	13.64	16.73	13.91
Fe_2O_3	5.34	7.10	5.09
CaO	7.50	6.69	8.05
MgO	3.59	1.41	3.70
SO_3	0.00	0.08	0.07
S^{2-}	0.00	0.00	0.01
Na_2O	1.24	0.48	1.14
K_2O	3.42	1.70	1.70
MnO	0.091	0.15	0.08
TiO_2	0.60	0.71	0.46
Summary:	100.36	100.40	100.19

Table 1. Results of chemical analysis

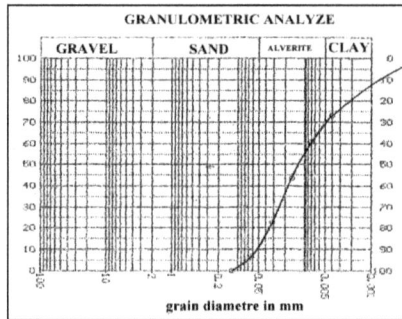

Figure 1. Granulometric test results for clay (I)

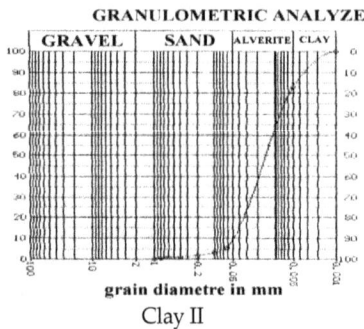

Clay II

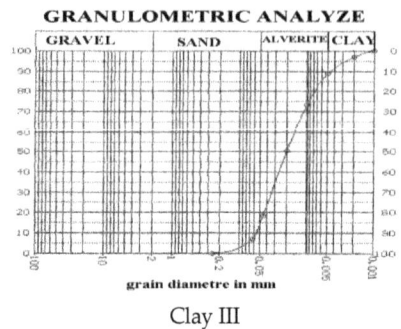

Clay III

Figure 2. Granulometric test results for Clay (II) and (III)

XRD examinations far all three clays were recorded on the Belgrade faculty of mining and geology, using the device PHILIPS PW 1710. DTA and TGA examinations for clay (I) was recorded in Belgrade ITNMS institute, using the device SDT Q600 (TA Instruments), while for clay (II) and clay (III) these examinations were done on the Belgrade chemistry faculty, using the device DERIVATOGRAPH-C (MOM Budapest) and DUPON.

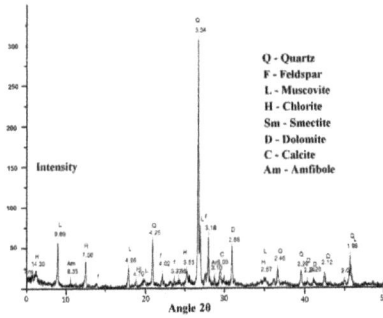

Clay I

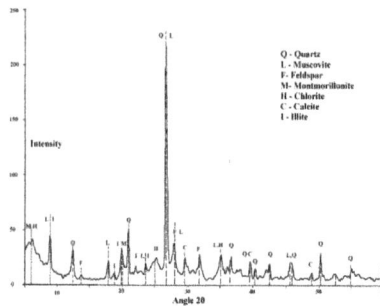

Clay II

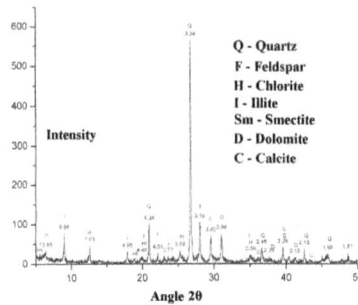

Clay III

Figure 3. XRD diagrams

Clay I   Clay II

Clay III

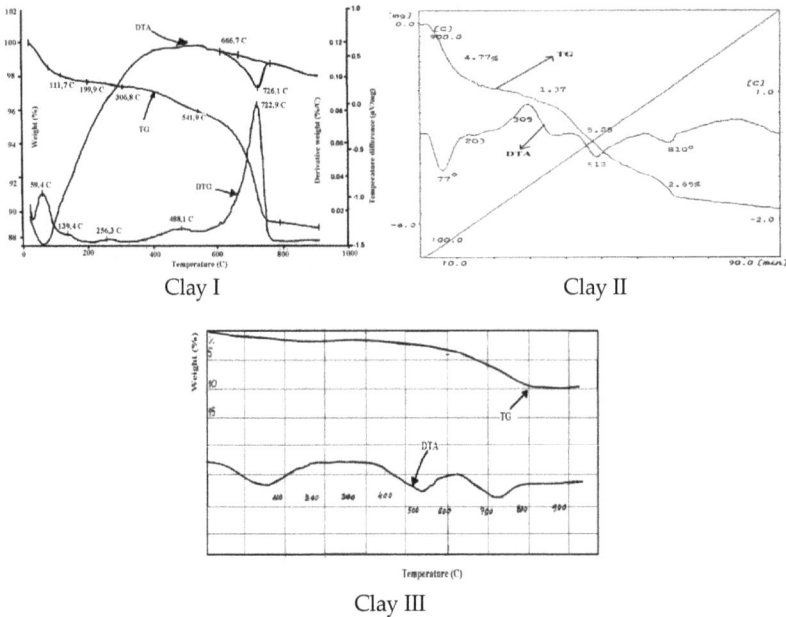

Figure 4. DTA/TG diagrams

Based on chemical test results it can be concluded that all three clays are representing usual masonry raw materials, with a relatively low content of aluminium oxide, a relatively small content of clay minerals and feldspars and increased carbonate content. Clay (III) has the highest SiO_2 and carbonate content. From fig. 1 and fig. 2 it can be seen that Clay (I) has the largest clay content of 30.18%, while in the case of Clay (II) and Clay (III) clay content were respectively 16.75%and 11.11%. From fig. 3 and fig. 4 it can be seen that the most common mineral in all three clays is quartz. Carbonate minerals: calcite and dolomite were present in all three clays too. Beside previously mentioned minerals clay (I) is consisted of mica, chlorite and a small amount of smectites, clay (II) is consisted of muscovite, montmorillonite, chlorite, and illite in traces, while clay (III) is consisted of illite, chlorite, and a small amount of smectites.

After initial clay characterization, the raw materials were subjected to further classical preparation. The raw material samples were first dried at 60°C and then milled down in a laboratory perforated rolls mill. After that, the clays were moisturized and milled in a laboratory differential mill, first at a gap of 3 mm and then of 1 mm. Laboratory samples of size 120x50x14 mm were formed in a laboratory extruder "Hendle" type 4, under a vacuum of 0.8 bar. These samples were used in further experimental work.

3.2. Drying experiments

Moropoulou [22] was investigating the influence of drying air, temperature (20 - 40°C), humidity (30-80%) and velocity (1 - 8 m/s) in order to develop a drying model which will include in its structure the drying air parameters. Mancuhan [3] was studying industrial drying of bricks in a tunnel dryer in order to find optimal drying air parameters which were necessary for rationalization an optimization of the drying process. On the base of previously mentioned studies and along with the years of industrial production experience range of drying air parameters: temperature (40-70⁰C), humidity (40-80%) and velocity (1-3 m/s) has been set up in this study as the boundaries of the planned drying experiment.

Drying kinetic curves were recorded, under the experimental conditions presented in Table 2, on the prepared heavy clay tiles (samples), by monitoring and recording the changes in weight and linear shrinkage of the clay tiles in a laboratory dryer, especially created for this purpose. Schematic view of the laboratory recirculation dryer is presented in Scheme 1.

3.2.1. Laboratory recirculation dryer

The laboratory recirculation dryer provides:

regulation of the drying air temperature within 0-125°C, with accuracy ± 0.2°C;
regulation of the relative humidity of the drying air within 20-100%, with an accuracy of 0.2%;
velocity regulation of the drying air within 0-3.5 m/s, with an accuracy of 1%;
monitoring and recording of the weight of the drying samples within 0-2000 g, with an accuracy of 0.01 g;
monitoring and recording the linear shrinkage within 0-23 mm with an accuracy of 0.2 mm; and
continuous time monitoring during drying.

Experiment	Air velocity, W / m/s	Air temperature, T / ⁰C	Air humidity, V / %
1	1	40	40
2	3	40	40
3	1	40	80
4	3	40	80
5	1	70	40
6	3	70	40
7	1	70	80
8	3	70	80

Table 2. Experimental conditions

Data acquisition, continuous time monitoring and recording of the temperature and relative humidity of the drying medium and the linear shrinkage of the drying samples were realized automatically, using PLC controllers and a standard Pentium IV computer.

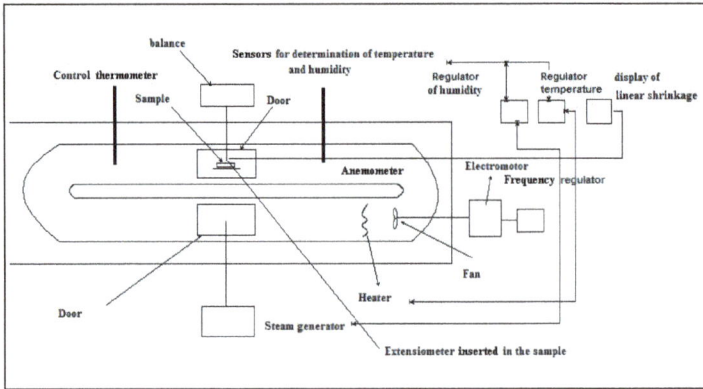

Scheme 1. Laboratory recirculation dryer

3.2.2. Interpretation

Four models for predicting the drying behavior (MR_{an}–t dependence) were obtained from previously described programs. Models A1 and B1 did not include shrinkage, while Models A2 and B2 did. Drying models results are presented in a form of table 3. Deviation of the predicted drying models form experimentally recorded data, presented in table 3, were given as a root mean square error (RMSE) calculated from equation (14). Typical graphical views of the experimental and predicted drying behaviour are presented in Fig. 5.

$$RMSE = \left[\frac{1}{N} \sum_{i=1}^{N} (MR_{exp,i} - MR_{pred,i})^2 \right]^{1/2} \tag{14}$$

Lower value of RSME is representing better agreement between model predicted and experimental drying behaviour.

Exp.	RSME											
	Clay (I)				Clay (II)				Clay (III)			
	Model				Model				Model			
	B2	A2	B1	A1	B2	A2	B1	A1	B2	A2	B1	A1
1	0.0281	0.0351	0.0517	0.0624	0.0241	0.0286	0.0586	0.0714	0.0166	0.0836	0.1079	0.1191
2	0.0111	0.0219	0.0694	0.0926	0.0194	0.0225	0.0523	0.0689	0.0186	0.0756	0.0995	0.1125
3	0.0143	0.0415	0.0731	0.0838	0.0174	0.0458	0.1104	0.1445	0.0524	0.0675	0.1159	0.1296
4	0.0128	0.0142	0.0461	0.0570	0.0219	0.0460	0.1002	0.1165	0.0378	0.0603	0.0944	0.1130
5	0.0121	0.0295	0.0611	0.0506	0.0212	0.0282	0.0856	0.1029	0.0474	0.0705	0.0935	0.1090
6	0.0073	0.0344	0.0689	0.0636	0.0232	0.0176	0.0532	0.0672	0.0173	0.0235	0.0841	0.1009
7	0.0125	0.0355	0.0467	0.0433	0.0188	0.0562	0.0996	0.1126	0.0171	0.0216	0.0709	0.0835
8	0.0201	0.0263	0.0730	0.0866	0.0247	0.0472	0.0942	0.1160	0.0242	0.0356	0.0820	0.0986

Table 3. Calculated RSME parameters

In all experiments value of RMSE had the lowest value for model B2, which means that model B2 had less deviation from experimental results than all other drying models.

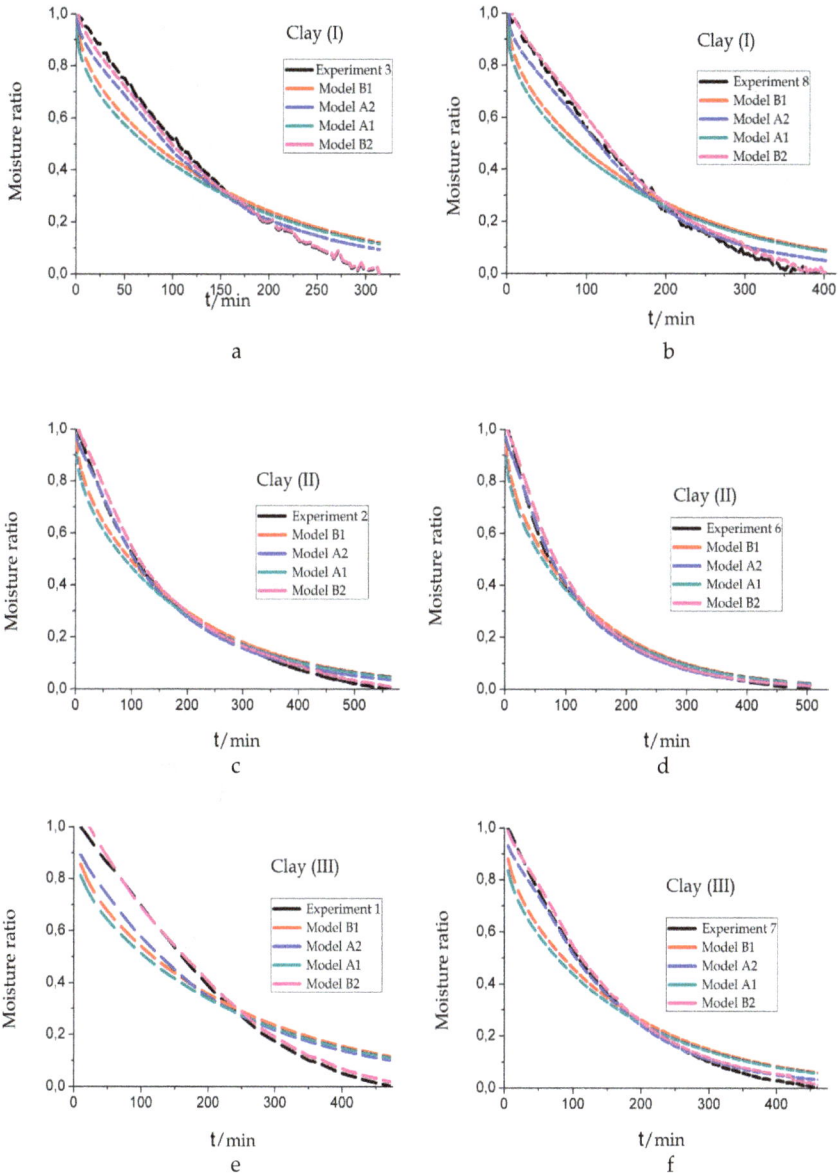

Figure 5. Experimental and calculated moisture ratio vs. drying time

The D_{eff} values obtained through the use of the described programs and from slope of equation (13) are presented in Table 4.

Exp.	$Deff \cdot 10^{10}$ / m²/s														
	Clay (I)					Clay (II)					Clay (III)				
	Model					Model					Model				
	A1	A2	B1	B2	SL	A1	A2	B1	B2	SL	A1	A2	B1	B2	SL
1	4.50	2.30	4.00	1.80	19.00	2.90	0.50	1.70	0.35	10.5	0.34	0.17	0.27	0.13	15.80
2	7.20	2.20	6.10	1.20	23.50	4.20	0.90	2.30	0.83	19.3	0.58	0.34	0.45	0.28	25.30
3	8.10	3.20	7.30	2.30	24.70	0.98	0.19	0.75	0.12	4.7	0.09	0.05	0.003	0.001	3.90
4	10.10	3.40	8.80	2.50	30.50	1.61	0.31	1.17	0.24	6.7	0.19	0.09	0.13	0.005	6.30
5	2.40	0.70	1.00	0.10	9.50	3.81	0.94	3.13	0.79	17.2	0.53	0.29	0.49	0.15	24.80
6	3.30	0.90	2.60	0.30	11.70	5.83	1.25	5.14	1.05	27.6	0.86	0.49	0.75	0.41	44.60
7	4.80	1.80	4.20	1.70	19.50	1.20	0.15	1.05	0.08	4.2	0.22	0.11	0.13	0.008	9,30
8	7.40	3.20	6.30	2.30	24.10	2.19	0,52	1.95	0.37	9.6	0.65	0.42	0.59	0.37	35.30

Table 4. Calculated values of effective diffusion coefficient

In all experiments, the value of effective diffusion coefficient D_{eff} determined by models, which has included the shrinkage correction in the calculation step (A2 and B2), was lower then the value of the same coefficient determined by other models. On analyzing experiments it can be seen that by increasing the drying air velocity from 1 to 3 m/s, the value of the effective diffusion coefficient also increases up to 38% for clay (II), 45% and 60% for clay (III). Diagrams have showed that the kinetic curves representing the models which neglect the shrinkage effect (A1 and B1) do not completely follow the configuration of the experimentally determined drying curves. Drying model B1 had less deviation from experimental results than drying model A1, in all experiments and for all three clays. It can be concluded that whatever the initial mineralogical composition of the clay model B1 is better than model A1. Deviations of these models from the experimental drying curves are higher at the beginning of the drying process and after some time in most case the deviations disappear. The moment of disappearance matches the moment from when the sample continues to dry further without shrinkage.

Drying kinetic curves of the model which includes shrinkage (A2 and B2) follow the configuration of the experimentally determined curves and their matching can be more than 98% as can be seen in experiments 3, 8 in case of clay (I) or in experiments 2, 6 in case of clay (II) or in experiments 1, 7 in case of clay (III). Drying model B2 had less deviation from experimental results than drying model A2, and all other drying models, in all experiments and for all three clays. It can be concluded that whatever the initial mineralogical composition of the clay model B2 is the best drying model. In the case of model B2, if minor deviations exist, they are at the beginning of the drying process and are most probably caused by the time interval which has to pass before stationary experimental conditions are fulfilled and the products are heated up to the temperature in the dryer. The intersection point of the experimental drying curves and modelled drying curves is characterized as the critical point. Critical point is a characteristic kinetic parameter which is important because it determines the moment after which the products no longer shrink.

From Table 4 it can be concluded that value of effective diffusion coefficient D_{eff} determined using the models which included the sample shrinkage correction is lower than the corresponding value determined using the models which neglected sample shrinkage or the slope model. The determined values of data of the D_{eff} from the slope model were higher than the data determined by other models. This result is in agreement with the D_{eff} determination and is representing additional proof that the models which included the shrinkage effect during drying have given more precise D_{eff} values. Effective diffusion coefficients for masonry clay products are in range of 10^{-7} up to 10^{-12} m²/s according to references [6,10]. This relatively large range for the D_{eff} values is connected with the different nature of the heavy clay and the different methods employed for their determination. The D_{eff} values presented in Tables 4 are lying below the previously mentioned range.

4. Conclusions

Calculation methods and computer programs specially designed for calculation of effective diffusion coefficient were developed. First calculation method was based on the mathematical calculation of the second Fick's law and Cranck diffusion equation. Second calculation method was based on the analytical solution of the Efremov differential diffusion equation with a boundary condition in the form of the flux. In both calculation methods, two program variations were designed to compute the effective diffusion coefficient. First program variation did not include shrinkage effect during drying into the computation algorithm while the second one has included it. Four models (A1, A2, B1 and B2) for predicting the drying behaviour were obtained as the result of cited program. This was the first time in the mathematical modelling of the drying of masonry clay that a shrinkage correction was entered into the calculation step. Drying diagram analysis have showed that irrespective of the nature the initial mineralogical composition of the clay, the drying curves representing the models which neglects the shrinkage effect (A1 and B1) did not fully follow the configuration of the experimentally determined kinetic curves, while in the case of the models which include shrinkage (A2 and B2), the resulting curves follows the experimental ones. From Figs. 1 - 6 it can be seen that the introduction of the shrinkage correction into equations (3) and (12) was entirely justified. Drying model B2 had less deviation from experimental results than drying model A2, and all other drying models, in all experiments and for all three clays. It can be concluded that whatever the initial mineralogical composition of the clay model B2 is the best drying model. The determined values of the effective diffusion coefficient were lower than the value that could be found in the literature. The values of the effective diffusion coefficient determined using the models which includes shrinkage were less than the values determined using the models which neglects shrinkage or the values obtained using the slope method. The intersection point of the experimental drying curves and the modelled drying curves is characterized as the critical point.

Author details

Miloš Vasić and Zagorka Radojević
Institute for Testing Materials -
IMS Institute Belgrade, Serbia

Željko Grbavčić
Faculty of Technology and Metallurgy of
the University of Belgrade, Belgrade, Serbia

Acknowledgement

This paper was realized under the project III 45008, which was financed by the Ministry of Science and Technological Development of the Republic of Serbia.

5. References

[1] Chemkhi S.; Zagrouba F. (2008). Development of a Darcy-flow model applied to simulate the drying of shrinkage media, *Braz. J. Chem. Eng.* Vol. 25, pp.503-514, ISSN 0104-6632

[2] Whitaker S. (1977). Simultaneous heat mass momentum transfer in porous media: A theory of drying, *Adv. Heat. Mass. transfer.* Vol. 13, pp. 110-203

[3] Cranck J. (1975). The mathematics of diffusion, (II edition), Oxford University press, London

[4] Efremov G. (2002). Drying kinetics derived from diffusion equation with flux-type boundary conditions, *Dry. Technol.* Vol. 20, pp. 55-66, ISSN: 0737-3937

[5] Vasić M.; Radojević Z. (2011). Establishing a method for determination of effective diffusion coefficient, *Proceedings of the 15th International Conference Modern Tecnologies, Quality and inovation, MODTECH 2011*, pp. 673-676, ISSN 2069-6739 Vadul lui Voda, Chisinau, Republic of Moldavia, May 25-27, 2011

[6] Chemkhi S.; Zagrouba F. (2005). Water diffusion coefficient in clay material from drying data, *Desalination.* Vol. 185, pp. 491-498, ISSN 0011-9164

[7] Mihoubi D.; Zagrouba F.; Amor Ben M. & Bellagi A. (2002). Drying of clay. I Material characteristics, *Dry. Technol.* Vol.20, pp. 465-487, ISSN: 0737-3937

[8] Zagrouba F.; Mihoubi D. & Bellagi A. (2002). Drying of clay. II Reheological Modelisation and simulation of physical phenomena, *Dry. Technol.* Vol.20, pp. 1895-1917, ISSN: 0737-3937

[9] Skansi D.; Tomas S. & Sokole M. (1994). Convection drying of porous material, *Ceram. Int.* Vol. 20, pp. 9-16, ISSN: 0272-8842

[10] Sander A.; Skansi D & Bolf N. (2003). Heat and mass transfer models in convection drying of clay slabs, *Ceram. Int.* Vol. 29, pp. 641-653, ISSN: 0272-8842

[11] Vasić M.; Radojević Z.; Arsenović M. & Grbavčić Ž. (2011). Determination of the effective diffusion coefficiente, *Ro. J. Of Mat.* Vol. 2, pp. 169-176, ISSN 1583-3186

[12] Zeden A.J.J. & Kerkhof P.J.A.M. (1996). Isothermal vapour and liquid transport inside clay during drying, *Dry. Technol.* Vol. 14, pp. 647-676 ISSN: 0737-3937

[13] Efremov, G. (1999). Development of the Generalized Calculation Methods for Non-stationary Heterogeneous Process in Chemical Technology and in a Furnish of Textile Materials, *PhD. Thesis*, MSTU, Moscow, Russia, (in Russian).

[14] Winitiziki S. (2008). A handy approximation for the error function and its inverse, avaliable at http://www.mendeley.com/research/a-handy-approximation-for-the-error-function-and-its-inverse-1/#page-1

[15] Winitiziki S. (2003). Uniform approximations for transcendental functions, *Computational Science and Its Applications—ICCSA, Proceedings, Part I International Conference,* Vol. 2667, DOI: 10.1007/3-540-44839-X, Montreal, Canada, May 18–21, 2003

[16] Katekawa M.E. & Silva M.A. (2006). A review of drying models including shrinkage Effects, *Dry. Technol.* Vol. 24 pp. 5-20 ISSN: 0737-3937

[17] Dissa A. O.; Desmorieux H.; Bathiebo J. & Koulidiati J. (2008). Convective drying of Amelie mango (Mangifera Indica L. Cv. Amelie) with correction for shrinkage,*J. Food Eng.* Vol. 88 pp. 429-437, ISSN: 0260-8774

[18] Ruiz-Lopez I.I.; Ruiz-Espinosa H.; Arellanes-Lozada P.; Barcenas-Pozos M.E. & Garcia-Alvarado M. (2011). Analytical model for variable moisture diffusivity estimation and drying simulation of shrinkable food products, *J. Food Eng, doi:10.1016/j.jfoodeng.2011.08.025*

[19] Silva W.P.; Pecker J.W.; Cleide M.D.P.S. & Gomes J.P. (2010). Determination of effective diffusivity and convective mass transfer coefficient for cylindrical solids via analytical solution and inverse method: Application to drying of rough rice",*J. Food Eng.* Vol. 98, pp 302-308, ISSN: 0260-8774

[20] Park K.J.; Ardito T.H.; Ito A.P.; Park K. J.B. & Oliveira R.A. (2007). Effective Diffusivity determination considering shrinkage by means of explicit Finite difference method, *Dry. Techn.* Vol. 25, pp. 1313-1319, ISSN: 0737-3937

[21] Silva W.P.; Pecker J.W.; Cleide M.D.P.S. & Diogo D.P.S. (2009). Determination of the effective diffusivity via minimization of the objective function by scanning: Application to drying of cowpea, *J. Food Eng.*, Vol. 95, pp 298-304, ISSN: 0260-8774

[22] Moropoulou A.; Karoglou M.; Giakoumaki A.; Krorida M.K.; Maroulis B.Z. & Saravacos D.G. (2005). Drying kinetic of some building materials, *Braz J Chem Eng.* Vol. 22, pp 203-208, ISSN 0104-6632

[23] Mancuhan E. (2009). Analzsis and Optimization of Drying of Green Brick in a Tunnel Dryer, *Dry Technol.* Vol. 27, pp. 707-713, ISSN: 0737-3937

Permissions

The contributors of this book come from diverse backgrounds, making this book a truly international effort. This book will bring forth new frontiers with its revolutionizing research information and detailed analysis of the nascent developments around the world.

We would like to thank Dr. Marta Valášková D.Sc and Dr. Gražyna Simha Martynkova, for lending their expertise to make the book truly unique. They have played a crucial role in the development of this book. Without their invaluable contribution this book wouldn't have been possible. They have made vital efforts to compile up to date information on the varied aspects of this subject to make this book a valuable addition to the collection of many professionals and students.

This book was conceptualized with the vision of imparting up-to-date information and advanced data in this field. To ensure the same, a matchless editorial board was set up. Every individual on the board went through rigorous rounds of assessment to prove their worth. After which they invested a large part of their time researching and compiling the most relevant data for our readers. Conferences and sessions were held from time to time between the editorial board and the contributing authors to present the data in the most comprehensible form. The editorial team has worked tirelessly to provide valuable and valid information to help people across the globe.

Every chapter published in this book has been scrutinized by our experts. Their significance has been extensively debated. The topics covered herein carry significant findings which will fuel the growth of the discipline. They may even be implemented as practical applications or may be referred to as a beginning point for another development. Chapters in this book were first published by InTech; hereby published with permission under the Creative Commons Attribution License or equivalent.

The editorial board has been involved in producing this book since its inception. They have spent rigorous hours researching and exploring the diverse topics which have resulted in the successful publishing of this book. They have passed on their knowledge of decades through this book. To expedite this challenging task, the publisher supported the team at every step. A small team of assistant editors was also appointed to further simplify the editing procedure and attain best results for the readers.

Our editorial team has been hand-picked from every corner of the world. Their multi-ethnicity adds dynamic inputs to the discussions which result in innovative outcomes. These outcomes are then further discussed with the researchers and contributors who give their valuable feedback and opinion regarding the same. The feedback is then collaborated with the researches and they are edited in a comprehensive manner to aid the understanding of the subject.

Apart from the editorial board, the designing team has also invested a significant amount of their time in understanding the subject and creating the most relevant covers. They scrutinized every image to scout for the most suitable representation of the subject and create an appropriate cover for the book.

The publishing team has been involved in this book since its early stages. They were actively engaged in every process, be it collecting the data, connecting with the contributors or procuring relevant information. The team has been an ardent support to the editorial, designing and production team. Their endless efforts to recruit the best for this project, has resulted in the accomplishment of this book. They are a veteran in the field of academics and their pool of knowledge is as vast as their experience in printing. Their expertise and guidance has proved useful at every step. Their uncompromising quality standards have made this book an exceptional effort. Their encouragement from time to time has been an inspiration for everyone.

The publisher and the editorial board hope that this book will prove to be a valuable piece of knowledge for researchers, students, practitioners and scholars across the globe.

List of Contributors

Shu Jiang
Energy & Geoscience Institute, University of Utah, Salt Lake City, USA

Oluwafemi Samuel Adelabu
Federal University of Technology, Ondo State, Nigeria

Károly Lázár
Centre for Energy Research, Institute of Isotopes, Hungarian Academy of Sciences, Budapest, Hungary

Zoltán Máthé
Mecsekérc Plc., Pécs, Hungary

Miloš René
Institute of Rock Structure and Mechanics, v.v.i., Academy of Sciences of the Czech Republic, Prague, Czech Republic

Burhan Davarcioglu
Department of Physics, Faculty of Arts and Sciences, Aksaray University, Aksaray, Turkey

Milan Gomboš
Slovak Academy of Sciences/Institute of Hydrology, Slovak Republic

Markoski Mile and Tatjana Mitkova
Department of Soil Science, Faculty of Agricultural Sciences and Food, Sc. Cyril and Methodius University in Skopje, Republic of Macedonia

Fabienne Trolard and Guilhem Bourrié
INRA, UMR1114, Environnement Méditerranéen et Modélisation des Agro-hydrosystèmes, Avignon, France
UAPV, UMR1114, Environnement Méditerranéen et Modélisation des Agro-hydrosystèmes, Avignon, France

Carla Eloize Carducci, Geraldo César de Oliveira and Nilton Curi
Department of Soil Science, Federal University of Lavras, Brazil

Eduardo da Costa Severiano
Federal Institute of Education, Science and Technology of Goiás State, Brazil

Walmes Marques Zeviani
Department of Statistics, Federal University of Paraná, Brazil

Marta Valášková and Gražyna Simha Martynková
Nanotechnology Centre, VŠB – Technical University of Ostrava, Ostrava-Poruba, Czech Republic
IT4Innovations Centre of Excellence, VŠB-Technical University of Ostrava, Ostrava-Poruba, Czech Republic

Masashi Ookawa
Department of Chemistry and Biochemistry, Numazu National College of Technology, Numazu, Japan

Hideo Hashizume
National Institute for Materials Science, Tsukuba, Japan

Zuzana Navrátilová and Roman Maršálek
Department of Chemistry, Faculty of Science, University of Ostrava, Ostrava, Czech Republic

Hisako Sato
Department of Chemistry, Graduate School of Science and Engineering, Ehime University, Matsuyama, Japan

Kenji Tamura
National Institute of Materials Science, Tsukuba, Japan

Akihiko Yamagishi
School of Medicine, Toho University, Ota-ku, Tokyo, Japan

Miloš Vasić and Zagorka Radojević
Institute for Testing Materials -IMS Institute Belgrade, Serbia

Željko Grbavčić
Faculty of Technology and Metallurgy of the University of Belgrade, Belgrade, Serbia